OXFORD MONOGRAPHS ON GEOLOGY AND GEOPHYSICS NO. 27

Series editors

# OXFORD MONOGRAPHS ON GEOLOGY AND GEOPHYSICS

# ATMOSPHERE–OCEAN INTERACTION

## SECOND EDITION

Eric B. Kraus

Joost A. Businger

OXFORD UNIVERSITY PRESS   New York
CLARENDON PRESS   Oxford
1994

Oxford University Press

Oxford   New York   Toronto
Delhi   Bombay   Calcutta   Madras   Karachi
Kuala Lumpur   Singapore   Hong Kong   Tokyo
Nairobi   Dar es Salaam   Cape Town
Melbourne   Auckland   Madrid

and associated companies in
Berlin   Ibadan

First edition © 1972 Oxford University Press
Copyright © 1994 by Oxford University Press, Inc.

Published by Oxford University Press, Inc.,
200 Madison Avenue, New York, New York 10016

Oxford is a registered trademark of Oxford University Press

Library of Congress Cataloging-in-Publication Data
Kraus, E. B. (Eric Bradshaw), 1913–
Atmosphere–ocean interaction / Eric B. Kraus, Joost A. Businger.—2nd ed.
p. cm.—(Oxford monographs on geology and geophysics; no. 27)
Includes bibliographical references and index.
ISBN 0-19-506618-9
1. Ocean–atmosphere interaction.
I. Businger, Joost Alois. II. Title. III. Series.
GC190.K72  1994   551.5—dc20   93–25300

1 3 5 7 9 8 6 4 2

Printed in the United States of America
on acid-free paper

# PREFACE

Following rather frequently voiced demands, we have completed this revised version of a book that was written by one of us (EBK) in 1972. The original edition had been out of print for many years. The first two chapters of this second edition are not very different from the original, but all the remaining chapters have been completely rewritten. The chapters "Radiation," "Turbulent Transfer near the Interface," and "Large-Scale Forcing by Sea Surface Buoyancy Fluxes" are entirely new. In a text of 325 pages, it is obviously not possible to account for all the relevant work that has been done during the past 23 years. We have tried to provide a sufficiently long list of references to permit students to search more deeply through the various particular fields.

We would like to express our gratitude to some of the persons who helped us in various ways. Howard Hanson and Henry Charnock reviewed the entire text and made valuable suggestions. Kristina Katsaros made useful suggestions about the parameterization of the radiation budget and the temperature at the sea surface. Mark Donelan provided up-to-date information on wave growth related to wave age. Robert A. Brown and Pierre Mourad helped us to understand some of the intricacies of secondary flows that are dealt with in Chapter 6. The satellite photograph of cloud streets that is presented there was given to us by Dieter Etling. Claes Rooth made substantial contributions to Chapters 7 and 8.

The hospitality of David Rogers, Scripps Oceanographic Institution, and of John Wallace, Joint Institute for the Study of the Atmosphere and Ocean, University of Washington, enabled us to meet at their respective departments and to work there together on the preparation of this text.

All figures in this monograph were prepared by the Graphics Department of the Natonal Center for Atmospheric Research under supervision of Justin Kitsutaka. Libby Bogard, University Corporation for Atmospheric Research spent much time and effort to administer a small National Science Foundation grant for travel and manuscript preparation. We are especially grateful to Marianne Kooiman for readying the manuscript and for editorial assistance.

*Pagosa Springs, Colorado*                                    E.B.K.

*Anacortes, Washington*                                       J.A.B.

# CONTENTS

# SYMBOLS

| | | | |
|---|---|---|---|
| $f$ | Coriolis parameter | | 1.2 |
| $F$ | fetch | | 5.3 |
| $F_e$ | exitance | | 3.1 |
| $F_i$ | irradiance, radiant flux density | (3.1) | 3.1 |
| $F_n$ | net irradiance | (3.3) | 3.1 |
| $F^*$ | black body irradiance | (3.10) | 3.1 |
| $Fr$ | Froude number | (1.74) | 1.5 |
| | | | |
| $g$ | acceleration of gravity | | 1.2 |
| $g'$ | reduced gravity | (1.26) | 1.2 |
| $g^*$ | restoring acceleration | (4.16) | 4.2 |
| $G(\ )$ | power spectrum | | 1.4 |
| $G_n$ | flux of trace constituent $C_n$ | | 5.4 |
| | | | |
| $h$ | height of boundary layer; depth of mixed layer | | 6.4 |
| $H$ | Henry's law constant | (2.18) | 2.1 |
| | | | |
| $\mathbf{i}$ | unit vector in $x$-direction | | 1.1 |
| $I$ | radiance | (3.1) | 3.1 |
| | | | |
| $\mathbf{j}$ | unit vector in $y$-direction | | 1.1 |
| | | | |
| $k$ | absorption coefficient | (3.13) | 3.1 |
| $k, k_*$ | von Karman constant | (5.16) | 5.1 |
| $k$ | thermal conductivity | (5.3) | 5.1 |
| $k, \mathbf{k}, k_j$ | wavenumber | | 1.4 |
| $\mathbf{k}$ | unit vector in $z$-direction | | 1.1 |
| $K$ | eddy transfer coefficient | (1.48) | 1.3 |
| | | | |
| $l$ | wavenumber in $x$-direction | | 1.4 |
| $l_m$ | mixing length | (6.15) | 6.3 |
| $L$ | horizontal length-scale | | 1.5 |
| $L$ | Obukhov length | (5.31) | 5.2 |
| $L$ | latent heat of evaporation | (2.6) | 2.1 |
| | | | |
| $m$ | molecular mass | | 2.1 |
| $m$ | wavenumber in $y$-direction | | 1.4 |
| $\mathbf{M}, M_i$ | wave momentum | | 4.1 |
| | | | |
| $n$ | wavenumber in $z$-direction | | 1.4 |
| $\mathbf{n}$ | unit vector normal to surface | | 1.2 |
| $N$ | Brunt–Vaisala frequency | (1.37) | 1.2 |
| $Nu$ | Nusselt number | (5.64) | 5.5 |

| | | | |
|---|---|---|---|
| $p$ | pressure | | 1.2 |
| $Pe$ | Péclet number | (1.78) | 1.5 |
| $Pr$ | Prandtl number | (1.79) | 1.5 |
| | | | |
| $q$ | specific humidity | | 2.2 |
| $q$ | velocity scale of turbulence | (1.66) | 1.4 |
| $q_*$ | specific humidity scale | (5.37) | 5.2 |
| $Q$ | heat flux | | 1.3 |
| | | | |
| $r, r_*$ | mixing ratio | | 2.2; 8.3 |
| $r$ | radius, polar coordinate | | 8.3 |
| $r$ | reflectance | (3.5) | 3.1 |
| $R$ | autocorrelation function | (1.49) | 1.4 |
| $R$ | gas constant for dry air | (2.21) | 2.2 |
| $R_1$ | universal gas constant | (2.5) | 2.1 |
| $R_v$ | gas constant for water vapour | (2.5) | 2.1 |
| $Ra$ | Rayleigh number | (1.81) | 1.5 |
| $Re$ | Reynolds number | (1.72) | 1.5 |
| $Ri$ | Richardson number | (5.33) | 5.2 |
| $Ri_f$ | flux Richardson number | | 5.5 |
| $Ro$ | Rossby number | (1.76) | 1.5 |
| $Rr$ | roughness Reynolds number | | 5.1 |
| | | | |
| $s$ | salinity | | 2.1 |
| $S$ | entropy | | 2.2 |
| $S_c$ | Schmidt number | (1.80) | 1.5 |
| $S'$ | indicator of instability | | 6.3 |
| | | | |
| $t$ | time | | 1.1 |
| $t_*$ | residence time | | 5.1 |
| $T$ | temperature | | 1.2 |
| $T$ | duration, period | | 1.4 |
| $T_v$ | virtual temperature | (2.24) | 2.2 |
| | | | |
| $u$ | optical thickness | | 3.1 |
| $u$ | turbulent component of velocity in $x$-direction | | 1.3 |
| $u_i$ | turbulent component of velocity in $x_i$-direction | | 1.3 |
| $u_*$ | friction velocity | (5.12) | 2.4; 5.1 |
| $U$ | mean component of velocity in $x$-direction, velocity-scale | | 1.3; 1.5 |
| $U_i, \mathbf{U}$ | mean component of velocity in $x_i$-direction, velocity vector | | 1.3 |
| $U_g, \mathbf{U}_g$ | geostrophic windspeed, geostrophic velocity vector | | 1.2 |

| | | | |
|---|---|---|---|
| $v$ | turbulent component of velocity in $y$-direction | | 1.3 |
| $v_n$ | transfer velocity | (5.57) | 5.4 |
| $V$ | mean component of velocity in $y$-direction | | 1.1 |
| $V$ | volume | | 1.2 |
| | | | |
| $w$ | turbulent component of velocity in $z$-direction | | 1.3 |
| $W$ | mean component of velocity in $z$-direction | | 1.3 |
| $W$ | work | | 2.3 |
| | | | |
| $x, y, z$ | Cartesian coordinates | | 1.1 |
| $z_0$ | surface roughness length | | 5.1 |
| | | | |
| $\alpha$ | coefficient of thermal expansion | (2.3) | 2.1 |
| $\alpha, \alpha_1$ | Kolmogorov constants | | 1.5 |
| | | | |
| $\beta$ | coefficient of saline contraction | (2.3) | 2.1 |
| $\beta$ | meridional variation of Coriolis parameter | (7.10) | 7.1 |
| | | | |
| $\gamma$ | lapse rate | | 1.3 |
| $\gamma$ | surface tension | | 2.3 |
| $\Gamma$ | adiabatic lapse rate | | 1.3 |
| | | | |
| $\delta$ | molecular diffusivity | | 1.3 |
| $\delta_{ij}$ | unit tensor | | 1.1 |
| | | | |
| $\epsilon$ | dissipation per unit mass | | 1.2 |
| $\epsilon$ | efficiency | | 8.3 |
| $\epsilon$ | emittance | (3.9) | 3.1 |
| $\epsilon_f$ | flux emittance | (3.20) | 3.3 |
| $\epsilon_{ijk}$ | alternating tensor | | 1.1 |
| | | | |
| $\zeta$ | departure of water surface from equilibrium level | | 4.2 |
| $\zeta$ | dimension-less height | (5.32) | 5.2 |
| | | | |
| $\eta$ | vorticity | | 1.1 |
| | | | |
| $\theta$ | angle of incidence; zenith angle | | 3.1 |
| $\theta$ | turbulent component of potential temperature | | 1.3 |
| $\theta_*$ | temperature scale | (5.36) | 5.2 |
| $\Theta$ | potential temperature | (2.28) | 2.2 |
| $\Theta_e$ | equivalent potential temperature | (2.31) | 2.2 |
| $\Theta_v$ | virtual potential temperature | | 2.2 |
| | | | |
| $\kappa$ | thermal diffusivity | | 1.2 |
| $\lambda$ | wavelength | | 3.1 |

| $\mu$ | chemical potential | | 2.1 |
|---|---|---|---|
| $\mu$ | ratio of diffusive capacities | | 4.2 |
| | | | |
| $\nu$ | kinematic viscosity | | 1.2 |
| $\xi$ | enthalpy | (2.7) | 2.1 |
| $\rho$ | density | | 1.2 |
| | | | |
| $\sigma$ | growth factor | (4.51) | 4.4 |
| $\sigma$ | Stefan–Boltzmann constant | (3.11) | 3.1 |
| $\sigma_{ij}$ | molecular momentum flux tensor | | 1.2 |
| | | | |
| $\tau, \tau_{ij}, \tau$ | stress magnitude; tensor; vector | | 1.2 |
| $\tau$ | transmittance | (3.8) | 3.1 |
| $\tau_f$ | flux transmittance | (3.25) | 3.3 |
| | | | |
| $\varphi$ | latitude | | 1.1 |
| $\varphi_m$ | dimension-less momentum gradient | (5.35) | 5.2 |
| $\varphi_q$ | dimension-less humidity gradient | (5.37) | 5.2 |
| $\varphi'$ | dimension-less temperature gradient | (5.36) | 5.2 |
| | | | |
| $\phi$ | distribution function | | 5.1 |
| $\Phi$ | gravitational potential | | 1.2 |
| | | | |
| $\Psi$ | velocity potential | | 1.2 |
| $\omega$ | frequency | | 1.4 |
| $\Omega$ | angular velocity of the earth | (1.13) | 1.2 |

# INTRODUCTION

We may as well start with the same quotation from *Genesis*: "And God...divided the waters which were under the firmament from the waters which were above the firmament," which introduced the first version of this monograph (Kraus, 1972). It was pointed out then that these initial conditions did not interdict dynamic changes. The laws of thermodynamics force water to pass across the interface that separates the two realms. And not only the waters...the sea surface acts like a membrane, which regulates the rate at which many substances circulate through the system that is our planet. It also transmits energy: with the exception of the tides, almost all the motion in the sea is driven directly or indirectly by atmospheric influences. In turn, latent heat from the oceans fuels a large part of the atmospheric circulation. In fact, it is hardly possible to interpret conditions in either medium without some knowledge of the role that is played by their common boundary.

The introduction to the first version also listed some of the conceptual and technological changes that had occurred during the 25 years following the publication of Sverdrup's (1945) *Oceanography for Meteorologists*. It was not only the science, but also the audience concerned with air-sea interactions, that had changed during those 25 years. Sverdrup addressed his book primarily to the many meteorologists who had been trained before and during World War II. He did not have to write for physical oceanographers, whose numbers where then so small that they were probably all known by him personally. By the early 1970s the number of physical oceanographers had increased substantially. The same applied to the number of books and journals that dealt with their subject. An attempt was made, therefore, in the first edition to cover not only oceanographic topics for meteorologists, but also some aspects of meteorology for oceanographers.

Another 25 years have almost gone by since then. It may be

appropriate to compare the rates of change during these two quarter-century periods. Computers and satellites were already fully operational well before 1970, as were methods to derive turbulent transports from observed high frequency fluctuation of the vertical velocity and the transported property. The decades since then were periods of incremental rather then fundamental innovation in our field. Computation has become faster and more powerful; satellite observations have become more comprehensive and their interpretation is now much more sophisticated. The upper ocean and the marine atmosphere have been investigated more intensively and by more people, who now can communicate more easily, not only among themselves but also with their instruments, than they could 25 years ago. All this permitted the mounting and management of numerous cooperative, national, and international observation programs, which have greatly increased the data base for air-sea interaction studies.

Most of our modern ideas about wind waves, wave-wave interactions, mixed-layers and radiative transfers above and below the interface had already been formulated in the nineteen fifties and sixties. These ideas have since then been refined, extended, and applied to particular situations. Much of this work was described during a series of meetings held during the 1970s under NATO auspices. Proceedings of these meetings have been edited and published by Favre and Hasselmann (1978), Kraus (1977) and Dobson et al. (1980). Accounts of more recent work, including many detailed observations made by scientists from the former Soviet Union, have been collected by Nihoul (1993).

The first edition of this book avoided discussion of interaction processes on the time scale of climate changes. To quote from the earlier introduction: "This omission does not indicate a lack of interest, but rather a feeling that the time [was] not opportune for a review of this topic." It was expected then that new numerical studies of the atmosphere–ocean system would make any review of "past speculations about climate changes obsolete in a very short time." Twenty-three years later, it seems to us that the time for such a review has still not arrived. Numerous efforts have been made indeed to simulate the coupled atmospheric and oceanic circulation. Some of these efforts have been described in a volume on *Coupled ocean-atmosphere models* edited by Nihoul (1985). It is probably fair to say that these efforts have not yet produced insights of a universal nature that would be acceptable generally to the scientific community. An exception to this pessimistic assessment of climate-scale interactions are the coupled processes associated with the El Niño/Southern Oscillation and with the thermohaline circulation in the Atlantic. Both of these topics have been the subject of concentrated research during the past 30 years and a consensus about their mechanism has begun to emerge. Some contradictions and uncertainties remain even there. They can be identified partly with different parameterizations of the transfer processes across the air-sea interface.

In the present book, we have tried to present a coherent, up-to-date account of processes that involve the transfer of energy, matter, and momentum between the atmosphere and the ocean. In Chapter 1, we explain the notation that is being used and recapitulate some of the relevant concepts of fluid dynamics. Though probably familiar to many readers, their presentation here avoids restatements in different places in the text. Chapter 2 deals with the state of matter near the interface. It follows the text of the first edition, but includes new material and references on sea ice and on bubbles and spray. Radiation is the topic of Chapter 3, which has been significantly revised for the present edition. Special attention has been given to the determination of the radiation balance at the sea surface by remote sensing.

Surface waves are a directly visible manifestation of the working of the wind upon the sea surface. The treatment in Chapter 4 begins with the classical model of harmonic, small-amplitude perturbations. We believe that this is needed to give new students of the subject some insight into the basic dynamics of a disturbed fluid interface. This is followed by an account of modern concepts of wave generation and wave spectra. Some relevant empirical formulas are also quoted. Surface wind waves have been the topic of several, specialized experimental programs. The relevant literature is too large to be summarized fully in this book. Readers who want to delve more deeply into this subject will find appropriate references in the text.

Turbulent transfers across the interface are a central feature of air-sea interaction studies. Chapter 5, which considers this topic, has been rewritten and expanded to deal with the problems of surface drift, wave influences on near-surface wind profiles and wind drag. A discussion of surface renewals has been introduced to give some insight into the structure of the near-surface molecular sublayer, and the transfer of trace gases across the interface. Chapter 6 now contains a description of coherent meso-scale boundary layer perturbations, which were not dealt with in the first edition of this book. The section on mixed-layers has also been expanded to include a more detailed account of atmospheric mixed-layers.

Common, large-scale features of oceanic and atmospheric dynamics have been considered thoroughly in Gill's (1982) monograph on this topic. Our treatment of some of the same material in Chapters 7 and 8 is somewhat less rigorous, but emphasizes more strongly the role of energy transfers between the two media. Conceptual models of the El Niño/Southern Oscillation phenomenon and of thermohaline circulation changes are discussed briefly in subsection 8.4.1.

It may be fitting to list some gaps in our present knowledge, which have become particularly obvious to us during the writing of this book. One area that needs a better understanding is the interaction between the wave field and the wind above. On the face of it, this seems a simple problem, but in spite of all that has been learned about waves since World War II, we still don't know quantitatively how they affect the downward flux of momentum, or the profile of the wind immediately above the surface. The vertical profile of wind-driven currents and their interaction with the wave field involves similar uncertainties. A different problem is raised by the necessity to parameterize transports by subgrid-scale coherent structures in large-scale circulation models. Better parmeterization of these transports would yield more realistic simulation of conditions both above and below the interface. Among other areas of quantitative uncertainty one could mention the passage of gases and other substances across the sea surface. Hopefully these lacunae will have been filled before somebody writes another book on this topic in another twenty five years. In the meantime they continue to challenge students of Atmosphere–Ocean Interaction.

# ATMOSPHERE–OCEAN INTERACTION

# 1

## BASIC CONCEPTS

### 1.1 Notation

Both Cartesian tensor and vector notation will be used in this text. The notation $x_i$ means the $i$-component of the vector $\mathbf{x} = (x_1, x_2, x_3)$. When used in the argument of a function [e.g., $\mathbf{f}(x_i)$], $x_i$ represents the whole vector, so that $\mathbf{f}(x_i)$ stands for $\mathbf{f}(x_1, x_2, x_3)$. Repeated indices indicate summations over all coordinate directions, ($u_i u_i = u_i^2 = u_1^2 + u_2^2 + u_3^2$). Two special and frequently used tensors are the *unit tensor* $\delta_{ij}$ and the *alternating tensor* $\epsilon_{ijk}$. The unit tensor has components equal to unity for $i = j$ and zero for $i \neq j$. The alternating tensor has components equal to $+1$ when the indices are in cyclical sequence 1, 2, 3 or 2, 3, 1 or 3, 1, 2; equal to $-1$ when the indices are not cyclical; and equal to zero when two indices are the same. The vorticity vector is defined by the relation

$$\eta_i \equiv \epsilon_{ijk} \frac{\partial U_k}{\partial x_j} \equiv \frac{\partial U_k}{\partial x_j} - \frac{\partial U_j}{\partial x_k}.$$

The symbol $\equiv$ is used throughout to represent a definition or identity.

Conditions near the sea surface are usually very anisotropic. It is often desirable to distinguish between the horizontal and vertical directions. We shall do so by using an $x, y, z$ coordinate system with the origin at mean sea level and the $z$-axis pointing upward. Unless otherwise specified, the $x$ and $y$ directions will be toward east and north. The vertical velocity will be denoted by $W$; the horizontal velocity by the vector $\mathbf{U}$ with components $U$ and $V$. Unity vectors in the $x, y, z$ directions are denoted by $\mathbf{i}, \mathbf{j}, \mathbf{k}$. The usual vector operation symbols will be used only to represent operations within the horizontal plane. For example,

$$\nabla \cdot \mathbf{U} \equiv \frac{\partial U}{\partial x} + \frac{\partial V}{\partial y}, \qquad \nabla \times \mathbf{U} \equiv \left( \frac{\partial V}{\partial x} - \frac{\partial U}{\partial y} \right) \mathbf{k} = \eta_z.$$

In a fluid one has to distinguish between local changes and changes that are experienced by an individual fluid element as it moves about. The former can be recorded by a fixed sensor and is represented by the partial time differential. The individual change could only be recorded by a sensor that would float with the element. It is denoted by the *total time differential*

$$\frac{d}{dt} \equiv \frac{\partial}{\partial t} + U_i \frac{\partial}{\partial x_i} \equiv \frac{\partial}{\partial t} + \mathbf{U} \cdot \nabla + W \frac{\partial}{\partial z}. \tag{1.1}$$

In a treatise that covers such a variety of topics, some use of the same symbols for different properties is inevitable. An attempt has been made to restrict this to properties that are so obviously different that confusion is unlikely. In the few cases that a symbol with two different meanings would appear in the same equation or on the same page, the symbol for one of the quantities is indexed with an asterisk. This has been noted in the list of the most-frequently used symbols in the appendix. Symbols that appear only locally in the text usually have not been included in the list.

## 1.2   Conservation equations

The physical processes on either side of the air–sea interface may be described by a set of equations that includes a formulation of conservation of mass (continuity equation), of conservation of momentum (equations of motion), of conservation of energy or enthalpy, and of the equation of state. The derivation of these equations can be found in Landau and Lifshitz (1959) or Hinze (1975). The conservation equations will be discussed in this section, and the equation of state will be discussed in detail in Chapter 2.

### 1.2.1   *Conservation of matter*

The *specific density, $\rho$,* is the amount of mass in a unit of volume. The continuity equation

$$\frac{\partial \rho}{\partial t} + \frac{\partial}{\partial x_i}(\rho U_i) = 0 \tag{1.2}$$

states that the rate with which this mass changes locally is equal to the divergence of $\rho U_i$, the mass flux per unit area, commonly called the *momentum* of the fluid.

Equation (1.2) can be written alternatively in the form

$$\frac{1}{\rho}\frac{d\rho}{dt} = -\left(\nabla \cdot \mathbf{U} + \frac{\partial W}{\partial z}\right). \tag{1.3}$$

In perturbations of relatively large horizontal extent (see (1.3)), the *hydrostatic equation*

$$dp = -g\rho \, dz$$

tends to be satisfied. This expression states that the decrease in pressure, $p$, with height is equal to the weight of the fluid per unit volume. The left-hand side of (1.3) may then be expressed in terms of the vertical velocity, $W = dz/dt$, and the square of the speed of sound, $c_s^2 = dp/d\rho$,

$$\frac{1}{\rho}\frac{d\rho}{dt} = \frac{1}{\rho}\frac{d\rho}{dp}\frac{dp}{dt} = -g\left(\frac{dp}{d\rho}\right)^{-1}\frac{dz}{dt} = -\frac{gW}{c_s^2} \equiv -\frac{W}{D_s}. \tag{1.4}$$

$D_s$ is called the *scale-depth* of the medium. In the atmosphere $D_s$ is about 10 km; in the sea, it would be of order 200 km and therefore much larger than the greatest ocean depth. On the right-hand side of (1.3) the term $\partial W/\partial z$ is of order $W/h$, where $h$ is the maximum vertical displacement. It follows that

$$\frac{1}{\rho}\frac{d\rho}{dt} \div \frac{\partial W}{\partial z} \approx \frac{h}{D_s}. \tag{1.5}$$

If this ratio is small, then the left-hand side of (1.3) is negligible compared to the individual terms on the right-hand side; that is, $\partial W/\partial z$ must be nearly balanced by the horizontal divergence term

$$\nabla \cdot \mathbf{U} + \frac{\partial W}{\partial z} \equiv \frac{\partial U_i}{\partial x_i} \approx 0. \tag{1.6}$$

The last relation describes continuity in an incompressible fluid. It is generally applicable to sea water in the upper ocean, and it is also valid in air, provided that vertical displacements remain much smaller than the scale-depth. The vertical displacements in the marine boundary layer, with which we are concerned here, tend to fall into that category.

If $\hat{\mathbf{U}}$ is the vertical average of the horizontal velocity in an incompressible fluid of limited depth $D$, then the vertical integral of (1.6), after division by $D$, can be expressed in the form

$$\nabla \cdot \hat{\mathbf{U}} + \frac{1}{D}\frac{dD}{dt} = 0. \tag{1.7}$$

The velocities of individual molecules affect the mass flow only through their mean, which is identical with the velocity of the fluid continuum. If the majority of molecules moves in a certain direction, then it is observable as a flux of mass. By contrast, any individual constituent in a mixture can diffuse through the space occupied by the fluid without a net flow of mass, as long as the other constituents move in the opposite direction. Concentrations of particular constituents in a mixture can also be affected by processes other than diffusion and advection. For example, spray in the marine atmosphere can evaporate. Fallout is another possible cause for a change of concentration.

Let $C_n$ stand here for the specific concentration or mass fraction of the $n$th constituent of a mixture. The mass per unit volume of this particular substance is $\rho_n = C_n \rho$.

$$\sum_n C_n = 1. \tag{1.8}$$

If there is no sedimentation, then the continuity equation for $C_n$ has the form

$$\frac{\partial}{\partial t}(\rho C_n) = -\frac{\partial}{\partial x_i}\rho\left(C_n U_i - \delta_n \frac{\partial C_n}{\partial x_i}\right) + S_n. \tag{1.9}$$

The flux of $C_n$, represented by the expression inside the brackets, has two parts: a bulk transport carried by the continuum velocity, $U_i$; and a flux produced by the random movement of molecules. In the absence of strong temperature or pressure gradients, this molecular diffusion flux is fully determined by the gradient of $C_n$ multiplied by the molecular diffusivity, $\delta_n$.

The source term, $S_n$, in (1.9) symbolizes the local creation of the $n$th constituent by internal phase changes. When there are no phase changes, the transformation of (1.9) with the aid of (1.2) yields, after division by $\rho$, and omission of the suffix $n$,

$$\frac{dC}{dt} = \frac{1}{\rho}\frac{\partial}{\partial x_i}\left(\rho\delta\frac{\partial C}{\partial x_i}\right) \approx \delta\frac{\partial^2 C}{\partial x_i^2}. \tag{1.10}$$

This is the conventional form of the diffusion equation in a moving fluid. It contains the implicit assumptions that the density is not changed by the diffusion process and that changes in the concentration of one constituent do not affect the concentrations of the other constituents. This assumption is a good approximation of the diffusion of salt in the upper ocean or of water vapour in cloud-free air. However, when we deal with mixtures of more or less equal parts or when there are any phase changes, the concentration of one constituent ceases to be independent of that of the others. The appropriate continuity equation then involves a summation over all substances in the mixture.

$$\sum_n \left[ \frac{\partial}{\partial t}(\rho C_n) + \frac{\partial}{\partial x_i}(\rho C_n U_i) \right] = \sum_n \delta_n \frac{\partial^2}{\partial x_i^2}(\rho C_n) = 0. \tag{1.11}$$

The last result follows from (1.2) and (1.8), which indicate that the left-hand side of (1.11) must be equal to zero.

### 1.2.2. *Conservation of momentum*

Momentum, being a three-dimensional vector, requires three equations for its specification. They can be written in the form

$$\frac{\partial \rho U_i}{\partial t} = -\rho \frac{\partial \Phi}{\partial x_i} - 2\epsilon_{ijk}\Omega_j \rho U_k - \frac{\partial}{\partial x_j}(\rho U_i U_j + p\delta_{ij} - \sigma_{ij}). \tag{1.12}$$

On the left-hand side of (1.12) is the local change of the momentum vector. The first term on the right-hand side represents the force of gravity. The local vertical is defined by the gradient of the *gravitational potential* $\Phi$. Its magnitude $\partial \Phi/\partial x_i = g = 9.81 \text{ m s}^{-2}$ can be considered constant in the present context. The following term represents the *Coriolis force,* an apparent inertial acceleration in a coordinate system that is fixed relative to the rotating earth. It is equal to the vectorial product of the earth's rotation vector $2\Omega_j$ with the momentum vector $\rho U_k$. The vector $\Omega_j$ is parallel to the polar axis and has a magnitude equal to the *earth's angular velocity*

$$\Omega = 2\pi/24 \text{ hr} = 0.76 \times 10^{-4} \text{ s}^{-1}. \tag{1.13}$$

The last term in (1.12) represents the convergence of the momentum flux. This flux is a tensor because the field of each of the three momentum components can be distorted or moved about by velocities which also have three components. As specified by the expression in brackets, the *momentum flux tensor* has three parts. The first one denotes the transport of the momentum, $\rho U_i$, by the continuum velocity vector, $U_j$. The following part involves the pressure, $p$, which may be interpreted as a transport of molecular momentum by the random molecular velocities. As the molecular motion is isotropic, without preferred direction, this contribution must be invariant to any rotation of the coordinate system, which is possible only if it can be represented by a scalar multiplied by the unit tensor $\delta_{ij}$. The third part of the momentum flux tensor represents the flux of momentum which is produced by the molecular random motions. These tend to equalize all continuum velocity gradients and differences. If $v$ denotes the kinematic viscosity,

$$\sigma_{ij} = \rho v \left( \frac{\partial U_i}{\partial x_j} + \frac{\partial U_j}{\partial x_i} \right) = \sigma_{ji}. \tag{1.14}$$

Air and sea water in natural conditions behave like incompressible fluids as far as viscous effects are concerned. With very good approximation, we therefore have

$$\frac{\partial}{\partial x_j} \sigma_{ij} = \rho v \left[ \frac{\partial^2 U_i}{\partial x_j^2} + \frac{\partial}{\partial x_i} \left( \frac{\partial U_j}{\partial x_j} \right) \right] = \rho v \frac{\partial^2 U_i}{\partial x_j^2}. \tag{1.15}$$

The velocity vector and the momentum flux tensor must both be continuous across the interface of two immiscible, real fluids such as air and water. The pressure must also be continuous. These requirements are known as the *kinematic* and the *dynamic boundary conditions.*

The classic *Navier–Stokes* form of the equations of motion can be obtained from (1.12) by elimination of the density differentials with the aid of (1.2)

$$\frac{dU_i}{dt} + 2\epsilon_{ijk}\Omega_j U_k = -\frac{1}{\rho}\frac{\partial p}{\partial x_i} - \frac{\partial \Phi}{\partial x_i} + v\frac{\partial^2 U_i}{\partial x_j^2}. \tag{1.16}$$

A slightly different form of (1.16) makes use of the identity

$$U_j \frac{\partial U_i}{\partial x_j} \equiv \frac{\partial}{\partial x_i} \left( \frac{1}{2} U_j^2 \right) + \epsilon_{ijk}\eta_j U_k. \tag{1.17}$$

The second term represents the gradient of the kinetic energy per unit mass. The vector product in the last term, sometimes called the *vortex force*, specifies an acceleration at right angles to the vectors of velocities and vorticity. In fluids of uniform density, (1.17) can be used to express (1.16) in the form

$$\frac{\partial U_i}{\partial t} + \frac{\partial}{\partial x_i} \left( \frac{1}{2} U_j^2 + \frac{p}{\rho} + \Phi \right) + \epsilon_{ijk}(\eta_j + 2\Omega_j)U_k - v\frac{\partial^2 U_i}{\partial x_j^2} = 0. \tag{1.18}$$

The term $(\eta_j + 2\Omega_j)$, called the *absolute vorticity,* measures the curl in a coordinate system that does not rotate with the earth around its axis.

In inviscid irrotational motion, the two last terms vanish in (1.18), which can then be integrated after the introduction of a velocity potential, $\psi$, defined by

$$U_i = \frac{\partial \psi}{\partial x_i}.$$  (1.19)

The result of this integration is known as *Bernoulli's equation*

$$\frac{\partial \psi}{\partial t} + \frac{1}{2} U_j^2 + \frac{p}{\rho} + \Phi = \text{const.}$$  (1.20)

If the motion is not irrotational but steady ($\partial/\partial t = 0$), we can obtain an analogous expression by integration along a streamline. The third term of (1.18), since it is normal to the velocity and therefore normal to the streamline, cannot contribute to such an integral. The term

$$\frac{1}{2} \rho U_j^2 + p$$  (1.21)

is called the *total pressure*. As it is the sum of a dynamic and a static component, it is the pressure that would be recorded by a transducer at the stagnation point of a streamline.

The effects of the earth's rotation and gravity tend to introduce a directional bias into all atmospheric motion patterns except those on a very small scale. It becomes more convenient in this case to replace the tensorial representation by vector notation in an $(x, y, z)$ coordinate system as defined in Section 1.1. If $\varphi$ now denotes the geographical latitude, then the components of the rotation vector $2\Omega_i$ in this system are 0; $f_c = 2\Omega \cos \varphi$; $f = 2\Omega \sin \varphi$. The time $4\pi/f$ is equal to the azimuthal period of a Foucault pendulum and is called the *pendulum day*.

In the new coordinate system (1.16) with consideration of (1.15) assumes the form

$$\rho \left[ \frac{d}{dt} - \nu \left( \nabla^2 + \frac{\partial^2}{\partial z^2} \right) \right] W - \rho f_c \mathbf{i} \cdot \mathbf{U} = -\frac{\partial p}{\partial z} - \rho g,$$

$$\rho \left[ \frac{d}{dt} - \nu \left( \nabla^2 + \frac{\partial^2}{\partial z^2} \right) \right] \mathbf{U} + \rho f \mathbf{n} \times \mathbf{U} = -\nabla p.$$  (1.22)

The symbols $\mathbf{i}$ and $\mathbf{n}$ denote unit vectors in the zonal and vertical directions. The contribution of the vertical velocity to the Coriolis force has been neglected in (1.22).

The density distribution can be represented by

$$\rho(x, z, t) = \bar{\rho}(z) + \rho'(x, z, t),$$  (1.23)

where $\bar{\rho}(z)$ is the horizontally averaged density, and $\rho'$ is the density fluctuation related to turbulence which is discussed in Section 1.3. The horizontally averaged density may slowly vary with $\mathbf{x}$ and $t$, as long as the averaging time, needed to exclude all turbulent fluctuations, is much shorter than the time variations of synoptic or mesoscale events. When the vertical accelerations in the fluid are small, the pressure is close to hydrostatic and is given by

$$\frac{\partial \bar{p}}{\partial z} = -g\bar{\rho}. \tag{1.24}$$

Therefore the right-hand side of (1.22) may be written as

$$\frac{\partial p}{\partial z} + \rho g = \frac{\partial p'}{\partial z} + \rho' g. \tag{1.25}$$

The frequently used Boussinesq approximation involves replacement of the right-hand side of the first equation (1.22) by (1.25) and neglect of the products of $\rho'$ with the acceleration terms on the left-hand side. This is permissible if $\rho'/\bar{\rho} \ll 1$. In the atmospheric boundary layer the relative density anomaly is typically of order $10^{-2}$, whereas in the ocean it is of the order $10^{-3}$.

A fluid element with a density that differs from its horizontal surroundings is forced vertically by a *buoyancy acceleration*

$$b' \equiv -g\frac{\rho'}{\bar{\rho}} \equiv -g' \approx -g\frac{\rho'}{\rho_r}, \tag{1.26}$$

where $g'$ is known as the *reduced gravity* and $\rho_r$ is a reference density, which may be the average density near the sea surface or an overall density average.

With the Boussinesq approximation and (1.25) and (1.26), the first equation (1.22) can be written as

$$\left[\frac{d}{dt} - \nu\left(\nabla^2 + \frac{\partial^2}{\partial z^2}\right)\right]W - f_c U = -\left(\frac{1}{\rho_r}\frac{\partial p'}{\partial z} + g'\right). \tag{1.27}$$

The right-hand side of (1.27) represents an unbalanced buoyancy force.

It is useful to introduce a fictitious velocity, $\mathbf{U}_g$, called the *geostrophic velocity,* for which the Coriolis acceleration exactly balances the horizontal pressure force

$$\nabla p + \rho f \mathbf{n} \times \mathbf{U}_g = 0. \tag{1.28}$$

Subtraction of (1.28) from the second equation (1.22) yields, after division by $\rho$,

$$\left[\frac{d}{dt} - \nu\left(\nabla^2 + \frac{\partial^2}{\partial z^2}\right)\right]\mathbf{U} = -f\mathbf{n} \times (\mathbf{U} - \mathbf{U}_g). \tag{1.29}$$

In analogy to (1.27) the right-hand side of (1.29) can be interpreted as an unbalanced Coriolis force. When $|(\mathbf{U} - \mathbf{U}_g)/\mathbf{U}_g| \ll 1$, the fluid is in quasi-geostrophic equilibrium.

In the special case of a fluid of limited depth, which moves with a vertically uniform velocity, (1.18) can be used to express (1.22) in the form

$$\left[\frac{\partial}{\partial t} + (\eta_z + f)\mathbf{n} \times \right]\mathbf{U} = -\nabla\left(\frac{1}{2}\mathbf{U} \cdot \mathbf{U} + \frac{p}{\rho}\right) + \nu\nabla^2\mathbf{U}.$$

It follows from vector calculus or from a cross-differentiation of the two component equations of this vector equation that its curl has the form

$$\frac{\partial}{\partial t}\eta_z + \mathbf{U} \cdot \nabla(\eta_z + f) + (\eta_z + f)\nabla \cdot \mathbf{U} = \nu\nabla^2\eta_z. \tag{1.30}$$

Forming the vertical average of (1.30) over a layer of depth $D$, one can use (1.7) and $\partial f/\partial t \equiv 0$, to derive the *vorticity equation*

$$\frac{d}{dt}(\hat{\eta}_z + f) - \frac{\hat{\eta}_z + f}{D}\frac{dD}{dt} = D\frac{d}{dt}\left(\frac{\hat{\eta}_z + f}{D}\right) = \nu\nabla^2\hat{\eta}_z. \tag{1.31}$$

Equation (1.31) indicates that the ratio of the vertically-averaged absolute vorticity $(\hat{\eta}_z + f)$, to the depth, $D$, of an individual fluid column remains constant, unless vorticity is diffused laterally into the column. The vorticity equation in this form is used widely, particularly by oceanographers.

### 1.2.3  *Conservation of energy*

Winds and currents are slow compared to the speed of sound. The kinetic energy of these bulk velocities is generally too small for their dynamic pressure (1.21) to produce significant density variations. This permits a separate treatment of the mechanical and thermal forms of energy in meteorology and oceanography. Conversion from one form into the other occurs mainly on a relatively large scale through the work of pressure during thermal expansion and on a very small scale through kinetic energy dissipation by viscosity.

A mechanical energy equation, which relates the local change of bulk kinetic energy per unit volume to the convergence of any kinetic energy flux, plus the dissipation and the work of pressure and external forces, can be derived analytically by scalar multiplication of (1.18) with the momentum vector $\rho U_i$. This eliminates the vortex and Coriolis accelerations which act at right angles to the velocity. After a simple transformation, one gets

$$\frac{\partial}{\partial t}\left(\frac{1}{2}\rho U_i^2\right) = -\frac{\partial}{\partial x_i}\left[U_i\left(\frac{1}{2}\rho U_j^2 + p\right) - U_j\sigma_{ij}\right] - \rho U_i\frac{\partial \Phi}{\partial x_i} + p\frac{\partial U_i}{\partial x_i} - \rho\epsilon. \quad (1.32)$$

The kinetic energy flux vector, expressed by the term in square brackets, has two parts. The first may be said to represent an energy flux produced by the total dynamic and static pressure. The second can be interpreted as a molecular flux of bulk kinetic energy. Viscosity transfers momentum and any momentum is always associated with kinetic energy. This viscous energy flux is relatively small, and it is usually neglected in air–sea interaction studies. The second term on the right-hand side of (1.32) represents the work of gravity during vertical displacements

$$\rho U_i\frac{\partial \Phi}{\partial x_i} = \rho Wg.$$

The third term represents the work of pressure during expansion, and the last term, the *dissipation rate per unit mass*, $\epsilon$, is derived from

$$\sigma_{ij}\frac{\partial U_j}{\partial x_i} = \sigma_{ij}\frac{\partial U_i}{\partial x_j} = \frac{1}{2}\rho v\left(\frac{\partial U_j}{\partial x_i} + \frac{\partial U_i}{\partial x_j}\right)^2 \equiv \rho\epsilon \quad (1.33)$$

where $\sigma_{ij}$ had been defined in (1.14).

At this stage it is convenient to introduce the concept of *available potential energy* (APE). Although the atmosphere and the oceans are stably stratified as a whole, we do find fluid masses of different densities at the same level. This causes the potential energy to be larger than it would be if the surfaces of equal density, $\rho_e$ were level. The available potential energy represents the excess over this theoretical minimum. Its specific or averaged value in a fixed closed volume, $V$, is given by

$$\text{APE} = \frac{1}{V}\int (\rho - \rho_e)gz \, dx \, dy \, dz. \quad (1.34)$$

The integration extends over the whole volume $V$.

A thermal energy balance must be based on the First Law of Thermodynamics.

$$T \, dS = c_p \, dT - \frac{1}{\rho} dp - \sum_n \mu_n \, dC_n, \tag{1.35}$$

where $S$ is the specific entropy; $c_p$ the specific heat at constant pressure; and $\mu_n$ is the specific chemical potential, which specifies the energy required to introduce a unit of mass of a new substance $n$ into the system.

The entropy of air or water can be changed only by molecular transport and by radiation. In adiabatic transformations [i.e., in processes during which the entropy remains constant ($dS = 0$) and the composition unchanged ($dC_n = 0$)], the First Law simply states that

$$\rho c_p \, dT = dp = c_s^2 \, d\rho, \tag{1.36}$$

where $c_s$ is the speed of sound as specified in Section 1.2.1. It follows from (1.4) and (1.26) that in a stratified fluid the buoyancy of an adiabatically displaced element will change at a rate

$$\frac{db'}{dt} = \frac{g}{\rho_r} \frac{d}{dt}(\bar{\rho} - \rho) = \left[ \frac{g}{\rho_r} \frac{\partial \bar{\rho}}{\partial z} + \left( \frac{g}{c_s^2} \right)^2 \right] W = -N^2 W. \tag{1.37}$$

The *Brunt–Vaisala frequency* or *buoyancy frequency*, $N$, which is defined by the last equality, corresponds to the frequency with which a displaced fluid element would oscillate about its equilibrium level in an inviscid fluid with upward density decrease ($\partial \bar{\rho}/\partial z < 0$). As indicated by (1.4) the length $c_s^2/g = D_s$ is much larger than the depth of the fluid layers with which we will be concerned here. The approximation

$$N^2 = -\frac{g}{\rho_r} \frac{\partial \bar{\rho}}{\partial z} \tag{1.38}$$

is therefore justified in the present context.

When the pressure remains constant, (1.35) can be rewritten to express the temperature change per unit mass and time by

$$c_p \frac{dT}{dt} = \frac{1}{\rho} \frac{\partial}{\partial x_i} \left( \rho c_p \kappa \frac{\partial T}{\partial x_i} \right) + \epsilon + \sum_n \mu_n \frac{dC_n}{dt} - \frac{\partial F_n}{\partial x_i}, \tag{1.39}$$

where $\kappa$ is the *thermal diffusivity*. The right-hand side of (1.39) represents the diabatic heating produced by the convergence of a molecular flux of sensible heat, by the frictional dissipation $\epsilon$, by phase changes, and by the convergence of a net radiation flux $F_n$.

In an incompressible fluid at rest, with no net absorption of radiation and without phase changes, (1.39) reduces to the classical equation of heat conduction

$$\frac{\partial T}{\partial t} = \kappa \frac{\partial^2 T}{\partial x_i^2}. \tag{1.40}$$

If the composition of a system varies in space or time, the molecular flux of sensible heat cannot be specified rigorously by the temperature gradient alone. In addition, it depends weakly on the concentration gradient. Equation (1.9) is affected similarly by a weak dependence of the diffusion flux $-\delta_n(\partial C_n/\partial x_i)$ on the temperature gradient. Relevant expressions can be found in Landau and Lifshitz (1959, p. 221). This coupling of the molecular fluxes plays an important role in the theory of irreversible processes. It probably has no practical effect on air–sea interactions, except possibly in the molecular sublayers of the interface itself, when the temperature difference between the sea and the air is large.

## 1.3  Turbulence and turbulent transport

The complexity of a turbulent flow is so formidable that even if we were able to describe its detailed structure, it would be impossible to comprehend. The study of turbulent flow is consequently directed toward statistical characteristics. We assume that the fluid motions can be separated into a slowly varying mean flow and a rapidly varying turbulent component. A discussion of the ramifications of this assumption can be found in Tennekes and Lumley (1972).

The averages of the variables will be taken over time, assuming that this will give a good representation of the *ensemble* average. The average is indicated by an overbar and the fluctuating component by a prime. Thus we have $T = \bar{T} + T'$; $q = \bar{q} + q'$, and so on. When there is no conflict with the notation it is possible to avoid the overbars and primes by writing the average in upper case and the fluctuating component, including waves, in lower case (e.g., $\tilde{u}_i = U_i + u_i$ and $\tilde{\theta} = \Theta + \theta$, where $\tilde{u}_i$ and $\tilde{\theta}$ represent the total values). We will use this notation from here on. In the presence of relatively large vertical displacements it is preferable to use the potential temperature $\theta$ instead of $T$, because of its invariant properties in a compressible medium (see subsection 2.2.4).

The averages of the deviations from the means are zero by definition:

$\overline{T'} = \overline{q'} = \overline{u_i} = 0$, and so on. This statement is based on the assumption that a mean exists that it is representative of the ensemble average and that it is relatively stationary in time. This is not always the case.

When the equation of continuity is averaged we get, with the specified notation,

$$\frac{\partial \bar{\rho}}{\partial t} + \frac{\partial}{\partial x_i} \overline{\rho \tilde{u}_i} = \frac{\partial \bar{\rho}}{\partial t} + \frac{\partial}{\partial x_i} \overline{(\bar{\rho} + \rho')(U_i + u_i)} = \frac{\partial \bar{\rho}}{\partial t} + \frac{\partial}{\partial x_i} (\bar{\rho} U_i + \overline{\rho' u_i}).$$

The last term represents the covariance of $\rho'$ and $u_i$. It can be different from zero only when there is a correlation between the density and velocity fluctuations. This happens in sound waves and, to a lesser extent, in convection when the density of rising fluid elements is systematically different from sinking elements. The relative amplitude of the density fluctuations $\rho'/\rho$ in the atmosphere or the sea is generally of order $10^{-2}$ or less. The order of the corresponding relative velocity fluctuations is usually not larger than unity. The averaged continuity equation reduces, therefore, with good approximation to the statement

$$\frac{\partial \bar{\rho}}{\partial t} + \frac{\partial}{\partial x_i} \bar{\rho} U_i \approx 0, \qquad (1.41)$$

which has the same structure as (1.2). Density fluctuations affect the geophysical motion only through buoyancy effects. This justifies use of the Boussinesq approximation, which neglects density fluctuations, except where they are multiplied by $g$ in the equations of motion.

A new element is introduced when we consider transient fluctuations of the fluid composition. The flux of a particular substance in the fluid mixture was specified by the expression in brackets of the right-hand side of (1.9). When this is averaged we get

$$\bar{\rho}\left( CU_i + \overline{cu_i} - \delta \frac{\partial C}{\partial x_i} \right).$$

It is customary in meterology and oceanography to speak of *advection* when we deal with a flux produced by the averaged velocities. The term *convection* is reserved for buoyancy-driven transport. In the free atmosphere and in the open ocean, the convective flux is much larger than the molecular diffusion. Like the latter, it can transfer matter when the mean mass flux $\rho U_i$ is zero or has a quite different direction.

Near the sea surface, vertical variations are much more pronounced then variations along the horizontal. If the latter are neglected and if $q$ is assumed to stand for the concentration of water vapour in the air (specific humidity) we can write the average of (1.9) in the form

$$\rho \frac{\partial \bar{q}}{\partial t} = -\frac{\partial}{\partial z} \left( \overline{q'w} - \delta \frac{\partial \bar{q}}{\partial z} \right) \equiv -\frac{\partial E}{\partial z}, \qquad (1.42)$$

where $E$ is the vertical transport of vapour mass per unit area and unit time.

When the vertical transport of sensible heat is treated in the same way, we get, in the absence of other heat sources,

$$c_p \rho \frac{\partial \bar{T}}{\partial t} = -c_p \frac{\partial}{\partial z} \rho \left( \overline{T'w} - \kappa \frac{\partial \bar{T}}{\partial z} \right) \equiv -\frac{\partial Q}{\partial z}. \qquad (1.43)$$

The momentum flux tensor was defined by the term in the brackets of (1.12). The contributions of the fluctuating velocities $\overline{\rho u_i u_j}$ to the momentum flux are known as the *Reynolds stress tensor*

$$\tau_{ij} \equiv -\overline{\rho u_i u_j}.$$

The vertical flux of horizontal momentum $\overline{\rho u w} = -\tau$ involves practically the entire frictional force that affects currents and winds in the boundary layers. To obtain an average of the second equation (1.22) we use (1.3) to represent it first in the momentum flux form (1.12). After minor rearrangement, and with neglect of the horizontal stress variations, averaging yields

$$\rho \frac{d\mathbf{U}}{dt} + \rho f \mathbf{n} \times \mathbf{U} = -\nabla p - \frac{\partial}{\partial z} \rho \left( \overline{uw} - v \frac{\partial \mathbf{U}}{\partial z} \right) \equiv -\nabla p + \frac{\partial \boldsymbol{\tau}}{\partial z}. \qquad (1.44)$$

Equation (1.44) is the equation of mean horizontal motion with the symbol $d/dt$ from here on denoting a differentiation following this mean motion.

An equation for the change of *turbulence kinetic energy* (TKE) can be obtained by expressing all physical variables in (1.32) as sums of their mean and fluctuating parts. When the results are averaged, we get an equation for the averaged local change of mechanical energy per unit mass. A second equation for the work done by the averaged velocities can be obtained by scalar multiplication of the averaged equations of motion with the mean veloctiy $U_i$. If this second equation is subtracted from the first, we get

$$\frac{1}{2}\frac{\partial}{\partial t}\overline{(u_j u_j)} = \overline{u_i u_j}\frac{\partial U_j}{\partial x_i} - \overline{g\rho' w}/\rho + \frac{\partial}{\partial x_i}\left[\frac{1}{2}U_i\overline{u_j u_j} + \overline{u_i\left(\frac{1}{2}u_j u_j + p'/\rho\right)}\right] - \epsilon.$$

(1.45)

This equation is known as the TKE equation, though the actual nature of the fluctuations has not been specified. In (1.45), it has been assumed that the molecular stress $\sigma_{ij}$ is negligibly small compared to the Reynolds stress. It also has been assumed that the dissipation acts only on the eddying motion and that its direct effect on the mean velocities is negligible:

$$\nu\overline{\left(\frac{\partial u_i}{\partial x_j} + \frac{\partial u_j}{\partial x_i}\right)\frac{\partial u_j}{\partial x_i}} \approx \epsilon, \qquad \nu\left(\frac{\partial U_i}{\partial x_j} + \frac{\partial U_j}{\partial x_i}\right)\frac{\partial U_j}{\partial x_i} \approx 0.$$

The mean velocities, of course, are affected indirectly through the formation of eddies that can be dissipated. If all eddy transports except those along the vertical are neglected, the TKE equation can be written as

$$\frac{1}{2}\frac{d}{dt}\overline{(u_j u_j)} = \overline{w\mathbf{u}}\cdot\frac{\partial \mathbf{U}}{\partial z} - \overline{g\rho' w}/\rho - \frac{\partial}{\partial z}\left(\overline{wp'}/\rho + \frac{1}{2}\overline{wu_j u_j}\right) - \epsilon. \qquad (1.46)$$

The first term on the right-hand side of (1.46) is the shear production term that is usually dominant near a boundary. The kinetic energy of a shear flow is always larger than the kinetic energy of a uniform flow with the same average momentum. Therefore, we may interpret the first term on the right-hand side of (1.46) as a generation of TKE by the reduction of the mean shear.

The term $-\overline{g\rho' w}/\rho$, which represents the work of the buoyancy force, can be either positive or negative depending on the APE. Approximating $\rho$ by $\rho_r$ and using (1.26), one can express this term in the form

$$-g\frac{\overline{\rho' w}}{\rho_r} = \overline{b' w}. \qquad (1.47)$$

An upward flux of buoyancy $(\overline{b' w} > 0)$ lowers the center of gravity of the fluid. The associated decrease of APE is compensated by an increase in TKE. The opposite happens when buoyancy is transported downward.

The energy flux associated with pressure fluctuations is indicated by the third term on the right-hand side of (1.46). The fourth term is a triple correlation, which may be interpreted as a transport of TKE that is carried by the turbulent eddies themselves. The last term represents the dissipation of TKE.

Equations similar to the TKE equation may be constructed for all second-order turbulent variables such as $\overline{u_i u_j}$ ($\overline{u_i u_i}$ is a special case of this), $\overline{\theta^2}$, $\overline{w\theta}$, $\overline{wq'}$, and so on. A more complete discussion of this subject matter can be found, e.g., in Busch (1973) and Businger (1982).

The difficulty of measuring turbulent vertical transports directly makes it desirable to express them parametrically as functions of the averaged or bulk quantities, which can be measured more easily. This may be accomplished by the introduction of eddy transfer coefficients, defined by the following relations

$$\overline{uw} = -K_m \frac{\partial U}{\partial z}, \qquad \overline{\theta w} = -K_h \frac{\partial \Theta}{\partial z}$$

$$\overline{q'w} = -K_e \frac{\partial \overline{q}}{\partial z},$$

$$(1.48)$$

where $K_m$ is the eddy transfer coefficient for momentum, or eddy viscosity, $K_h$ the eddy transfer coefficient for sensible heat, or eddy thermal diffusivity, $K_e$ the eddy transfer coefficient for water vapour, or eddy diffusivity for water vapour, and $\Theta$ the potential temperature as defined by (2.28).

The eddy coefficients have the same dimension as the molecular diffusivity or viscosity, but their value is generally very much larger and they are not intrinsic properties of the fluid. Unlike the molecular coefficients, they vary with the location, the state of the fluid, the stability and the averaging period. When the mean gradient becomes zero, the representation of transports by the expressions (1.48) involves infinitely large values of the eddy coefficients, which makes no practical sense. The concept of a parametric representation of transports by eddy coefficients and mean gradients is inapplicable under these circumstances.

The preceding discussion has been ambiguous on at least one important point: Nothing was said about the length of the averaging interval, which defines the local mean value of the temperature, velocity, and so on. This is particularly important when we try to combine a deterministic treatment of slow changes with a statistical parametric treatment of fast fluctuations. If the averaging interval is too long, then the averaged equations cannot serve for prediction. If it is too short, then the value of the means becomes biased by individual, short fluctuations.

In some cases, the choice of an averaging interval is obvious. For example, the sea-surface height varies over a period of hours with the tides and over a period of seconds with the passage of wind waves. An equation for tides can be based on records of the sea-surface height averaged over 20-minute intervals. This procedure is possible because wind waves and tides have such different frequencies and because they usually do not modulate each other to any significant extent. In general, the para-meterization of the high-frequency fluctuations by averaging is legitimate only if the averaging period can be associated with a 'gap' in the frequency spectrum (Lumley and Panofsky, 1964). A definition of such spectra will follow in the next section.

## 1.4 Statistical description of fluctuating quantities

In Section 1.3, averages and deviations from averages were introduced, as well as variances and covariances. Here we will summarize several standard statistical techniques, used to digest the data from these fluctuat-ing variables, and present a few applications of these techniques. More detailed information on the subject can be found in textbooks on statistics (e.g., Bendat and Piersol, 1986).

### 1.4.1 *Correlation functions and spectra*

In the following discussion the symbol $f$ is used for a continuously fluctuating variable, such as temperature. The time-averaged product of values of $f$ that are separated by a specified time interval, $\delta$, is the lag or *autocorrelation* of the record

$$R(\delta) = \overline{f(t)f(t + \delta)} = \lim_{T \to \infty} \frac{1}{T} \int_{-T/2}^{T/2} f(t)f(t + \delta) \, dt, \qquad (1.49)$$

where $T$ is the averaging time. Setting the time lag $\delta = 0$, we get the mean square value

$$R(0) = \overline{f^2}.$$

With the exception of some very special cases,

$$R(\infty) = (\bar{f})^2.$$

The *variance*, $V(f)$, is defined by the difference

$$V(f) = \overline{f^2} - (\bar{f})^2 = R(0) - R(\infty). \qquad (1.50)$$

In practice, averages often have to be computed over relatively short-time intervals. It is customary to consider such limited records as *stationary* if the averages over finite time intervals and the corresponding autocorrelations do not vary 'significantly' with the starting time $t$. Significance here implies that the variations in the averages are no larger than might be expected as a result of instrumental and statistical sampling variations.

If the values obtained by the operation (1.49) are identical at different locations, then the statistical distribution of the variable $f$ is said to be spatially homogeneous. Time and space averages are interchangeable if a process is both *statistically stationary* and *homogeneous*. Such processes are often called 'ergodic', though this term has a somewhat narrower meaning in statistical mechanics.

A stationary time series can be expressed symbolically by the Fourier series

$$f(t) = \sum_n A_n \exp(-i\omega_n t) = \sum_n A_n^* \exp(i\omega_n t), \qquad (1.51)$$

where $i = \sqrt{-1}$ and the symbols $\omega_n$ represent a numerable set of angular frequencies. The real part of the series indicates a wave field and the imaginary part represents the random time series generated by turbulence. If the length of the fundamental period is denoted by $T$,

$$\omega_n = \frac{2\pi n}{T} \qquad (1.52)$$

The complex conjugate amplitudes $A$ and $A^*$ can be obtained from the record of $f$ by the integral

$$A_n = \frac{1}{T} \int_{T/2}^{-T/2} f(t) \exp(+i\omega_n t)\, dt$$

$$A_n^* = \frac{1}{T} \int_{T/2}^{-T/2} f(t) \exp(-i\omega_n t)\, dt. \qquad (1.53)$$

For $\omega_0 = 0$, introduction of expression (1.51) into (1.49) yields, because of the periodicity and the orthogonality of the trigonometric functions,

$$\bar{f} = A_0 = A_0^* \qquad \overline{f^2} = R(0) = A_0^2 + \frac{1}{2} \sum_n A_n A_n^*. \qquad (1.54)$$

A plot of the spectrum of the amplitudes as a function of frequency consists in this case of a set of discrete normal lines, spaced at equal intervals $\omega_1$ along the $\omega$ axis. The line-spacing is inversely proportional to the duration of the fundamental period.

Infinitely long periods are in principle required for the exact specification of all nonconservative processes (i.e., processes in which the production of entropy plays some role). With $T \to \infty$, it follows from (1.52) that $\omega_1 \to 0$. In that case the series (1.51) has infinitesimal intervals between adjacent frequencies and the resulting amplitude spectrum is said to be continuous. Usually it is possible, with some not very stringent assumptions, to represent such a nonperiodic record by the *Fourier integral*

$$f(t) = \frac{1}{\sqrt{2\pi}} \int_{-\infty}^{\infty} a(\omega) \exp(-i\omega t)\, d\omega$$

$$= \frac{1}{\sqrt{2\pi}} \int_{-\infty}^{\infty} a^*(\omega) \exp(i\omega t)\, d\omega \tag{1.55}$$

with the inverse

$$a(\omega) = \frac{1}{\sqrt{2\pi}} \int_{-\infty}^{\infty} f(t) \exp(i\omega t)\, dt = a^*(-\omega). \tag{1.55'}$$

Expressions (1.51) or (1.55) can specify a particular record, but this is not what is needed in general. Our recording period is never infinite, and our observations are always subject to error. The same process will generate somewhat different records when it is observed on different occasions. When analyzed, these records will all yield different values for the coefficients $A_n$ or $a(\omega)$. Different starting times and the resulting phase shifts can easily cause a complete reversal of sign; arithmetic means are meaningless under these circumstances. We want a statistic that is common to all record samples. Such a statistic can be obtained from the mean square or from the variance of the coefficients that are drawn from different samples. The distribution of the products $AA^*$ or $aa^*$ as a function of frequency rather than of the set of amplitudes, provides the most meaningful way of summarizing the frequency characteristics of a record. These products are closely connected with the distribution among various frequencies of the energy of a process, or of its power, or intensity. In fact, it is this power spectrum that is seen when white light is decomposed experimentally into a continuous band of colours.

Power spectra describe a statistical property of data collections. They do not provide any information about the chronological history or phase in the realm of time. The mathematical tools and numerical methods used to decompose a record into its spectral components are described in many textbooks, and appropriate programs are available in most computing centers. Procedural details, therefore, need not occupy us here.

The *power spectral density*, $G(\omega)$, multiplied by $d\omega$ is a measure of the contribution made by perturbations with frequencies between $\omega$ and $\omega + d\omega$ to the mean square value $\overline{f^2}$. It can be defined by either

$$\lim_{\Delta\omega\to 0} \sum_{\omega}^{\omega+\Delta\omega} (AA^*)_n \to G(\omega)\, d\omega,$$

or

$$a(\omega)a^*(\omega) = G(\omega).$$

(1.56)

Although $G(\omega)$ can be computed directly from a Fourier expansion of the basic data, it is sometimes more convenient to derive first the autocorrelation function (1.49). For stationary processes, the two functions are related by the *Fourier transform*

$$G(\omega) = \int_{-\infty}^{\infty} R(\delta) \exp(-i\omega\delta)\, d\delta = 2\int_{0}^{\infty} R(\delta) \cos(\omega\delta)\, d\delta.$$

(1.57)

The second equality is explained by the fact that the autocorrelation of a stationary process is always an even function: $R(\delta) = R(-\delta)$.

The mean square value of $f$ is given by

$$\overline{f^2} = R(0) = \int_{-\infty}^{\infty} G(\omega)\, d\omega.$$

(1.58)

Transports are caused by the joint variations of two properties: temperature and velocity in the flux of heat, or two velocity components in the Reynolds stress. The *cross-correlation* of two functions $f(t)$ and $g(t)$ is specified in analogy to the definition of the autocorrelation in (1.49) by

$$R_{fg}(\delta) = \overline{f(t)g(t+\delta)} = \lim_{T\to\infty} \frac{1}{T}\int_{-T/2}^{T/2} f(t)g(t+\delta)\, dt.$$

(1.59)

The two functions $f$ and $g$ are said to be statistically uncorrelated if $R_{fg}(\delta) = \overline{f}\,\overline{g}$ for any time interval, $\delta$. The *covariance* of $f$ and $g$ is defined by

$$CV(f, g) = R_{fg}(0) - \overline{f}\,\overline{g}.$$

The cross-correlation, unlike the autocorrelation, does not necessarily have a maximum for $\delta = 0$, and it need not be an even function. When $R_{fg}(-\delta) = -R_{fg}(\delta)$ for all $\delta$ and when the covariance is zero, the functions $f$ and $g$ are said to be in *quadrature*.

In analogy to (1.57), the *cross-spectral density* of two time series can be obtained by the Fourier transform of the cross-correlation function. Because the cross-correlation is not necessarily even, this generally yields a complex number

$$\int R_{fg}(\delta) \exp\left(-i\omega\delta\right) d\delta = C_{fg}(\omega) - iQ_{fg}(\omega). \tag{1.60}$$

The real part $C_{fg}(\omega)$ is called the *co-spectral density* function or simply the *co-spectrum*; the imaginary part $iQ_{fg}(\omega)$ is the *quadrature spectrum*.

A useful technique, to test whether or not a time series is stationary, is the integration of the variance or covariance spectrum over wavelength or frequency, starting with small wavelengths. The resulting curve is called an *ogive*. An example is given in Fig. 1.1, where Oncley (1989) has calculated the co-spectrum and the corresponding ogive from the time series of $u$ and $w$. It is clear that in this case for periods $>200$ seconds the ogive curve levels off and the co-spectrum is confined to shorter wavelengths. This confirms the stationary character of the time series. It also specifies the time required to obtain a complete co-spectrum.

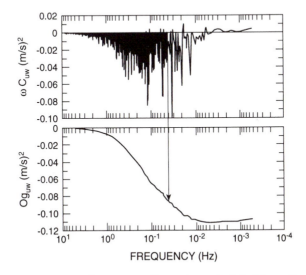

**Fig. 1.1.** The cospectrum and corresponding ogive calculated from time series of $u$ and $w$ measured on Oct. 12, 1987, from 07:16–07:48 MDT, at 4-m height near Carpenter, Wyoming. $U = 2.75\,\mathrm{m\,s^{-1}}$ and $L = 14.8\,\mathrm{m}$ [see (5.31)]. Courtesy S. Oncley.

When a property varies not only in time but also in space, $f = f(x_j, t)$, it can be represented by the multiple Fourier integral

$$f(x_j, t) = (2\pi)^{-2} \int a(k_j, \omega) \exp\left[i(k_j x_j - \omega t)\right] dl\, dm\, dn\, d\omega. \qquad (1.61)$$

The symbol $k_j$ denotes a radius vector in a wavenumber space. Its components, designated here by the letters $l$, $m$, $n$, are a measure of the number of times the pattern repeats itself in $2\pi$ units of distance in the three directions of geometric space. The integral extends over all time and over the whole three-dimensional wavenumber domain.

In this case the autocorrelation function is defined by

$$R(\xi_j, \delta) = \overline{[f(x_j, t)f(x_j + \xi_j, t + \delta)]}, \qquad (1.62)$$

where $\xi_j$ is a displacement vector and the average extends over the whole domain of space and time. A statistical specification of the function $f(x_j, t)$ by the four-dimensional wavenumber frequency spectrum $G(k_j, \omega)$ can then be obtained from the multiple Fourier transform of $R(\xi_j, \delta)$. In analogy to (1.58), we can also derive the mean square value of $f$ in space and time from the integral of $G(k_j, \omega)$ over the whole wavenumber domain and frequency domains.

For many purposes it is convenient to specify the wavenumber space with polar instead of Cartesian coordinates. This permits us to assess the contribution from wave components with different zenith angles $\theta$ and azimuths $\alpha$

$$G(k_j, \omega)\, dl\, dm\, dn = G(k, \theta, \alpha, \omega)k^2 \sin\theta\, dk\, d\theta\, d\alpha, \qquad (1.63)$$

or, in the two-dimensional case,

$$G(\mathbf{k}, \omega)\, d\mathbf{k} = G(\mathbf{k}, \alpha, \omega)\mathbf{k}\, dk\, d\alpha. \qquad (1.64)$$

The determination of the wavenumber-frequency spectrum $G(k_j, \omega)$ requires continuous observations in space and time, which are very rarely available. Methods to obtain wavenumber spectra with a limited array of discreetly spaced sensors have been discussed by Barber (1963). The need for comprehensive measurements is reduced when there is a known dynamic or kinematic relationship between the various wave numbers and frequencies. Fortunately, this is rather common. Two cases are of particular interest here: waves and isotropic turbulence.

Waves or wave fields are characterized by a physical relationship between the wavenumber, the frequency, and one of several external parameters. The existance of such a 'dispersion relation' generally permits us to omit without loss of information either the magnitude of the wavenumber vector or the frequency from our statistical description.

### 1.4.2 *Isotropic turbulence*

*Isotropic turbulence* is a state of complete randomness in which the statistical feature of the motion exhibits no preference for any direction. This symmetry permits us to specify the wavenumber spectrum as the function of a single parameter. A relation with locally observable frequencies can be established if the fluctuations are imbedded in some mean wind or current and if they retain their individual character while being swept past a fixed observation point. With $U$ denoting the velocity of the advective stream and $l$ a streamwise wavenumber, we have

$$l = \frac{\omega}{U},\tag{1.65}$$

where $\omega$ is a locally established cyclical frequency. The hypothesis of a fluctuation field, that remains 'frozen' over periods that are long in comparison to the advection time scale $L/U = 2\pi/lU$, is known by its originator as *Taylor's hypothesis.*

A detailed review of isotropic turbulence theory has been published by Hinze (1975). The topic is also discussed by Phillips (1977). Even when the large-scale flow is anisotropic, it is usually permissible to apply the results of isotropic theory to those features of the actual flow that are determined by the fine-scale structure. When this is done, we speak of *local isotropy*; the adjective 'local' here being applicable to wavenumber space rather than to the space of our senses.

There is no correlation between the different velocity components in isotropic turbulence. Therefore it cannot contribute to the momentum flux. The turbulent energy, since it is partitioned equally between the three directions of space, can be expressed as the integral of a spectrum function $G(k)$ which depends only on the length of the radius vector in the wavenumber domain

$$\frac{1}{2}\overline{u_i^2} = \frac{1}{2}q^2 = \int_0^\infty G(k)\,dk\tag{1.66}$$

where $q$ is the root–mean–square turbulent velocity. The dissipation associated with isotropic turbulence is

$$\epsilon = 2\nu \int_0^\infty k^2 G(k)\, dk. \tag{1.67}$$

This follows immediately from (1.33), if all the velocity components are represented by a set of harmonics with wavenumber $k$.

The three-dimensional wavenumber spectrum $G(k)$ can be obtained from simultaneous observations at a large number—ideally an infinitely large number—of spatially distributed sensors. Alternatively, under favourable conditions with an overall advection velocity, we can use Taylor's hypothesis (1.65) to derive $G(k_1)$ from a streamwise or *downstream power spectrum* $G_1(k_1)$. This downstream spectrum, which can be obtained from observations at a single point, can be interpreted geometrically as a 'one-dimensional cut' of $G(k)$. In locally isotropic motion, the relation between the two spectra is (Hinze, 1975, p. 208)

$$2G(k_1) = k_1^2 \frac{\partial^2 G_1}{\partial k_1^2} - k_1 \frac{\partial G_1}{\partial k_1}.$$

Introduction of this term into (1.67) gives

$$\epsilon = 15\nu \int_0^\infty k_1^2 G_1(k_1)\, dk_1. \tag{1.68}$$

Furthermore, the relation between the transverse spectra $G_2$ and $G_3$ and the downstream spectrum is given by

$$G_2(k_1) = G_3(k_1) = \frac{1}{2}\left[ G_1(k_1) - k_1 \frac{\partial G_1(k_1)}{\partial k_1}\right].$$

In (1.67) and (1.68) the spectrum functions are weighted by the square of the wavenumber. This indicates that the dissipation will be associated mainly with the higher harmonics, or in a physical sense with the smaller eddies, which justifies the statement after (1.45) that its direct effect on

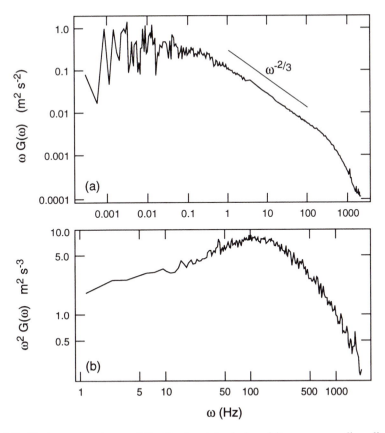

**Fig. 1.2.** Variance spectrum of the horizontal wind and its corresponding dissipation spectrum measured with a hot wire anemometer. Using Taylor's hypothesis $G(k)$ is transformed to $G(\omega)$ in (1.66) and (1.67) because the observations provide frequencies rather than wavenumbers. The observations were made near Carpenter, Wyoming, during the fall of 1990. $U = 10 \, \text{m s}^{-1}$; $u_* = 0.62 \, \text{m s}^{-1}$; and $z/L = -0.01$. Courtesy S. Oncley.

mean winds and currents is negligible. The separation between the part of the spectrum in which most of the energy resides and the high-wavenumber region where dissipation occurs is illustrated by Fig. 1.2.

## 1.5 Scaling techniques and similarity relations

For many practical applications it is useful to formulate the equations in nondimensional form. If this is done correctly it reduces the number of variables in the equations and it allows us to evaluate the relative magnitude of the terms. It also allows us to develop similarity relations for geophysical processes.

One technique is to take one of the governing or derived equations and divide it by one of its terms. The result is an equation in nondimensional form. If we then choose appropriate scales for the variables, we automatically obtain nondimensional groups that represent similarity characteristics for the flow. This is sometimes straightforward, but it often requires a careful selection of possible scales.

In the following discussion we will apply these rules to several equations. Instead of doing this systematically, we shall select a few examples that are useful for the following chapters. Scaling parameters will also be introduced in later chapters, when appropriate.

Dimensional scaling factors for the equations of motion may be introduced by

$$x, y = (x^*, y^*)L,$$

$$z = z^*D,$$

$$t = t^*\omega^{-1}, \tag{1.69}$$

$$U, V, \mathbf{U} = (U^*, V^*, \mathbf{U}^*)U,$$

$$W = W^*UDL^{-1},$$

where $L$ and $D$ are horizontal and vertical length-scales, respectively, $\omega$ is a characteristic frequency, and $U$ is a velocity scale. The quantity $D/L$ is known as the *aspect ratio*. The starred variables are nondimensional and of order unity or smaller. Starred symbols will also be used later to denote vector operations with nondimensional coordinates.

When $d/dt$ is written in its explicit form in (1.27), we get, after introduction of the new variables from (1.69) and division by $-b'$,

$$\frac{D}{L}\left[\frac{\omega U}{b'}\frac{\partial}{\partial t^*} + \frac{U^2}{Lb'}\left(\mathbf{U}^* \cdot \nabla^* + W^*\frac{\partial}{\partial z^*}\right) - \frac{\nu U}{D^2 b'}\left(\frac{D^2}{L^2}\nabla^{*2} + \frac{\partial}{\partial z^{*2}}\right)\right]W^*$$

$$-\frac{f_c U}{b'}U^* = \left(1 + \frac{1}{\rho_r b'}\frac{\partial p'}{\partial z}\right). \tag{1.70}$$

Here $b'$ is to be interpreted as $b' = -g\rho^{-1}\Delta\rho$ or $b' = g\Theta^{-1}\Delta\Theta$, where $\Theta$ is the potential temperature as defined in (2.28); $\Delta\rho$ and $\Delta\Theta$ are intervals of density and potential temperature relevant to the scales under consideration. It can be noted that $f_c U/b'$ is invariably very small. Even on the equator, where $f_c$ has a maximum, it would require a windspeed of about $70\,\mathrm{m\,s^{-1}}$ or $140$ knots to balance a reduced gravitational acceleration as low as $g/1000$. The term in square brackets on the left-hand side of (1.70) is multiplied by the aspect ratio, $D/L$. The vertical thickness of the atmosphere or the oceans is much smaller than their horizontal extent. They are relatively thin fluid shells. The aspect ratio, therefore, is very small in any motion pattern that involves large horizontal displacements. In particular, in the atmospheric cyclones and anticyclones of our weather charts or in large-scale oceanic circulations, $D/L$ is about $10^{-3}$ or at most $10^{-2}$. The right-hand side of (1.70) which represents the relative departure from hydrostatic equilibrium, is then nearly zero, which means that the motion in these large-scale circulations is very nearly hydrostatic.

When the aspect ratio approaches unity, the departure from hydrostatic equilibrium is determined by the relative magnitude of the terms in square brackets of (1.70). These terms involve three time scales. In addition to the characteristic period of local transients, supposed to be of order $\omega^{-1}$, we have an advective time scale, $L/U$, which measures approximately the time needed for a motion pattern of horizontal scale $L$ to be advected past a fixed observation point. The time needed for viscous effects to penetrate a depth, $D$, is of order $D^2/\nu$ and defines a viscous time scale.

The motion is quasi-steady when the period of local transients is large compared to the two other time scales. Whether viscous or inertial forces predominate is then indicated by the ratio

$$\frac{U}{L} \div \frac{\nu}{D^2} = \frac{UL}{\nu}\left(\frac{D}{L}\right)^2. \tag{1.71}$$

where

$$\frac{UL}{\nu} = Re, \tag{1.72}$$

the *Reynolds number*. Viscosity cannot affect the motion significantly if that ratio is at least one order larger than unity. In small-scale isotropic motion pattern $D/L \sim 1$. Viscous forces can thus be neglected compared to inertial forces if $Re \gg 1$. On the other hand, we see that, if the motion pattern is anisotropic with a small aspect ratio (i.e., if the motion is more or less horizontal), then molecular viscosity may begin to play a significant role at much higher Reynolds numbers.

If the characteristic time scale of the viscous effects is much shorter than either of the two other time scales, then we can only have steady laminar motion. Friction and buoyancy then balance each other. This happens, for example, in the fall of cloud drops or in the rise of small bubbles through the sea. The aspect ratio is unity for a sphere, and the balance of forces is indicated by the dimension-less factor in front of the third bracketed term in (1.70). It can be seen that the appropriate velocity scale

$$U \propto \frac{b'D^2}{\nu}.$$                                           (1.73)

With $U$ the terminal velocity and $D$ the radius of the spheres, the proportionality factor was computed by Stokes more than a century ago to be equal to 2/9.

The dimensionless factor in front of the second bracketed term in (1.70) is known as the *Froude number*

$$Fr \equiv -\frac{U^2}{Lb'}.$$                                          (1.74)

It measures the ratio of inertial to gravitational accelerations. These can balance each other, for example, in the flow around large raindrops that have reached their terminal velocity. In this case $Fr = 1$. Froude numbers of order unity are also approached in gravity waves just before they break, or in violent cumulus convection.

When the perturbation time-scale is short compared to both the advective and the viscous time scale, and when the resulting accelerations are equal to the buoyancy, we have, from the first bracketed term of (1.70),

$$-\frac{b'}{\omega} = U \equiv c,$$

where $c$ is the phase speed of small amplitude gravity waves at the boundary of two infinitely deep inviscid fluids.

Equation (1.29) can be treated in a similar way. When $d/dt$ is again written in its explicit form and nondimensional starred variables are introduced from (1.69), we get, after division by the Coriolis acceleration $fU$,

$$\left[ \frac{\omega}{f}\frac{\partial}{\partial t^*} + \frac{U}{Lf}\left( \mathbf{U}^* \cdot \nabla^* + W^*\frac{\partial}{\partial z} \right) - \frac{\nu}{D^2 f}\left( \frac{D^2}{L^2}\nabla^{*2} + \frac{\partial}{\partial z^{*2}} \right) \right]\mathbf{U}^*$$

$$= (\mathbf{U}^* - \mathbf{U}_g^*) \times \mathbf{n}. \quad (1.75)$$

It is convenient to choose the magnitude of the geostrophic wind as a scale velocity ($U = U_g$). The nondimensional unbalanced pressure force on the right side of (1.75) varies then usually between zero and unity

$$0 < \frac{|\mathbf{n} \times (\mathbf{U} - \mathbf{U}_g)|}{U_g} = |\mathbf{n} \times (\mathbf{U}^* - \mathbf{U}_g^*)| < 1.$$

It is zero when the motion is geostrophic. It does not usually exceed unity because it is very rare for the actual wind to be either opposed to or to be twice as large as the geostrophic wind. Any finite value has to be balanced by one or several of the scaling factors on the left-hand side of (1.75). The Coriolis force has no significant influence on the motion pattern when any of these factors is larger than unity.

The ratio of the inertial to the Coriolis force, as measured by the scaling factor of the second bracketed term in (1.75), is known as the *Rossby number*,

$$Ro \equiv \frac{U}{fL}. \tag{1.76}$$

When multiplied by $2\pi$, this is equal to the ratio of the half-pendulum day to the advective time-scale. When $Ro$ is small, the motion is approximately geostrophic. When it becomes comparable in magnitude to the right-hand term in (1.75), with the other scaling factors remaining comparatively small, we have steady, nonviscous motion in which pressure, Coriolis, and inertial acceleration balance each other. The resultant velocity is known as the *gradient wind*. Such a balance may be approached, for example, in a stationary low pressure system above the ground friction layer.

The ratio of the inertial to the frictional forces is the same in the horizontal as along the vertical and is given by (1.71). The scaling factor of the last bracketed term in (1.75) measures the relative magnitude of the frictional to the Coriolis force. This ratio is known as the *Ekman number*, *Ek*,

$$\frac{v}{D^2 f} = \frac{Ro}{Re} \left(\frac{D}{L}\right)^2 = Ek. \tag{1.77}$$

It is seen again from (1.77) that friction may affect motion even at relatively large Reynolds numbers, when the vertical scale, $D$, is small. This is the case, for example, near a horizontal boundary where major vertical displacements are inevitably suppressed. The layer where this happens, and where friction partly balances the pressure and Coriolis force, is known as the *Ekman layer.*

Turning now to the first term in (1.75), we can consider the following cases. When $\omega \sim 0$, we have approximately steady motion as described in the previous two paragraphs. A perturbation time scale that is of the same order as the advective time scale, $\omega^{-1} \sim L/U$, is typical for synoptic scale perturbations in the atmosphere. These perturbations are characterized by $Ro \ll 1$ and $Re \gg 1$. They cannot cause large departures from geostrophic equilibrium. When $\omega = f$, we have the case of the so-called Coriolis or inertial oscillations with periods equal to half a Foucault pendulum day.

Perturbations with an even shorter time scale or higher frequency $\omega > f$ cannot be balanced by the horizontal pressure gradient and by the Coriolis force alone. From (1.70) it can be seen that they cannot be balanced by buoyancy either when

$$\frac{U\omega}{b'} > 1.$$

The external forcing functions become unimportant in this case and we must have an internal balance between the terms on the left-hand side of (1.75). It will be seen below that this is typical for the inertial and dissipation range of small-scale turbulence.

Next we turn our attention to the conservation of energy as expressed by (1.39). When $d/dt$ is written in its explicit form and the ratio of the advection term and the thermal diffusion term is taken, the scaling leads us to the dimensionless ratio

$$\frac{UL}{\kappa} \equiv Pe, \tag{1.78}$$

which is known as the *Peclet number.* This number is similar to the Reynolds number; the kinematic viscosity, $\nu$, has been replaced by the thermal diffusivity, $\kappa$. Molecular diffusion of heat is negligible when $Pe \gg 1$. The thermal diffusivity of air near the surface varies from $0.189 \, \text{cm}^2 \, \text{s}^{-1}$ at $0°C$ to $0.284 \, \text{cm}^2 \, \text{s}^{-1}$ at $30°C$, and of water at atmospheric pressure from $0.0013$ to $0.0015 \, \text{cm}^2 \, \text{s}^{-1}$ over the same temperature range. It follows that, except for extremely small-scale phenomena with low velocities, the Peclet number in nature is always large.

When more than one equation is needed to describe a phenomenon, the scaling becomes more complex. There is some similarity between the conservation equations if the kinematic viscosity and the diffusivities are of the same order of magnitude. In the case of the equations of motion and conservation of energy this means that $Re$ and $Pe$ are of the same order of magnitude. The ratio

$$\frac{Pe}{Re} = \frac{v}{\kappa} \equiv Pr \tag{1.79}$$

is the *Prandtl number,* which is an important similarity parameter.

When $Pr = 1$, the viscous time scale is equal to the time scale of thermal diffusion, $D^2/\kappa$, and similarity exists between viscous dissipation and thermal diffusion. For air, $Pr \simeq 0.7$ and for water $Pr \simeq 13$ at 0°C and $\simeq 7$ at 20°C.

Similarly, in the case of the equations of motion and the conservation of matter the similarity parameter is

$$\frac{v}{\delta} \equiv Sc, \tag{1.80}$$

the *Schmidt number.* In sea water $Sc$ for salinity varies from 2500 to 800 for temperatures between 0 and 20°C. There is clearly very little similarity between the diffusion of sea salt and the viscous effects on the fluid.

A good example of the interaction of the equations of motion and conservation of energy is the onset of convection in an unstably stratified fluid, where the density difference is caused by a temperature difference. The buoyancy force, $b'$, is being diminished by thermal diffusion and counteracted by viscous force. The interplay results in a dimensionless number, the *Rayleigh number*

$$Ra = \frac{b'D^3}{\kappa v}, \tag{1.81}$$

where $D$ is the depth of the unstable layer. When $Ra$ is below the critical value—1708—convection is suppressed (Chandraseckar, 1961). When $Ra$ increases above the critical value, laminar convection cells may develop. Further increase of $Ra$ produce irregular chaotic motions, and finally fully developed turbulence. In the convective boundary layer $Ra$ is frequently very large, and fully developed turbulence exists. Yet, coherent structures that resemble laminar flow patterns, can be found often even in this intensely turbulent environment.

Convection may also be caused by density differences due to variations in salinity. In this case, the buoyancy force is counteracted by diffusion of salt and viscosity. Therefore the thermal diffusivity $\kappa$ in (1.81) should be replaced by the diffusivity $\delta$. This may be accomplished by replacing $Ra$ by $Ra\,Pr^{-1}\,Sc$.

In conclusion, we present a similarity argument related to the various scales in a fully turbulent flow. When $Re$ is very large ($10^6$–$10^9$), as is usually the case in the boundary layers on both sides of the air–sea interface, there is a large range of eddy sizes. The large eddies in the turbulent flow have their own large Reynolds number, and viscosity plays an insignificant role in their evolution. Furthermore, it is clear from the equation of turbulence kinetic energy (1.46) that both the shear production term and the buoyancy term mainly produce large turbulent eddies. On the other hand, dissipation of turbulence becomes important when the Reynold's number of the eddies is of the order 1. Therefore, there must be a range of eddy sizes where neither the input nor the dissipation of energy plays a significant role, but where energy cascades from larger to smaller eddies. In the spectrum of turbulence this range is called the *inertial subrange*. This line of reasoning led Kolmogorov (1941) to a famous similarity argument.

If $k_0$ and $k_d$ are the characteristic wavenumbers of the energy containing eddies and the energy dissipating eddies, respectively, then the inertial subrange for $k$ is given by

$$k_0 \ll k \ll k_d. \tag{1.82}$$

In this range the three-dimensional variance spectrum, $G(k)$, of turbulence kinetic energy is entirely determined by the rate of dissipation, $\epsilon$, and the wavenumber, $k$. This leads on dimensional grounds to

$$G(k) = \alpha\epsilon^{2/3}k^{-5/3}. \tag{1.83}$$

Similarly, the one-dimensional longitudinal or downstream spectrum may be written as

$$G_1(k_1) = \alpha_1\epsilon^{2/3}k_1^{-5/3}. \tag{1.84}$$

In (1.83) and (1.84), $\alpha$ and $\alpha_1$ are constants of proportionality, the *Kolmogorov constants*. The last spectrum is most easily obtained from measurements. This means that, once the constant $\alpha_1$ is known, it is possible to determine $\epsilon$ from the spectrum in the inertial subrange. In order to determine $\alpha_1$, it is necessary to determine $\epsilon$ with (1.68), which requires the total spectrum, especially the high frequency end or dissipation range, to be measured. Such measurements are difficult and the results vary. Champagne et al. (1977) report $\alpha_1 = 0.5$, whereas Williams and Paulson (1977) find $\alpha_1 = 0.54$. Recent measurements by Fairall et al. (1990) resulted in $\alpha_1 = 0.59$, which is in agreement with the value reported by Dyer and Hicks (1982). The range of uncertainty is still rather large and further data analysis is needed.

# 2

# THE STATE OF MATTER
# NEAR THE INTERFACE

A thermodynamic phase specifies a substance or mixture of substances that occupies a limited volume with a characteristic temperature, $T$, a definite pressure, $p$, at its boundary, and a composition that can be described at any moment by the masses of its constituents. A well-mixed sample of air or of sea water in contact with our instruments, therefore, is defined as a phase. Different phases, brought together in a thermodynamic system, tend to change until an equilibrium has been established. This involves an exchange of energy and matter between the originally different phases. If the process is isolated from external influences, then it results in an increase of entropy. In fact, entropy is always generated by mixing (i.e., by the transport of a conservative property down its own gradient). This can be said to include the transport of momentum by viscosity. Entropy can be diminished locally when gradients are sharpened by a flux in the opposite direction. In nature this can be brought about by a coupling of the transports of different properties.

The potential for change that exists in a physical system is the subject of the thermodynamics of irreversible processes. The ocean and the atmosphere are open systems which can exchange matter through their boundaries. Though the system that they form together is closed—at least on the time scales with which we are concerned—it is not isolated, being subject to heat exchange with the surrounding universe. The inhomogeneity and variable temperature of both oceans and atmospheres cause further complications. Although equilibrium thermodynamics alone cannot provide information about the rate or the mechanism of energy transformations, it does provide, in the First Law, a constraint that must always be obeyed. It also allows us to predict the general direction of development in limited regions that only interact slowly with their surroundings.

## 2.1  Sea water

### 2.1.1  *The equation of state*

The properties or 'coordinates' which specify a thermodynamic phase are not independent. Anyone of them can be expressed as a function of all the others by an equation of state. In a liquid, the state is strongly affected by molecular interaction. Pure water generally is not a random aggregation of separate $H_2O$ molecules but consists partly of small groups that hang together. This polymerization increases with decreasing temperature.

   In the real ocean, the density of sea water can be affected by plankton, organic debris, dust, and especially bubbles near the surface. If these suspensions are disregarded, one can express the equation of state in the canonical form $v \equiv v(T, s, p)$; where $v$ denotes a specific volume of uncontaminated sea water, $T$ the temperature, $s$ the salinity and $p$ the pressure. Instead of $v$ it is often convenient to use the density $\rho \equiv 1/v$ as state variable.

   In SI units, $\rho$ is specified in kilograms per cubic meter $(kg\,m^{-3})$ and $p$ in Pascal $(1\,Pa = 1\,kg\,m^{-1}\,s^{-2} = 10^{-5}\,bars)$. A pressure of one bar is approximately equivalent to the atmospheric surface pressure or to the pressure exerted by a 10-m water column. Oceanographers usually specify the temperature, $T$, in degrees Celsius $(°C)$ and the salinity, $s$, in grams of dissolved material per kilogram of sea water [i.e., in parts per thousands (ppt) or in almost identical 'practical salinity units' (psu)]. Sea water in the world ocean is characterized by a salinity range of 33–37 ppt and a surface density range of 1022–1028 $kg\,m^{-3}$. The narrow width of this range makes it convenient to express the numerical value of the density more briefly by

$$\rho - 1000 \equiv \sigma(T, s, p). \tag{2.1}$$

In many older oceanographic textbooks, $\sigma$ is defined in nondimensional terms by

$$\sigma = 1000\left(\frac{\rho}{\rho_m} - 1\right). \tag{2.2}$$

As the maximum density of pure water $\rho_m = 999.975\,kg\,m^{-3}$, these two definitions differ numerically by a small amount, which is insignificant in the present context.

   The equation of state for a gas, like air, can be derived analytically from the principles of statistical mechanics. This is not possible for liquids. For sea water at standard atmospheric surface pressure $(p = p_0 = 1013.2\,mb = 101{,}320\,Pa)$, Millero et al. (1976) have derived explicit empirical equations for the evaluation of (2.1). A contour plot of $\sigma$, derived

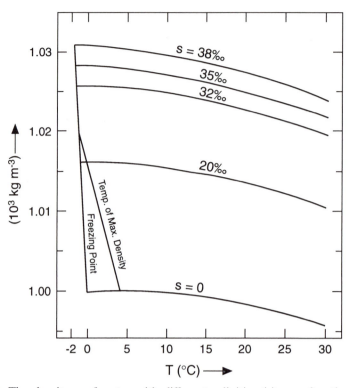

**Fig. 2.1.** The density $\rho$ of water with different salinities ($s$) as a function of the temperature $T$.

from these values, is shown in Fig. 2.1. The figure illustrates that, near freezing point, the density of sea water becomes relatively insensitive to temperatue variations. In that case, small salinity differences can play a major dynamic role.

Coefficients of *thermal expansion* and *saline contraction,* as defined by

$$\alpha = -\rho^{-1}\left(\frac{\partial \rho}{\partial T}\right) \quad \text{and} \quad \beta = \rho^{-1}\left(\frac{\partial \rho}{\partial s}\right), \tag{2.3}$$

are listed in Tables 2.1 and 2.2.

**Table 2.1.** Coefficients of Thermal Expansion ($10^6/°C$) of Sea Water at Atmospheric Pressure

| Temperature (°C) | −2 | 0 | 5 | 10 | 15 | 20 | 25 | 30 |
|---|---|---|---|---|---|---|---|---|
| Fresh water | −106 | −68 | 16 | 88 | 151 | 206 | 256 | 302 |
| Salt water (35 ppt) | 23 | 52 | 114 | 167 | 214 | 256 | 296 | 332 |

**Table 2.2.** Coefficients of Saline Contraction ($10^6$ ppt) of Sea Water at Atmospheric Pressure

| Temperature (°C) | −2 | 0 | 5 | 10 | 15 | 20 | 25 | 30 |
|---|---|---|---|---|---|---|---|---|
| ($s = 35$ ppt) | 795 | 788 | 774 | 762 | 752 | 744 | 737 | 732 |

Useful linear approximations for mixing studies at temperatures above 5°C and atmospheric pressure, have the form

$$\alpha \approx 77.5 + 8.70T \quad \text{and} \quad \beta \approx 779.1 - 1.66T. \qquad (2.4)$$

The error resulting from these approximations is smaller than 5 per cent and hence smaller than that likely to be caused by inaccurate data and inexact parameterizations of sea surface fluxes.

If the temperature is kept invariant, realistic sea surface pressure and salinity changes would cause the value of the expansion coefficients to vary by less than $10^{-5}$. These changes are therefore negligible for most air–sea interaction studies. However, the effect of pressure changes cannot be neglected during deep convection or any large vertical displacements. The resulting problems have been studied and summarized by McDougall (1987). Empirical expressions for the thermal expansion coefficient as a function of pressure have been published by Bryden (1973).

### 2.1.2   Latent heat and saturation vapour pressure of pure water

Geophysicists are concerned with water in both its liquid and in its gaseous and solid phases. The behavior of water vapour near the sea surface approximates that of an ideal gas with an equation of state

$$e = \frac{\rho_v R_1 T}{m_v} \equiv \rho_v R_v T. \qquad (2.5)$$

The symbol $e$ in meteorology traditionally represents the pressure of water vapour, $\rho_v$ is the vapour density, and $T$ is here the absolute temperature (K). The universal gas constant $R_1 = 8.31436 \text{ m}^2 \text{ s}^{-2} \text{ K}^{-1}$ and the gas constant for water vapour $R_v$ equals $R_1$ divided by the molecular mass of water $m_v = 18.02$. A liquid and its vapour are in equilibrium when the average number of molecules that move in either direction across their interface is equal. The saturation vapour pressure over water, $e_s$, is defined as the pressure of a vapour in equilibrium with a plane water surface. The vapour pressure over ice is similarly defined.

The saturation vapour pressure increases with temperature. The *Clausius–Clapeyron equation*

$$L = T\left(\frac{1}{\rho_v} - \frac{1}{\rho_w}\right)\frac{de_s}{dT} \approx \frac{T}{\rho_v}\frac{de_s}{dT} = \frac{R_v T^2}{e_s}\frac{de_s}{dT} \tag{2.6}$$

relates the *latent heat of evaporation, L,* to the amount of work required to expand a unit mass of water substance against pressure from the specific volume of liquid water, $1/\rho_w$, into the volume of the gaseous phase, $1/\rho_v$. Equation (2.6) can be evaluated if one knows the variations of the latent heat with temperature. To compute this, one considers that the specific heat content or *enthalpy, $\xi_v$,* of a vapour is equal to the sensible heat of the liquid before evaporation, plus the latent heat at the temperature of evaporation, plus the change in sensible heat which occurs when the vapour is heated to its actual temperature. The actually observable value of the vapour enthalpy does not provide information about the temperature at which the evaporation occurred and therefore cannot be affected by it. If $T$ is the actual temperature and $T_0$ a suitable lower reference temperature, it follows that

$$\xi_v = cT_0 + L(T_0) + c_{pv}(T - T_0) = cT + L(T). \tag{2.7}$$

The specific heat of pure liquid water $(c = 1.0 \text{ cal g}^{-1} \text{ K}^{-1} = 4187 \text{ J kg}^{-1} \text{ K}^{-1})$ and the specific heat of water vapour at constant pressure $(c_{pv} = 4R_v = 1846 \text{ J kg}^{-1} \text{ K}^{-1})$ can be considered constant in meteorological and oceanographic conditions. Setting $T_0 = 0°C = 273.16 \text{ K}$ one gets

$$L(T) = L(T_0) - \frac{c - c_{pv}}{T - T_0} = 2.500895 - 2344(T - T_0) \quad (\text{J kg}^{-1}). \tag{2.8}$$

With this expression for $L(T)$ (2.6) can be integrated to give the saturation vapour pressure as a function of temperature. The listing in the *Smithsonian Meteorological Tables* (1971) is based on such a computation, but it also takes the rather insignificant deviations from perfect gas behavior into account. If the temperature dependence of latent heat is neglected $(L(T) = L(T_0))$, then integration of (2.6) yields

$$e_s(T) \approx e_s(T_0) \exp\left[\frac{L(T_0)}{R_v T_0}\left(1 - \frac{T_0}{T}\right)\right]$$

$$= 610.8 \exp\left[19.85\left(1 - \frac{T_0}{T}\right)\right] \quad \text{(Pa)}. \quad (2.9)$$

At temperatures below 30°C, saturation vapour pressure values derived from this equation differ by less than 2 per cent from the exact figure.

The latent heat of sublimation is very nearly constant

$$L_i = 2.835 \times 10^6 \quad (\text{J kg}^{-1}) \tag{2.10}$$

and the saturation vapour pressure over pure ice, $e_{si}$, can be expressed with very good approximation by

$$e_{si} = 610.8 \exp\left[22.47\left(1 - \frac{T_0}{T}\right)\right] \quad \text{(Pa)}. \tag{2.11}$$

The latent heat of fusion is given simply by the difference between (2.10) and (2.8).

When water is in contact with a saturated vapour, additional evaporation requires an 'external' heat supply. When the vapour is not saturated, this heat can be provided partly by cooling (i.e., sensible heat reduction of the evaporating liquid). The rate at which this cooling occurs depends on the tendency of water molecules to pass spontaneously from the liquid into the vapour phase and vice versa. This is measured by the chemical potential difference between the two phases. The chemical potential of the $n$th constituent in a mixture of ideal gases is

$$\mu_n = \mu_n^0 + R_n T \ln(p_n), \tag{2.12}$$

where $\mu_n^0$ is a function of the temperature only; $p_n$ is the partial pressure; $R_n = R_1/m_n$ the gas constant of gas $n$; and $m_n$ the molecular mass. When pure water is in equilibrium with water vapour at the same temperature, molecules in either phase have the same tendency to pass across the interface into the other phase. This means that the chemical potentials of the two phases must be equal. It follows that the chemical potential, $\mu_v$, of an unsaturated vapour with partial pressure, $e$, differs from that of a saturated vapour, $\mu_{sv}$, or from that of liquid water, $\mu_w$, at the same temperature by

$$\mu_v - \mu_{sv} = \mu_v - \mu_w = R_v T \ln\left(\frac{e}{e_s}\right). \tag{2.13}$$

The ratio $e/e_s$ is the relative humidity. In an unsaturated vapour, the difference $\mu_v - \mu_w$, known also as the *affinity of vaporization*, is always negative. It specifies the energy involved in an isothermal expansion of the vapour during which the pressure drops from $e_s$ to $e$. Alternatively, it can be said that the energy needed for evaporation is provided by the liquid water and, consequently, the water cools.

### 2.1.3  Colligative properties

The saturation vapour pressure over saline water, $e_s(s)$, is lower than over fresh water. This is one of the 'colligative properties' of a solution; others being the freezing point depression, boiling point elevation, and osmotic pressure. Though sea water does not obey exactly the theory of colligative properties in a dilute solution, one can use Raoult's law to equate the ratio $e_s(s)/e_s$ to the mole fraction of $H_2O$ in sea water. One gram of sea water contains $(1-s)/m_w$ moles of $H_2O$ and $cs/m_s$ moles of various salts. The mean molecular mass of the dissolved salts is denoted by $m_s$ and $c$ is the dimension-less *van't Hoff factor*, which accounts for the fraction of dissociated (ionized) salt molecules. It follows that

$$\frac{e_s(s)}{e_s} = \left(\frac{1-s}{m_w}\right)\Big/\left(\frac{1-s}{m_w} + \frac{cs}{m_s}\right) \approx 1 - \frac{m_w}{m_s}cs. \tag{2.14}$$

For the salts in sea water $cm_w/m_s = 0.527$. With $s = 35 \times 10^{-3}$, (2.14) therefore yields a vapour pressure reduction of about 2 per cent. The affinity of vaporization over the sea can be obtained from (2.13) by the substitution of $e_s(s)$ for $e_s$.

Salts in solution lower the freezing temperature $T_f$. In sea water, according to Millero (1978)

$$T_f = -0.0527s + 1.71523 \times 10^{-3}s^{3/2} - 2.155 \times 10^{-4}s^2 - 7.53 \times 10^{-8}p \tag{2.15}$$

with units $T_f$ (°C), $s$ (psu) $\approx s$ (ppt), and $p$ (Pa). Maykut (1985) states that the linear approximation

$$T_f = -0.055s \qquad (2.15')$$

is sufficiently accurate for most purposes.

For pure water the temperature of maximum density $T_m = 4°C$. This is therefore the temperature of the deeper layers in all major fresh water lakes of the temperate zones. The value of $T_m$ decreases linearly with increasing salinity. The density maximum disappears in salt water with $s > 24.7$ ppt. The density of sea water therefore increases monotonically as it is cooled down to its freezing point. The temperatures $T_f$ and $T_m$ are plotted in Fig. 2.1.

The *partial osmotic pressure* caused by the salinity of sea water is directly proportional to the freezing point depression $T_f$ multiplied by the ratio $(273 + T)/273$ as indicated by Sverdrup et al. (1942). The contribution of dissolved atmospheric gases to the osmotic pressure is discussed in the following section.

### 2.1.4   *Atmospheric gases in solution*

The state of the atmosphere–ocean system is determined by the relative amounts of various gases that are found in the atmosphere or are dissolved in the ocean. This changing distribution has determined the climate of the globe along with biological and ecological evolution.

Sea water contains all of the atmospheric gases in solution. Concepts involved in quantitative considerations of this phenomenon include the *concentration, C,* or *mixing ratio, r,* the *solubility, $C^*$,* and the *saturation* or *equilibrium concentration, $C^s$.*

The osmotic pressure, $p'_n$, of the $n$th component in a dilute solution is adequately determined by

$$p'_n = \frac{R_1 TC_n}{m_n} = R_n TC_n. \qquad (2.16)$$

This equation has the same structure as the equation of state in a mixture of ideal gases, $R_1$ is the universal gas constant, $m_n$ the molecular mass and $R_n = R_1/m_n$ the gas constant for the gas $n$. The concentration $C_n$ in this case has the dimension of density.

In the oceanographic literature, the concentration is often represented by values of $C_n/m_n$, expressed in units of kg-moles per cubic meter or equivalently in gram-moles per liter of sea water. It is then proportional to the number of molecules of the gas $n$ that are found within a unit volume of sea water. It can alternatively be expressed in nondimensional units of moles per kilogram:

$$C(\text{moles kg}^{-1}) = C(\text{moles m}^{-3})/\rho_w = r \times \text{mole-number},$$

where $\rho_w$ is the density of sea water and $r$ is the mixing ratio, that is the ratio of the mass of solute to the mass of sea water in an arbitrary volume. In yet another way, the concentration is sometimes specified in terms of milliliters or standard cubic centimeters of dissolved gas per liter of sea water. The volume of the dissolved gas is defined in this case as the volume it would occupy in the gas phase at a temperature of 0°C ($T_0 = 273.2$ K) under pressure of one standard atmosphere ($p_0 = 1013.2$ mb $= 1.0132 \times 10^5$ Pa). The concentration so specified is related to the other specifications by

$$C(\text{ml l}^{-1}) = 1000C(\text{moles m}^{-3})R_1 T_0/p_0.$$

The solubility $C_n^*$ represents the greatest possible amount of the gas that can be dissolved in a unit volume of liquid (i.e., its maximum possible concentration). It is determined by an equation of the form

$$C_n^* = \text{const.} \frac{p_0(n)}{P_n}. \tag{2.17}$$

The symbol $p_0(n)$ denotes an arbitrary reference pressure in the ambient gas phase. It is customary to choose $p_0(n)$ equal to $p_0$ (i.e., the solubility of oxygen is determined by the maximum amount that could be dissolved if the solution were in contact with a pure oxygen atmosphere of standard pressure). The symbol $P_n$ is the saturation vapour pressure of the gas $n$ that would be in equilibrium with its condensed phase. For oxygen at 0°C, $P_n$ equals about 36,000 standard atmospheres. The saturation pressure $P_n$ decreases with the ambient temperature. This causes the solubility of all atmospheric gases to increase with falling temperatures and hence with depth and latitude.

For gases that are not dissociated in sea water, the proportionality constant in (2.17) depends only on the nature of the solvent. Because the saturation vapour pressure, $P_n$, of the inert gases decreases with increasing molecular weight, their solubility, $C_n^*$, increases correspondingly. This relation does not apply to dissociated compounds. The solubility of all gases decreases with increasing salinity because the presence of heavy salt molecules automatically lowers their mole fraction in the solution.

In the atmosphere, the partial pressure, $p_n$, of each component gas is necessarily smaller than $p_0$. The ratio $p_n/p_0$ equals the fraction of the substance $n$ in a unit volume of surface air. The standard value of the equilibrium or saturation concentration of $n$ in surface sea water is therefore

$$C_n^s = C_n^* \frac{p_n}{p_0} = \text{const.} \frac{p_n}{P_n} \equiv H \frac{p_n}{P_n}. \tag{2.18}$$

In a submerged air bubble, the partial pressure $p_n$ is larger than in surface air. The equilibrium concentration increases therefore with depth. The proportionality factor, $H$, in (2.18) is often called the 'Henry's law constant', although it is generally not really a constant since it varies with temperature. It represents again the fraction of the dissolved gas $n$ that occurs in nondissociated or non-ionised form. It equals unity for gases like nitrogen, but becomes relatively small for gases like carbon dioxide that undergo dissociation.

The actual concentration, $C_n$, of a dissolved, nondissociated gas is generally not in equilibrium with its partial pressure in the ambient air. It therefore has a tendency to go into solution or vice versa according to whether $C_n$ is larger or smaller than $C_n^s$. The tendency is measured by the chemical potential difference and is given, analogously to (2.13), by

$$\mu_n - \mu_n^s = R_n T \ln \left( \frac{C_n}{C_n^s} \right), \tag{2.19}$$

where $\mu_n$ is the actual chemical potential of the $n$th constituent of a gas mixture in solution and $\mu_n^s$ is the potential in equilibrium with the ambient partial pressure $p_n$.

The solubilities listed in Table 2.3 were obtained from Tables 3.1 and 3.2 in Broecker and Peng (1982), who also provide caveats and original sources for these numbers.

The compounds carbon dioxide ($CO_2$), nitrous oxide ($N_2O$), and freon 11 ($Cl_3FC$), which are listed in the last three rows, all affect the climate of the atmosphere–ocean system. The volume partition coefficients in the last column of Table 2.3 relate the number of molecules in a volume of air, at standard pressure, to the equilibrium number of the same species of molecules in an equal volume of water. For example, for every helium atom in a liter of surface sea water at 24°C, one could expect to find 135 He atoms in a liter of air above. In the case of oxygen, this ratio is reduced to 1:42; for carbon dioxide it becomes 2:3.

Table 2.3 demonstrates the lowering of the solubility and equilibrium concentration as the water gets warmer. For oxygen, this is also illustrated by Fig. 2.2. The decrease of oxygen available to fish and other organisms

**Table 2.3.** Solubility and Saturation Concentration of Atmospheric Gases in Sea Water ($s = 35$ ppt) Under Pressure of 1013.2 mb

| Gas | Mole Wt. | Atmospheric Volume Fraction | Sea Water Concentrations (ml l$^{-1}$) | | | | Vol. Part Coef. 24°C |
| | | | Solubility $C_n^*$ | | Saturation $C_n^s$ | | |
| | | | 0°C | 24°C | 0°C | 24°C | |
|---|---|---|---|---|---|---|---|
| N$_2$ | 28.0 | 0.781 | 18.3 | 11.8 | 14.29 | 9.22 | 85 |
| O$_2$ | 32.0 | 0.209 | 38.7 | 23.7 | 8.09 | 4.95 | 42 |
| A | 39.9 | 0.0093 | 42.1 | 26 | 0.39 | 0.24 | 39 |
| He | 4.0 | $52.0 \times 10^{-7}$ | 7.8 | 7.4 | $4.1 \times 10^{-5}$ | $3.8 \times 10^{-5}$ | 135 |
| Ne | 20.2 | $180.0 \times 10^{-7}$ | 10.1 | 8.6 | $18.0 \times 10^{-5}$ | $15.0 \times 10^{-5}$ | 116 |
| Kr | 83.7 | $10.8 \times 10^{-7}$ | 85.6 | 46.2 | $9.2 \times 10^{-5}$ | $5.0 \times 10^{-5}$ | 22 |
| Xe | 131.0 | $0.9 \times 10^{-7}$ | 192 | 99 | $1.7 \times 10^{-5}$ | $0.9 \times 10^{-5}$ | 10 |
| Rn | 222.0 | * | 406 | 186 | — | — | 5.4 |
| CO$_2$ | 44.0 | $3350.0 \times 10^{-7}$* | 1437 | 666 | 0.48 | 0.22 | 1.5 |
| N$_2$O | 44.0 | $3.0 \times 10^{-7}$* | 1071 | 476 | $32.0 \times 10^{-5}$ | $14.0 \times 10^{-5}$ | 2.1 |
| Cl$_3$FC | 137.0 | * | — | 677 | — | — | 1.5 |

* denotes variable amount.

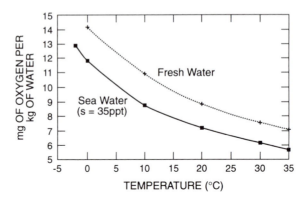

**Fig. 2.2.** Sea water and fresh water oxygen concentrations in equilibrium with a standard atmosphere at a pressure of 1013 mb.

is one of the detrimental effects of thermal pollution in rivers and coastal waters. Ocean surface waters are often supersaturated with oxygen, either because of rapid warming, or because of bubbles that went into solution at a pressure greater than $p_0$. Free oxygen is produced internally in the sea by photosynthesis. It is removed by the decay of organic matter and by respiration.

Carbon dioxide has a much higher boiling point and a lower saturation pressure than oxygen or nitrogen. It can also interact chemically with sea water. In contrast to other atmospheric gases, these factors combined cause $CO_2$ to be about sixty times more abundant in the ocean than in the atmosphere. A comprehensive description of the whole carbon dioxide system in the sea is given by Riley and Skirrow (1965). More detailed accounts of the carbon dioxide content of surface ocean waters and of recent man-made influences can be found in Broecker and Peng (loc. cit.).

Another gas that has attracted some attention is DMS—a sulphur compound that is produced organically in sea water. After passage into the atmosphere it tends to form small droplets of sulphuric acid which then can act as nuclei for the condensation of water vapour. The process may have some influence on atmospheric precipitation regimes. Data on the solubility of DMS can be found in Dacey et al. (1984).

### 2.1.5 Molecular transport coefficients

Values for the *kinematic viscosity, thermal diffusivity* and of the *diffusivity of sea salts* at sea-level pressure are listed in Table 2.4.

Only the coefficients of viscosity were obtained from experiments with actual sea water. The other properties tabulated under sea water are from measurements with sodium chloride solutions.

The viscosity of water increases as the temperature is lowered, while the coefficients of thermal diffusivity and diffusivity decrease. The ratio of the kinematic viscosity to the thermal diffusivity is specified by the Prandtl number $Pr = v/\kappa$. For sea water with a temperature of 10°C, $Pr \approx 10$. The thermal diffusivity is in turn more than 100 times larger than the diffusivity of sea salts. These differences are the cause of double-diffusion, and they may also allow anomalies in the vertical distribution of salt near the surface to persist while vertical velocity and temperature contrasts are being smoothed out.

**Table 2.4.** Kinematic Molecular Transport Coefficients ($10^{-7}\, m^2\, s^{-1}$) in Water at Standard Atmospheric Pressure.*

|  | Pure Water | | Sea Water (s = 35 ppt) | |
|---|---|---|---|---|
| Temperature (°C) | 0 | 20 | 0 | 20 |
| Viscosity ($v$) | 17.87 | 10.04 | 18.26 | 10.49 |
| Thermal diffusivity ($\kappa$) | 1.34 | 1.43 | 1.39 | 1.49 |
| Diffusivity ($\delta$) of NaCl | 0.0074 | 0.0141 | 0.0068 | 0.0129 |

* From American Inst. of Physics Handbook, 1963 edition.

**Table 2.5.** Kinematic Diffusivities $(10^{-9}\,\mathrm{m^2\,s^{-1}})$ of Atmospheric Gases in Sea Water*

| Gas | Diffusion Coefficients | |
|---|---|---|
| | 0°C | 24°C |
| Nitrogen ($N_2$) | 1.1 | 2.1 |
| Oxygen ($O_2$) | 1.2 | 2.3 |
| Argon | 0.8 | 1.5 |
| Helium | 2.0 | 4.0 |
| Neon | 1.4 | 2.8 |
| Krypton | 0.7 | 1.4 |
| Xenon | 0.7 | 1.4 |
| Radon | 0.7 | 1.4 |
| Carbon Dioxide | 1.0 | 1.9 |
| Nitrous Oxide | 1.0 | 2.0 |

* From Broecker and Peng (1982).

Table 2.5 gives diffusivity values for the gases that were listed in Table 2.3. These values are all of the same order of magnitude, comparable to the values for NaCl in sea water (Table 2.4). The coefficients for nitrogen and oxygen are known to be possibly incorrect by as much as 15 per cent. Values for the other gases may involve even larger uncertainties.

## 2.2   Moist air

### 2.2.1   *The equation of state*

In this section we will consider air as a mixture of ideal gases with water vapour as the only variable constituent. Spray, rain or fog drops, salt particles, and other suspended particulate matter do not affect the density significantly with the possible exception of water spouts and hurricanes. The equation of state for a mixture of gases with partial pressures, $p_n$, and specific concentrations, $C_n$, is

$$p = \sum_n p_n = R_1 T \sum_n \frac{C_n}{m_n}. \tag{2.20}$$

The volume fractions of the various constituents of dry air are practically constant. They have been listed in Table 2.3 together with their individual molecular mass. The mean molecular mass of the dry air mixture $m_a = 28.97$, which makes the gas constant for dry air

$$R = \frac{R_1}{m_a} = 287.04 \text{ J kg}^{-1} \text{ K}^{-1}. \tag{2.21}$$

Let $\rho$ without suffix denote the density of moist air. The specific concentration of water vapour is known as *specific humidity* and is commonly denoted by $q$. In moist air $q = \rho_v/\rho$. The specific concentration of dry air is $(1 - q)$. The molecular mass ratio $m_w/m_a = 0.622$. With this symbolism the equation of state for moist air can be written in the form

$$p = \rho R_1 T \left( \frac{1-q}{m_a} + \frac{q}{m_w} \right) = \rho RT \left[ 1 + \left( \frac{m_a}{m_w} - 1 \right) q \right] \approx \rho RT \left( 1 + \frac{3}{5} q \right). \tag{2.22}$$

Alternatively, the vapour content of air is often specified in meteorology by the humidity mixing ratio, $r$. This is defined by the amount of vapour that is mixed with a unit mass of dry air, that is, by the ratio of vapour density to that of dry air

$$r \equiv \frac{\rho_v}{\rho_a} = \frac{q}{1-q} = \frac{m_w}{m_a} \frac{e}{p-e}. \tag{2.23}$$

In terms of this quantity, the equation of state for moist air reads

$$p \approx \rho RT \left( 1 + \frac{3}{5} \frac{r}{1+r} \right) \approx \rho RT \left( 1 + \frac{3}{8} \frac{e}{p} \right). \tag{2.22'}$$

The equation of state (2.22) or (2.22'), shows that moist air is less dense than dry air of the same pressure and temperature. The so-called *virtual temperature*

$$T_v \approx T \left( 1 + \frac{3}{5} q \right) \approx T \left( 1 + \frac{3}{8} \frac{e}{p} \right) \tag{2.24}$$

indicates the temperature to which dry air would have to be heated, without change in pressure, to make its density equal to that of the moist air. Values of $T_v$ in the Smithsonian Meteorological Tables are based on a

**Table 2.6.** Molecular Transport Coefficients ($10^{-4}\,\mathrm{m^2\,s^{-1}}$) in Air at Standard Surface Pressure.

| Temperature (°C) | −10 | 0 | 10 | 20 | 30 |
|---|---|---|---|---|---|
| Kinematic viscosity ($v$) | 0.1259 | 0.1346 | 0.1437 | 0.1529 | 0.1623 |
| Thermal diffusivity ($\kappa$) | 0.177 | 0.189 | 0.202 | 0.215 | 0.228 |
| Diffusivity of vapour ($\delta$) | 0.211 | 0.226 | 0.241 | 0.257 | 0.273 |

slightly more exact definition that takes deviations from ideal gas behavior into account. Over the warm tropical and subtropical oceans, the difference between the actual and the virtual temperature can be as large as 5°C. Density fluctuations under these conditions are largely a function of variable moisture content.

### 2.2.2  Molecular transport coefficients

The coefficient of diffusion in an ideal gas is proportional to the free path length and therefore inversely proportional to the density. As in the case of the sea, molecular transports in the atmosphere are of interest mainly in the immediate vicinity of the interface. The relevant coefficients and their variation with temperature are listed in Table 2.6. Unlike water, gases become less viscous with decreasing temperature. At 20°C, the kinematic viscosity of air is about fifteen times larger than that of water. The diffusivity of water vapour is of the same order as the conductivity, causing moisture and heat to be transported down their gradients at about the same rate.

### 2.2.3  Isobaric mixing and fog formation

Apart from an arbitrary additive constant, the specific enthalpies of dry air and water vapour are

$$\xi_a = c_p T = \frac{7}{2} RT = 1004.84T \quad (\mathrm{J\,kg^{-1}}) \tag{2.25}$$

$$\xi_v = c_{pv} T = 4R_v T = 1846.40T \quad (\mathrm{J\,kg^{-1}}).$$

The specific enthalpy of moist air is therefore also a linear function of $T$.

$$\xi_m = (1 - q)\xi_a + q\xi_v = c_p T(1 + 0.84q) + \mathrm{const.} \tag{2.26}$$

When two air parcels are mixed together, the total amount of moisture

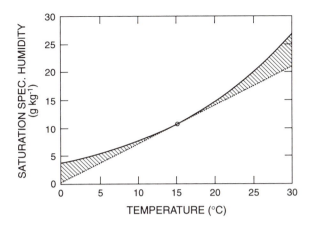

**Fig. 2.3.** Marine fog formation. The curved solid line represents the saturation specific humidity $q_s(T)$ at a standard atmospheric pressure of 1013 mb. The straight dotted line is the tangent to this curve at 15°C. Sea fog can be generated at this sea temperature by vertical mixing, if atmospheric values of $T$ and $q$ are represented by points in the shaded area.

is conserved provided there is no precipitation. In the absence of external heating and pressure changes, the total enthalpy is also conserved. On a $(T, q)$ plane, the mixing of two air parcels characterized by $T_1, q_1$ and $T_2, q_2$ produces therefore a mixture with a temperature and specific humidity that lies somewhere along the straight line that joins the two initial conditions. It follows from the curvature of the saturation specific humidity, $q_s$, in Fig. 2.3 that the mixing of two unsaturated air parcels can produce a supersaturated mixture, as long as the straight mixing line intersects the $q_s$ curve at two points. Fog can be produced in this way.

Air in immediate contact with the sea surface is always nearly saturated. The generation of sea fog, through mixing of this surface air with the free unsaturated atmosphere above, has been studied in some detail by Saunders (1964). This type of fog is common when relatively warm air flows over a cold water surface (steam fog), though it can occur also in the opposite case. Whether fog actually does form depends on the position of the final mixture along the 'mixing' line, that is, on the ratio of mass that goes into the mixture from the surface and from above. This in turn is a function of the rate of evaporation and heat conduction from the sea.

The fact that the equilibrium pressure over salt water is 2 per cent lower than over pure water and that therefore the actual sea surface specific humidity, $q_s(s)$, lies slightly below the $q_s(0)$ curve, can significantly alter the intersection between that curve and the mixing line in Fig. 2.3.

This fact therefore has some bearing on an assessment of the conditions under which fog may occur. When such an assessment is made, it must also be considered that the stipulated absence of external heating is not always justifiable. Particularly when the atmosphere is warm and cloudless, long-wave radiation tends to cool the air, and this further favors fog formation. The mixing argument alone may account for sea fogs that form when cold air moves over warmer water.

### 2.2.4   Adiabatic and pseudo-adiabatic changes of state

The isobaric mixing argument is not valid when the mixing of two air masses involves significant pressure variations. In that case the enthalpy is not conserved. What does remain constant, in the absence of phase changes, is the *potential temperature*, $\Theta$, which is defined as the temperature that would be acquired by the air when brought adiabatically to a standard pressure, $p_0$, of 1000 mb or $10^5$ Pa. There can be no conduction, absorption, or viscous dissipation during adiabatic changes. Division of the First Law of Thermodynamics (1.36) by the absolute temperature, $T$, and subsequent integration, yields with consideration of the equation of state (2.22) for $q = 0$

$$c_p \ln \left( \frac{T}{\Theta} \right) - R \ln \left( \frac{p}{p_0} \right) = 0. \tag{2.27}$$

As $p_0 = 1000$ mb and $R/c_p = 2/7$, this is equivalent to

$$\Theta = T \left( \frac{1000}{p} \right)^{2/7}. \tag{2.28}$$

The *entropy* of a dry air parcel is fully specified by the logarithm of the potential temperature of the parcel. It cannot vary during adiabatic processes when the heat content remains unchanged; but when air parcels are mixed, the potential temperature of the mixture becomes equal to the arithmetic mean of $\Theta$ in the original parcels. Entropy, being proportional to $\ln (\Theta)$, is increased in this case.

If $\Theta$ is vertically constant though a layer of limited depth ($\partial\Theta/\partial z = 0$), differentiation of (2.27), with the assumption of hydrostatic equilibrium, indicates that the lapse rate of the actual temperature, $T$, in the same layer is linear

$$\left( \frac{\partial T}{\partial z} \right)_{\Theta = \text{const.}} = \frac{RT}{c_p p} \frac{\partial p}{\partial z} = \frac{\rho}{c_p} \frac{\partial p}{\partial z} = -\frac{g}{c_p} = -9.8 \times 10^{-3} \, °\text{C} \, \text{m}^{-1}. \tag{2.29}$$

This relation is applicable only when the ambient Froude number is relatively small. It is not satisfied in the presence of vertical accelerations that amount to a significant fraction of $g$. In the atmosphere this only happens during very violent convection. Ordinarily, air columns are always unstable when the lapse rate exceeds the adiabatic value of $g/c_p$. The Brunt–Vaisala frequency (1.38) is imaginary in that case ($N^2 < 0$).

Although relatively large pressure changes may be experienced by an air parcel that spirals in horizontally toward the center of a hurricane or a tornado, it is more common for such pressure changes to be the result of vertical displacements. However, when an air parcel changes its height by less than 20 m, the corresponding adiabatic temperature change remains less than 0.2°C. This justifies the approximation of the potential by the actual temperature in the discussion of many marine surface layer problems.

To compute the temperature changes that result from pressure changes in moist air, one applies (1.35) to one unit of dry air plus $r$ units of vapour. The enthalpy of dry air $\xi_a = c_p T + \text{const}$. For unsaturated water vapour the chemical potential was given by (2.13) and the enthalpy by (2.7). When these expressions are introduced into (1.35) we obtain, after division by $T$ and subsequent integration,

$$c_p \ln (\Theta) + r \left[ c \ln (T) + \frac{L}{T} + R \ln \left( \frac{e}{e_s} \right) \right] \approx c_p \left[ \ln (\Theta) + \frac{rL}{c_p T} \right]$$

$$\approx S + \text{const.,} \quad (2.30)$$

where $S$ is the entropy of $(1 + r)$ mass units of moist air. The exponential of the last term in brackets is known as the *equivalent potential temperature*

$$\Theta_e \equiv \Theta \exp \left( \frac{rL}{c_p T} \right). \qquad (2.31)$$

It represents a conservative property which is not changed by expansion, even if this involves condensation and precipitation. An unsaturated air column is *potentially unstable* (also called *convectively unstable*) if it is characterized by an upward decreasing equivalent-potential temperature ($\partial\Theta_e/\partial z < 0$). If such a column is lifted bodily until it is saturated, it will become unstable (i.e., its density at the top will be larger than at the bottom). The lower tropical troposphere over the oceans is almost everywhere potentially unstable.

## 2.3    The liquid–gas interface

### 2.3.1    *Laminar sublayers*

Turbulence is always suppressed by molecular viscosity in a thin layer along a material surface. It is generally assumed that such molecular sublayers also characterize conditions at the sea surface much of the time. In Chapter 5 these molecular sublayers will be discussed in more detail.

Turbulence energy can be transmitted directly from air to water by viscous and pressure forces across the interface. However, within each molecular sublayer, the motion can be expected to remain locally laminar and parallel to the interface. Transport normal to the interface can be produced there only by molecular motion. It has been indicated in Table 2.5 that various dissolved gases have different molecular diffusivities in sea water. Their flux rates through the laminar sublayer will therefore differ accordingly, as will be discussed in 5.4.

Evaporation causes the interface to act as a source of water vapour in the atmosphere and of salinity in the oceans. Vapour diffuses rather easily through air, but the molecular diffusivity of salt in water was shown in Table 2.4 to be rather small. This can be expected to make the sublayer anomalously saline and dense. It may cause spray drops to leave the sea surface with a higher salinity than bulk sea water. It also may cause a downward convection of this saline water, especially when the surface temperature drops at night. This process can contribute to the presence of highly saline water at the bottom of the surface mixed-layer in subtropical regions.

### 2.3.2    *Surface tension*

Surface tension results from molecular forces near the interface. It tends to reduce the area of free surfaces or interfaces and therefore causes these to behave like elastic skins or membranes stretched over a frame. This affects the shape of spray droplets and suspended air bubbles, the surface spreading of contaminations, and the dynamics of capillary waves.

The very existence of the liquid state is due to van der Waals forces of attraction between closely packed molecules. In the interior of a liquid, these attractive forces act isotropically in all directions. Their resultant is therefore zero. On the surface, however, this resultant has an unblanced component that is directed toward the interior of the liquid and so tends to pull the surface molecules inward. Any stretching of a boundary surface or interface has to bring molecules from the interior to the surface. It therefore requires work to overcome this inward pull. The work needed to extend an element of surface $dA$ by an amount $\delta(dA)$ can be represented by

$$\delta W = \Gamma_s \delta(dA) = \gamma \delta(dA). \tag{2.32}$$

The quantity $\Gamma_s$ represents the *specific surface free energy*. It measures the work per unit surface (Joule per meter squared) that is needed to form an interface. If a thermodynamical system includes free surfaces or interfaces, the surface free energy is part of its internal energy. The *surface tension*, $\gamma$, measures the force that is needed to bring molecules from the interior to the surface; therefore, it is expressed in Newtons per meter ($= 10^3$ dyne cm$^{-1}$). Although the surface tension and the specific surface free energy are different concepts, they have the same dimension (kilograms per second squared) and are also numerically equal except for fluids with a very high viscosity.

The elastic properties of boundary surfaces between two fluids cause an increased pressure, $\Delta p$, at the concave side of any curved part of the interface. The work, $\delta W$, which has to be done against this pressure difference when a surface element, $dA$, is displaced in the direction of its normal by an infinitesimal distance, $\delta \zeta$, is

$$\delta W = \Delta p \, \delta \zeta \, dA. \tag{2.33}$$

It can be deduced from analytical geometry that deformations of a curved surface generally involve changes in its area that locally are of the order

$$\delta(dA) = \delta \zeta \left( \frac{1}{r_1} + \frac{1}{r_2} \right) dA = -\nabla^2 \zeta \, \delta \zeta \, dA,$$

where $r_1$ and $r_2$ are the radii of surface curvature in two orthogonal planes normal to the interface. The last approximation is justified if the interface can be represented by a surface $z = \zeta(x, y)$ that does not deviate much from the plane $z = 0$. Combination of the last three equations yields Laplace's formula

$$\Delta p = \gamma \left( \frac{1}{r_1} + \frac{1}{r_2} \right) = -\gamma \nabla^2 \zeta. \tag{2.34}$$

When $r_1$ and $r_2$ are positive, $\Delta p$ is also positive, thus causing a pressure increment on the concave side of the boundary. The pressure difference is zero when the interface is a plane ($r_1 = r_2 = \infty$).

The shape of bubbles and spray drops depends partly on their size.

Buoyancy and other dynamic forces affect them at a rate proportional to the contained volume. Surface tension becomes dominant if that volume is relatively small. It then forces the bubble or drop to contract into a sphere ($r_1 = r_2 \equiv r$), which is the smallest surface to contain a given volume. Under these conditions

$$\Delta p = \frac{2\gamma}{r}. \tag{2.34'}$$

The surface tension at an air–water boundary is larger than that at the boundary of air with most other fluids that occur in the natural environment. It decreases with temperature. This is due to a weakening of molecular attraction by increased agitation. On the other hand, the presence of salts in solution increases the mean value of the intermolecular attraction and the surface tension of sea water is therefore slightly greater than that of fresh water at the same temperature. Sverdrup et al. (1942) quote a study by Fleming and Revelle, who suggested an empirical expression of the form

$$\gamma(\text{seawater}) = 75.63 - 0.144T + 0.221s \quad (10^{-3}\,\text{N m}^{-1}), \tag{2.35}$$

where $T$ is specified in °C and $s$ in ppt.

### 2.3.3 *Contamination*

The water found at the sea surface is always contaminated by all sorts of organic material, by airborne dust from the continents, and, increasingly, also by human waste and debris. The surface tension of contaminated water is usually smaller than that of sea water that has been brought up freshly from the depths of the ocean. On the other hand, adhesion of surface-active materials can stiffen an air–water interface. In particular it can cause air bubbles with a radius smaller than $100\,\mu$m to move like a rigid body through the surrounding water (Wu, 1992).

The spreading of a drop of insoluble oil on a water surface involves the buoyancy of oil on water as well as the surface interaction between the three substances: oil, water, and air. This is illustrated schematically in Fig. 2.4. The rim of an oil lens is pulled outward if the air–water surface

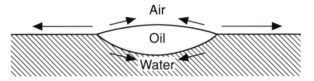

**Fig. 2.4.** Surface tensions at the rim of an oil lens (schematic).

tension is larger than the sum of the air–oil and oil–water surface tensions. The spreading of an oil slick depends therefore on the sign of the *spreading coefficient*

$$Sc = \gamma(\text{air/water}) - \gamma(\text{air/oil}) - \gamma(\text{oil/water}). \qquad (2.36)$$

Spreading occurs if $Sc$ is positive; if it is negative, then buoyancy effects due to the density difference between oil and water may still push the rim of the oil patch outward, but the surface tension now pulls it inward. For example, at 20°C the air–water surface tension is $72.8 \times 10^{-3}\,N\,m^{-1}$; the surface tensions of two different types of oil at the same temperature are (in $10^{-3}\,N\,m^{-1}$):

|  | Air–oil | Oil–water |
|---|---|---|
| Olive oil | $32.0 \times 10^{-3}\,N\,m^{-1}$ | 20.6 |
| Paraffin oil | 26.5 | 48.4 |

It follows from (2.36) that olive oil will spread on clean water while paraffin tends to form lense-shaped globules like fat on a hot soup. Heavy mineral oil tends to behave like paraffin in this regard.

If $Sc > 0$ the oil continues to spread until it becomes distributed as a thin film or as a monolayer on the water. A film may be defined as a sheet of material still thick enough to have two independent interfaces, one with the water and one with the air. Each of these will have its own characteristic surface tension. Films can show interference colours, provided that their thickness is larger than the wavelength of the incident light.

Monomolecular layers or *monolayers* have a thickness equal to the effective diameter of a single molecule and are formed by surface-active materials. These materials are made up of molecules with both a hydrophilic and a hydrophobic part. Many insoluble substances, such as fatty acids and various alcohols, fall into this category. They will spread on the water surface and tend to form a layer one molecule in thickness with the hydrophillic —COOH or —OH groups oriented toward the water phase and the hydrophobic hydrocarbon chains sticking out into the air. Monolayers always reduce the surface tension.

A closely packed monolayer of some substances (e.g., cetylic alcohol) can prevent the escape of the water molecules into the vapour phase. Depending on weather conditions, this can reduce the evaporation rate from small tanks by more than 50 per cent. The operational use of this phenomenon for water conservation is limited, however, by the difficulty of maintaining closely packed unbroken monolayers on large natural water surfaces. At the sea surface in particular, monolayers and organic films, although never absent from large areas, are always patchy and broken up by waves. Their effect on evaporation is therefore probably insignificant. However, they may have some effect on surface temperature fluctuations and hence on convection in the atmosphere above and in the sea below. They also can influence the deformation of the interface.

Contaminations do not necessarily slow down the local exchange rates through the interface. They can have the opposite effect. For example, it appears that the passage of $CO_2$ molecules through a molecular sublayer into the water is accelerated when suitable catalysts are present as contaminants. It has been shown by Quinn and Otto (1971) that this can augment the exchange rate through the interface by a considerable amount.

## 2.4  Bubbles and spray

The air–sea boundary is not a simple continuous surface. Bubbles in the water and drops in the air extend the area of interaction between the two fluids. In a hurricane, the state of the matter near the air–sea interface has been described as being too thick to breathe and too thin to swim in. The present section deals briefly with the formation, terminal velocity, and size distribution of air bubbles in the water and spray droplets in the marine atmosphere. Global environmental effects are listed in subsection 2.4.7.

### 2.4.1  *Generation of bubbles and spray droplets*

Bubbles are formed most frequently from air that has been entrapped mechanically by breaking waves. Little seems to be known quantitatively about the actual process that produces a collection of small bubbles from the bulk air entrained in a breaking wave. To a lesser extent, bubbles can be generated also by the impact of rain, snow, or dust on the sea surface, by reduction of solubility caused by spring heating, and by various organic processes.

Spray can be formed by the mechanical disruption of waves in strong winds. The spray plumes or spindrift torn from the wave crests in this way are made up of relatively large drops. More numerous and generally not so large are drops produced by the bursting of bubbles at the surface. The physics of this generating process has been described in detail by Blanchard (1963, 1983). The surface free energy of a collapsing bubble is converted into the kinetic energy of a vertical jet, which rises from the centre of the cavity. This jet becomes unstable, and breaks up into 1–10 drops. The diameter of these jet drops tends to be about one-tenth that of the generating bubble. Drops are also produced by the shattering of the film that separates the bubble from the atmosphere before bursting. These film droplets are much more numerous than are the jet drops. They are also much smaller. Their diameter is generally less than $4\,\mu$m. Their chemical composition may also be different. Bubbles with a larger surface area produce fewer jet drops, but many more film droplets than do smaller bubbles. The bursting of a large bubble with diameter of about 6 mm can yield up to 1000 such film droplets, while a 2 mm bubble gives rise to 100 at most and a 0.3 mm bubble appears to produce no film droplets at all. Later investigations summarized by Wu (1992) suggest that the number of droplets per bursting bubble is generally somewhat lower and that it can be approximated by power laws of the form

$$n_j = \mathrm{mod}\,(7.0e^{-D/3}) \quad \text{and} \quad n_f = \mathrm{mod}\,(1.07D^{2.15}), \tag{2.37}$$

where $n_j$, $n_f$ are the respective numbers of jet and film droplets that are produced by a bursting bubble of diameter $D$.

The jet drops seem to be the main source of the so-called giant salt nuclei in the atmosphere. The ejection speed of these drops is strikingly fast, but decreases with increasing bubble diameter from nearly $100\,\mathrm{m\,s}^{-1}$ for a bubble diameter of about 0.07 mm to $10\,\mathrm{m\,s}^{-1}$ for bubbles with a diameter of 1 mm. These ejection speeds may be reduced by contamination. The height to which drops rise depends on the initial velocity and on the frictional retardation. The latter is so large for small drops that in spite of an injection speed of about $50\,\mathrm{m\,s}^{-1}$, drops from bubbles with a diameter of 0.1 mm do not rise more than 1 cm in still air. This height increases to a maximum of about 18 cm for bubbles in the 1–3 mm range and decreases again for drops from larger bubbles because of the slower ejection speed

### 2.4.2   Equilibrium pressure in air bubbles and spray droplets

Surface tension usually causes the matter within bubbles and droplets to be under a pressure that exceeds the pressure of the surrounding medium. Air molecules that pass from a bubble into water or vice versa must work with or against this excess pressure. To evaluate this process let $p = p_0 + g\rho_w z$ denote the ambient pressure on a bubble at depth $z$. Following (2.34'), the pressure within a spherical bubble is $p + \Delta p = p + 2\gamma/r$. As a first approximation it is assumed that the relative proportions or ratios of partial pressures of the different atmospheric constituents are not changed when air is entrapped in a bubble or goes into solution. The chemical potential of air molecules in solutions, which are in equilibrium with bulk air at pressure $p$, is denoted by $\mu^s(\infty)$. The equilibrium potential relative to the air inside the bubble is specified by $\mu^s(r)$. It follows then from (2.12), (2.34'), and the equation of state for air that

$$\mu^s(r) - \mu^s(\infty) = RT[\ln (p + \Delta p) - \ln (p)] \approx RT\frac{\Delta p}{p} = \frac{2\gamma}{\rho_a r}. \qquad (2.38)$$

where $\rho_a$ is the density of air at the ambient temperature $T$ and pressure $p$. Equation (2.38) indicates that, to be in equilibrium, air molecules in water surrounding a bubble must have a higher concentration and greater osmotic pressure than dissolved air under a plain water surface.

Similar considerations apply to airborne drops of pure water. The ambient vapour pressure must be larger to be in equilibrium with a drop than with a plane water surface. If $e_s$ denotes the saturation vapour pressure over a plane pure water surface at temperature $T$, $\rho_v$ is the corresponding vapour density and $e_s(r)$ is the equilibrium vapour pressure over a drop of radius $r$, one gets from (2.13) and (2.34')

$$R_v T \ln \left[ \frac{e_s(r)}{e_s} \right] = \frac{2\gamma}{\rho_v r}. \qquad (2.39)$$

Sea spray drops are not made of pure water. Fractionation during their formation and evaporation in the air can cause their salinity to be very different from that of bulk sea water. Once in the air, the mass of salt in the drop tends to remain constant, but the amount of water and the density are liable to change. If the salt mass in the drop is denoted by $M_s$, one can use (2.14) to express the ratio of the equilibrium vapour pressures over salt water $e_s(s, r)$ and fresh water $e_s(r)$ drops by

$$\frac{e_s(s, r)}{e_s(r)} = \frac{\frac{4}{3}\pi r^3 \rho_w - M_s}{\frac{4}{3}\pi r^3 \rho_w - M_s + c\frac{m_w}{m_s} M_s}. \qquad (2.40)$$

Elimination of $e_s(r)$ between (2.39) and (2.40) yields, after some simplification and neglect of terms that are small in a relatively dilute solution,

$$e_s(s, r) \approx e_s\left(1 + \frac{a}{r} - \frac{b}{r^3}\right) \qquad (2.41)$$

with

$$a = \frac{2\gamma}{\rho_v R_v T} \approx \frac{3.3 \times 10^{-7}}{T} \quad (\text{m})$$

$$b = \frac{3cm_w M_s}{4\pi m_s \rho_w} \approx 1.26 \times 10^{-7} \quad (\text{m}^3). \tag{2.42}$$

The temperature in the first expression (2.42) has to be specified in degrees Kelvin. As the saturation pressure over a plane fresh water surface is a tabulated function of the temperature only, (2.41) and (2.42) can be used to derive the approximate equilibrium vapour pressure over salt water droplets as a function of their salt content, radius, and ambient temperature. In an unsaturated environment $[e < e_s(s, r)]$, the droplets lose water by evaporation. As their radius shrinks, the magnitude of the last term in (2.41) increases faster than does that of the second term. This causes a decrease in the value of $e_s(s, r)$ that tends to continue until a new equilibrium with the ambient air vapour pressure has been reached. By adjusting their size, spray droplets can survive in this way over a wide range of environmental conditions.

### 2.4.3  Terminal velocities of gas bubbles and spray droplets

Most bubbles and spray droplets are small enough for the Reynolds number, associated with their movement through the ambient fluid, to be unity or less. Inertial accelerations are then negligible and the motion through a fluid at rest is affected only by the buoyancy and the viscosity. The resultant terminal velocity must satisfy the dimensional argument of (1.73). Conditions are somewhat complicated by the fact that the transport of a clean fluid sphere through another fluid creates both an external and an internal circulation. At the surface, the velocities must be continuous and the viscous stress equal and opposite on both sides. Batchelor (1967) gives a derivation of the appropriate formula. With subscripts 'w' and 'a' referring to water and air and $\rho_a/\rho_w \ll 1$ the terminal velocity of a small, clean rising bubble of radius $r$

$$w_b = \frac{1}{3} \frac{gr^2}{v_w} \frac{\rho_w - \rho_a}{\rho_w} \frac{\rho_w v_w + \rho_a v_a}{\rho_w v_w + 3/2\rho_a v_a} \approx \frac{1}{3} \frac{gr^2}{v_w}. \tag{2.43}$$

The terminal fall velocity of a small drop is

$$w_d = \frac{1}{3} \frac{gr^2}{v_a} \frac{\rho_a - \rho_w}{\rho_a} \frac{\rho_a v_a + \rho_w v_w}{\rho_a v_a + 3/2\rho_w v_w} \approx -\frac{2}{9} \frac{\rho_w}{\rho_a} \frac{gr^2}{v_a}. \tag{2.44}$$

Typical Reynolds numbers associated with these motions are

$$Re = 2rw_b/v_w \approx 2/3r^3gv_w^{-2} \qquad \approx 2.9 \times 10^{12}r^3 \quad \text{(bubbles)}$$
$$= 2rw_d/v_a \approx 4/9(\rho_w/\rho_a)r^3gv_a^{-2} \approx 15.5 \times 10^{12}r^3 \quad \text{(droplets)},$$

with $r$ in meters. This means that $Re$ will exceed unity for bubbles with $r > 70\ \mu$m and for drops with $r > 40\ \mu$m. The temperature dependence of the sea water viscosity, $v_w$, causes small bubbles to rise about twice as fast in equatorial as in polar waters. The corresponding ratio of the Reynolds numbers is about $4:1$.

The terminal velocity of larger drops or bubbles is not adequately determined by (2.43) or (2.44). An equation for the vertical velocity of bubbles with a radius $r > 80\ \mu$m by Thorpe (1982) has the same form as (2.43) divided by a parameter

$$A = 72\{1 - 2/[1 + (1 + 0.0912gr^3v_w^{-2})^{1/2}]\}.$$

Bubbles with a radius $r > 1$ mm tend to be deformed and their shapes pulsate. According to Levich (1962), their rise velocity becomes independent of the radius and approaches a value of about 30 cm s$^{-1}$. The motion of deformed drops has been investigated by Taylor and Acrivos (1964) and by Matunobu (1966).

Air bubbles that rise through sea water rapidly collect surface-active material, including algae and bacteria, on their surface. This can cause them to behave dynamically like rigid spheres. The terminal velocity of these 'dirty' bubbles is somewhat slower than that of 'clean' bubbles with the same radius. Equation (2.43) is not applicable even to small bubbles under these circumstances. Empirical formulas for the rise velocity of dirty bubbles can be found in Thorpe's (1982) and Wu's (1992) reviews. Wu's paper also has a graph for the rise velocity of both clean and dirty bubbles as a function of their diameter.

### 2.4.4 The size and flux spectra of air bubbles in bubble clouds

Below breaking waves, clouds of bubbles are formed. Because of their relatively long lifetime, these bubble clouds overlap when the wind speed is sufficiently high. They then tend to form a continuous layer, analogous to low stratus or stratocumulus clouds in the atmosphere. The maximal depth to which bubble clouds are entrained was observed by Thorpe (1982) to be proportional to the wave height. Laboratory experiments by Hwang et al. (1990) confirmed the existence of a linear relationship between the bubble entrainment depth and the significant wave height $H_{1/3}$ [see (4.55)].

The size spectrum of bubbles in any individual cloud is a function of depth and time. Change in the total ambient pressure with depth compresses bubbles or permits them to expand during their rise to the surface. The bubbles also grow or shrink according to whether their internal pressure is smaller or larger than the partial pressure of dissolved air in the water. If that partial pressure is in equilibrium with the bulk atmospheric pressure at the sea surface, a bubble with a diameter of 50 $\mu$m would go into complete solution within about 300 s, at a depth of 1 cm below the surface. At a depth of 1 m, it would dissolve in about 100 s. The chemical composition of bubbles also varies during their lifetime because oxygen dissolves at a faster rate than nitrogen.

Thorpe (1982) described the resulting transients in bubble radius and composition with a set of rather long and complicated semi-empirical equations. He concluded that bubbles tend to change in volume by diffusive processes about twenty times faster than by vertical displacements and compressibility. Except for the top few centimeters, the size distribution of bubbles found in the ocean seems to be determined primarily by gas exchanges with the surrounding water.

Bubble populations below the surface were measured with a bubble trap by Kolovayev (1976) in the open subtropical Atlantic. Johnson and Cooke (1979) photographically determined the number of bubbles within 17 $\mu$m diameter bands at three different depths. The use of active sonar for the observation of bubble clouds was reviewed by Thorpe (1982). Farmer and Lemon (1984) used an ingenious method to investigate bubble layer clouds from the subsurface attenuation of wind noise. This permitted them to derive vertically integrated spectra in quasi-stationary bubble layers at relatively high wind speeds. Bubble population densities were specified in this study by a power law.

$$N(a) = Ka^{-p}, \qquad (2.45)$$

where $N(a)$ is the number of bubbles with radius $a$ and where $K$ and $p$ may be functions of windspeed.

Other parametric models of bubble spectra were developed by Thorpe (1982) and by Crawford and Farmer (1987). On the basis of earlier studies, Wu (1992) represented the normalized probability density $\Pi(D)$, of bubbles with diameter $D$ in the near-surface ocean, by power laws of the form

$$
\begin{aligned}
&\Pi(D) = 6.25 \times 10^{-10} D^4 && D < 70\ \mu\text{m} \\
&\Pi(D) = 0.015 && 70\ \mu\text{m} < D < 100\ \mu\text{m} && (2.46) \\
&\Pi(D) = 1.5 \times 10^6 D^{-4} && D > 100\ \mu\text{m}.
\end{aligned}
$$

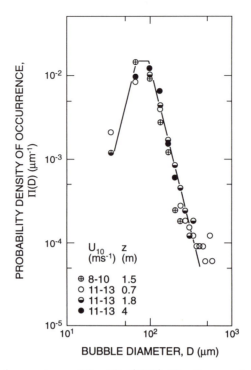

**Fig. 2.5.** Bubble size spectrum. After Wu (1992). The lines are based on (2.46) and the data points are from Johnson and Cooke (1979).

This distribution is represented by the solid lines in Fig. 2.5. Juxtaposition with Johnson and Cooke's data suggests that the normalized distribution (2.46) is not affected by changing wind velocities.

To obtain bubble flux spectra, Wu (1992) multiplied the size spectrum specified by (2.46) with the appropriate rise velocities. Assuming that turbulence and orbital wave motions advect bubbles upward and downward at the same rate, he used terminal velocities for these computations (see Section 2.4.3).

### 2.4.5  *Sea surface bubble spectra and whitecap coverage as a function of windspeed*

The rising bubbles that reach the sea surface after waves have broken affect the reflectivity and emissivity of the sea surface in many parts of the electromagnetic radiation spectrum. In the visible range, they can be seen as whitecaps. The persistence of an individual whitecap depends on the duration of the upward bubble flux and hence upon the depth to which air from a breaking wave has been entrained. It is also affected by the longevity of bubbles on the surface. The great majority of whitecap bubbles have diameters smaller than $100\,\mu$m and cannot be seen individually. The preponderance of these very small bubbles makes the area over which bubbles actually rise appreciably larger and more persistent than the visible whitecap area. In the latter, bubbles may be seen to rise to the surface for a minute or more, but they continue to give the water locally a cloudy appearance long after the whitecap as such has disappeared.

The main bubble population at the surface is related to the average fraction $A(W)$ of sea surface area that is covered by whitecaps. This coverage depends on the surface windspeed, $U$, on the wind gustiness, and on the sea–air temperature difference. Monahan and O'Muircheartaigh (1980) developed the empirical relationship

$$A(W) = 3.84 \times 10^{-6} U_{10}^{3.41}, \tag{2.47}$$

where $U_{10}$ is the windspeed at a 10-m height. Wu (1988) suggested the formula

$$A(W) = 0.2 u_*^3, \tag{2.48}$$

where $u_*(\mathrm{m\,s^{-1}})$ is the friction velocity (see subsection 5.1.2). Expressions (2.47) and (2.48) are not very different if one considers that the friction velocity is equal to the windspeed multiplied by the square root of the drag coefficient. This parameter increases slowly with increasing windspeed (see Fig. 5.6). It will be shown in Chapter 6 that kinetic energy is transferred from air to water at a rate proportional to the third power of the friction velocity $u_*$. The structure of the relation (2.48) is in keeping with this argument.

According to Wu (1992), the total bubble population $N_b$ at the sea surface is also proportional to $u_*^3$ and is given by

$$N_b = 2.9 \times 10^6 u_*^3. \tag{2.49}$$

A spectrum of bubbles at the sea surface can be obtained from the product of $N_b$ with probability densities (2.46).

Relations (2.46)–(2.49) cease to be valid at very low windspeeds. Thorpe (1982) reports that no bubbles were observed at windspeeds below $2.5\,\mathrm{m\,s^{-1}}$. He also cautions that use of these equations for an assessment of the global effect of whitecaps on gas exchange from existing wind statistics is not warranted at this stage. The model would need considerable improvements and tightening before global deductions could be made with any confidence.

### 2.4.6  *The size and flux spectra of spray droplets*

Spray was first measured photographically by Monahan (1968). The impact left by spray droplets of different sizes on oil-covered plates was sampled by Preobrazhenskii (1973). Another impaction method was used by de Leeuw (1987), who also published a critical review (de Leeuw, 1989) of the different observational methods that had been used. Wu (1990) synthesized all the quoted investigations including his own data. He also gave detailed figures and suggested parameterizations of droplet spectra as a function of size, windspeed, and elevation above the sea surface.

Spray consists mainly of the relatively large jet drops, which are ejected vertically. The vertical distribution of these drops is a function of size and windspeed. De Leeuw (1987) found that the drop concentration had a maximum some distance above the sea surface at low windspeeds. It is conceivable that this can be explained by the finite ejection speed of the jet drops. At higher windspeeds the drop concentration decreases monotonically with height. The average diameter of spray drops seems to be largest some 2 m above the surface, as indicated by Fig. 2.6.

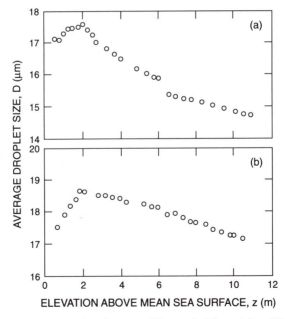

**Fig. 2.6.** Average spray droplet sizes at different heights. After Wu (1990). Data points based on de Leeuw (1987). Wind speeds: (a) 5.5 m s$^{-1}$; (b) 13.0 m s$^{-1}$.

The surface flux, $F_d(r)$, of spray drops with radius $r$ is a function of the whitecap coverage $A(W)$ or of the total bubble population $N_b$ at the sea surface. This flux might in principle be obtained from products of (2.37), (2.46), and (2.49). It seems rather doubtful whether these estimates and parameterizations are good enough individually to warrant such a deduction. The subject has been reviewed by Monahan (1989). He derived $F_d$ on the basis of physical arguments from

$$\frac{\delta F_d}{\delta r} = A(W)\tau^{-1}\frac{\delta N}{\delta r}, \tag{2.50}$$

where $\tau$ is a time constant that describes the exponential decay rate of individual whitecaps and $N$ is the number of aerosol particles with radius $r$ that are produced per unit whitecap area. Monahan also gives a rather complicated empirical formula for $F_d$, which is based on whitecap simulation tank experiments.

### 2.4.7 Environmental effects of bubbles and spray

The preponderant role of the southern hemisphere in bubble and spray production is illustrated by Fig. 2.7, which had been prepared by Blanchard (1963) before systematic satellite observations had become available. The small secondary peak in the total June–August whitecap

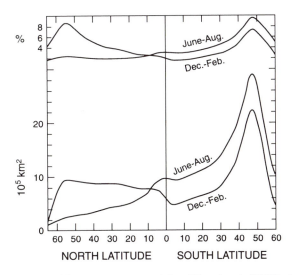

**Fig. 2.7.** Mean ocean whitecap coverage. After Blanchard (1963). Upper curves: whitecap coverage in percentage of area. Lower curves: total whitecap area in $10^5$ km$^2$ within 10° zonal strips, centered on indicated latitude.

area just north of the equator is mainly due to persistent, strong monsoon winds over the Arabian Sea.

In general, spray or bubbles do not have a significant effect on the density and local dynamics of the surrounding medium. In very roiled waters, bubbles can change the average density by up to 0.5 ppt. In sea water at a temperature of 20°C this would be equivalent to the density change produced by a 2°C increase in temperature. Such a density decrease could occur only very close to the surface under conditions of very high windspeed. Its effect on the vertical stability and on the generation of turbulence kinetic energy (1.46), is then overshadowed completely by the mechanical stirring due to the wind stress.

Similar arguments apply to the marine atmosphere. Spray increases the density of the air–water mixture and the virtual temperature is therefore reduced. In the tropics, at 27°C, a spray load of $3 \, \mathrm{g \, m^{-3}}$ would have the same effect on the air density as a 1°C decrease in temperature. The spray loads observed by de Leeuw (1989), Marks and Monahan (1989), and others above open ocean waters in winds up to $15 \, \mathrm{m \, s^{-1}}$ were considerably smaller. The effect of spray on density is negligible under these circumstances. However, the spray load is roughly proportional to the third or a higher power of the windspeed. One can expect that the effect of spray on air density may become noticeable at very high windspeeds. Spray loads have never been measured quantitatively under these conditions.

The large-scale chemical effects of bubbles and spray are much more significant than their dynamical consequences. Wind-stirred, whitecap-covered waters are often supersaturated with oxygen and nitrogen and other atmospheric gases. Likewise, the air above the sea surface can be supersaturated with moisture. Entrained air bubbles can influence chemical reactions that occur in the sea, including those in living organisms. Salt-water spray makes the marine atmosphere far more corrosive than the atmosphere over land. It is probably a major source of airborne condensation nuclei and therefore affects the physics of precipitation and the character of the global climate. Spray also contributes to the electrification of the atmosphere by carrying significant amounts of positive charge upward from the sea surface (Blanchard, 1963).

Sea spray is a dominant factor in the global salt circulation. Eriksson (1959) estimated that some 1000 million tons of salt are transferred annually from the oceans into the atmosphere. Though this figure may seem large per se, the contribution of sea spray to the total volume of water that is circulated through the atmosphere is probably insignificant. Assuming a mean oceanic salinity of 35 ppt, an annual flux of $10^{12} \, \mathrm{kg}$ of salt from the global sea surface area of $4.01 \times 10^{14} \, \mathrm{m^2}$ must be associated with an upward water flux of $0.07 \, \mathrm{kg \, m^{-2} \, year^{-1}}$. This corresponds approximately to the annual removal of 0.07 mm of water from the sea surface, which is at least 1000 times smaller than global sea surface evaporation estimates. Eriksson's estimate is not necessarily exact, but the figure cannot be much larger without causing an unduly large deposition of salt on the continental surfaces.

As bubbles rise to the surface they collect any material, dissolved or particulate, that is surface-active. When the bubbles burst, their skin is skimmed off by the resulting jet and filmdrops, which are highly enriched in this surface-active material. Marine aerosols contain therefore relatively large amounts of organic and inorganic substances. The deposition of this material on the surface of the continents is part of the general geochemical circulation. Blanchard (1989) estimates that the enrichment factor for some bacteria can exceed several hundred in jet drops that are ejected from bubbles that have risen only a few centimeters through a bacterial suspension. It has been suggested that the resulting spray-borne transport of bacteria from the sea surface may cause an occasional health hazard for people in coastal areas.

## 2.5  Sea ice

Sea ice is a heterogeneous variable substance; this limits the generality of quantitative statements about its nature. The present section deals with the formation and physical properties of sea ice. It has been based largely on a paper by Maykut (1985), which also provides an extensive bibliography from which additional information can be obtained.

### 2.5.1  *Formation and growth*

The effect of salinity on the freezing temperature of sea water was specified by (2.12) and (2.12′). If sea water is cooled at the surface, its density increases until the freezing point has been reached. Convection therefore tends to distribute the cooling throughout the depth of a water column with uniform salinity. When the feezing point has been reached, some supercooling may occur before freezing begins. The degree of supercooling probably amounts to only a few hundredths of a degree in the open ocean, but it can reach 0.2–0.4°C in protected calm water (Doronin and Khelsin, 1975).

Ice begins to form at or near the surface as a cloud of crystals with diameters of no more than 3–4 mm, which is called *frazil ice*. The individual crystals tend to have the shape of small platelets or vertically floating needles. Continuing freezing increases the number of crystals in the mixture and leads to the formation of so-called *grease ice*. The floating crystals give the water a milky appearance and a mushy consistency, which tends to suppress high-frequency surface waves. The surface freezing increases the salinity and therefore the density of the surrounding waters, which helps to maintain convection below.

The viscosity of the frazil or grease ice soup increases with the number of crystals per unit volume. The transition to solid cover begins when the ice fraction exceeds 30–40 per cent. In the presence of waves, *pancake ice* tends to form. It consists of discreet plates, with diameters on the order of 1 m. Under favorable conditions these plates then consolidate into an unbroken 'new ice' cover. Continuing growth occurs on the underside of the ice sheet, mainly by crystal growth at the ice-water interface, but also by frazil ice accretion. The individual ice crystals are hexagonal prisms. Molecular forces cause growth on their base plates to be slower than growth on the six prism sides. The growing crystals tend to align themselves with the prism sides parallel to the ice-water interface. These crystals then grow or accumulate into a layer of thin 1–3-cm long ice platelets, that are oriented normal to the ice water interface.

The accretion of frazil ice to the underside of a solid ice appears to be more important in Antarctic waters than it is in the Arctic Basin. Large concentrations of algae have been found in new ice formed by frazil ice accretion. This suggests that organic cells are being scavenged by the rising crystals from the surrounding medium but may also serve as a nucleating agent in supercooled water. Suspended silt may also be scavenged in this way.

Within the pancake ice, or within the new ice sheet into which it may consolidate, are many air bubbles and cavities filled with sea water. When freezing is rapid these cavities tend to be larger than those in ice that was formed during a more gradual drop of temperature. Continuing cooling will cause incremental freezing within the cavities. This further increases the salinity in the entrapped brine. With any further decrease in temperature, some of the salt in the cells may begin to crystallize. The resulting structured substance, formed of water ice, salt crystals, and brine together is called an *eutectic mixture.* Crystallization begins to occur in a pure solution of sodium chloride at a temperature of $-21.2°C$. A solution of calcium chloride begins to crystallize at $-55°C$. In sea water, where many salts are present simultaneously, these temperatures are somewhat reduced; in the case of NaCl, to about $-23°C$.

The rapidity of the freezing process affects the size of cavities as well as the actual salt content of the new ice. When the temperature suddenly drops to low values, brine and salt crystals may accumulate even at the surface. This can make the ice surface wet down to temperatures of $-40°C$ and greatly increase the friction on sled runners or skis. The poor thermal conductivity of ice slows the freezing rate at the bottom of thick ice sheets, causing the salinity to decrease with increasing depth. The vertical salinity distribution in a sheet of newly-formed sea ice was found by Malmgren (1927) to decrease from 6.74 ppt at the surface to 3.17 ppt at about a 1-m depth.

Conditions are reversed in old or perennial sea ice. After a few years of exposure in the polar environment, this becomes nearly salt-free at the surface and can be used as a source of drinking water. At that stage, the ice sheet thickness tends to approach a climatologically stationary state in which growth at the under side is balanced by surface melting and evaporation during summer. In the present climate, old ice in the Arctic Basin has an equilibrium thickness of about 3–4 m and an age of about 7–10 years. It follows that ice molecules migrate upward through the perennial ice sheet at a rate of 30–60 cm year$^{-1}$ until they melt at the surface or pass into the atmosphere.

It has been pointed out by Untersteiner (1968) that this upward migration of ice particles must be associated with a corresponding downward flux of salt. Repeated thawing and freezing provides a mechanism for this downward flux. The seasonal temperature rise melts ice surrounding the brine-filled cavities. As the melting goes on, the cavities grow larger in size and joint, allowing the highly saline trapped solution to trickle down. This process is favoured by the hydrostatic head of melt water at the surface and also by converging horizontal large-scale movements which can pack and press the ice together into pressure ridges and hummocks. The pressure produced on the internal cavities during freezing also assists in the expulsion of brine.

Surface ablation in the Southern Ocean is less important than it is in the Arctic Basin. The large reduction or disappearance of the antarctic ice pack during summer is caused presumably by melting at the ice-water interface. Some of the necessary heat can be supplied by the absorption of short-wave radiation in open leads or near the ice edge. The remainder probably comes from mixing with warmer water below the pycnocline.

### 2.5.2 *Physical properties of sea ice*

Of particular concern for the study of atmosphere-ocean interaction are the effects of sea ice upon light transmission and heat exchange between the two media. The resistance of an ice pack to deformation—its inertia and large scale horizontal viscosity—affects the transmission of momentum and vorticity to the ocean waters below.

At 0°C the density of pure ice is 916.76 kg m$^{-3}$; that of pure liquid water is 999.87 kg m$^{-3}$, which is about 9 per cent higher. Because of its air and salt content, the density of sea ice may be either above or below that of pure ice. Values between 0.92 and 0.86 have been reported. Pure ice contracts when being cooled at a practically constant rate. Its coefficient of thermal expansion $\alpha_i = -\partial \ln (\rho_i)/\partial T = 1.7 \times 10^{-4}$. On the other hand, lowering of the temperature in sea ice may cause incremental freezing of entrapped brine, which can force expansion of the ice mass as a whole. The resulting expansion can be rather large at relatively high subfreezing temperatures. Sea ice with a salinity of 4 ppt only tends to contract at temperatures below $-10°C$. Contraction only begins to occur at still lower temperatures if the salt content is higher. The expansion of an ice sheet during freezing can contribute to the formation of pressure ridges and the expulsion of brine.

Values of $\alpha_i$ as a function of temperature (°C) and salinity (ppt) have been tabulated by Sverdrup et al. (1942) on the basis of observations by Malmgren (1927). At temperatures $-1 > T > -14$, the empirical equation

$$\alpha_i = 0.383 - 0.093T - 45.953sT^{-2} \quad (°C^{-1}) \qquad (2.51)$$

expresses these values with a mean square error less than $10^{-5} °C^{-1}$. Equation (2.51) indicates that the effect of salinity diminishes at low temperatures.

Variations in the specific heat of sea ice are related to those of the expansion coefficient, because they too are affected by freezing of entrapped brine or by melting of ice surrounding brine-filled cells. The amounts of heat involved in these processes can be large. The specific heat of sea ice with a salinity of 8 ppt and a temperature of $-2°C$ was computed by Malmgren to be $45420 \, J \, kg^{-1} °C^{-1}$. That is more than 10 times the specific heat of liquid water. At $-8°C$ the same ice would have a specific heat of only $4232 \, J \, kg^{-1} °C^{-1}$ and at $-20°C$ this drops to $2514 \, J \, kg^{-1} °C^{-1}$. At very low temperatures, when most of the salts have crystallized out, the specific heat of sea ice approaches that of pure ice, which is about $2010 \, J \, kg^{-1} °C^{-1}$. An empirical formula for the specific heat of sea ice, $c_i$, as a function of temperature and salinity was developed by Untersteiner (1961). A slight modification of that formula by Ono (1967), which is valid for $T > -8°C$, has the form

$$c_i = 2114 + 7.5T + 18050sT^{-2} \quad (J \, kg^{-1} °C^{-1}). \qquad (2.52)$$

At 0°C, the heat of fusion of pure water $L_0 = 333691 \, J \, kg^{-1}$. It becomes less at lower temperatures. In sea ice the latent heat of fusion, $L_i$, is not constant because the inclusion of salt and brine permits melting to take place whenever the temperature rises, no matter how cold it may have been. Data by Malmgren (1927) for values of $L_i$ at initial temperature of $-1$ and $-2°C$ at different salinites have been reproduced in the Smithsonian Meteorological Tables (1971). The table is accompanied by an empirical equation for $L_i$. Other empirical relations were developed by Ono (1967) and by Maykut (1985). The last one has the form

$$L_i/L_0 = 1 - s_i(0.000532 - 0.04919/T_i). \qquad (2.53)$$

Sea ice with a salinity of 10 ppt has a latent heat of fusion that is only about half that of pure ice at $-1°C$.

The value of $L_i$ indicates the amount of heat that must be conducted through an ice cover to permit melting or freezing at the ice-water interface. This conductive heat flux, $Q_i$, can be expressed in the form

$$Q_i = -k_i \, \partial T_i/\partial z.$$

Compared to pure ice, the conductivity, $k_i$, of sea ice is lowered by the inclusion of brine, air bubbles and salt suspensions. These effects were considered theoretically by Schwerdtfeger (1963), who also published tables of $k_i$ as a function of temperature, salinity, and ice density. A parametric equation by Untersteiner (1961) represents the conductivity by

$$k_i = 20.77 + 0.13s_i/T_i \quad (W\,m^{-1}\,°C^{-1}). \qquad (2.54)$$

It can be seen that ice with a relatively high temperature and salinity has both a large heat capacity and a small heat conductivity. An ice cover of this type has therefore a large thermal inertia and resistance to change. As the porosity of the ice decreases with depth, we may expect its conductivity to increase. This is confirmed by observations as shown in Table 2.7.

The radiant energy transfer across sea ice is affected by the surface albedo and by the extinction of solar radiation within the ice sheet or ice pack. Albedo can vary from 0.80 to 0.85 for snow-covered polar ice to 0.30 to 0.50 for melting first year ice. The albedo decreases with increasing wavelength. This decrease becomes very pronounced for ice sheets that have much liquid water near the surface.

Ice is a translucent material. It generally permits transmission of some

**Table 2.7.** Heat Conductivity of Sea Ice $(W\,m^{-1}\,°C^{-1})$ at Different Depths (Malmgren, 1927)

| Depth (m) | 0.0 | 0.25 | 0.75 | 1.25 | 2.0 |
|---|---|---|---|---|---|
| Conductivity | 0.7 | 1.4 | 1.9 | 2.1 | 2.1 |

light to the water below. The remainder, which is absorbed, can affect brine concentration and biological activity within an ice sheet. Ice is most transparent for greenish light with wavelengths between 0.40 and 0.55 $\mu$m. Absorption increases with increasing wavelength as well as with the amount of liquid water, algae, and other contaminants that the ice contains. Red and infrared radiation with wavelengths larger than 0.7 $\mu$m can be absorbed even by a relatively thin layer of uncontaminated ice and therefore cannot reach the water below. The amount of all transmitted light is much reduced by snow which has a much higher extinction coefficient than solid ice. Thick layers of algae that bloom in spring at the ice underside also intercept light transmission very effectively, particularly in the 0.45–0.55 $\mu$m band.

Detailed spectral data for the albedo and absorption associated with different types of ice cover have been published by Grenfell and Maykut (1977). A summary of their studies and graphs of spectral distributions can be found in Maykut (1985). The effect of short-wave radiation on the heat budget of pack ice with open leads has been studied by Maykut and Perovich (1987).

# 3

# RADIATION

The earth receives virtually all of its energy from the sun in the form of electromagnetic radiation. This radiation is absorbed, reflected, and scattered by the earth's surface, the ocean, and the atmosphere. The absorbed radiation is transformed into heat and other forms of energy, and eventually it is returned to space as low-temperature terrestrial radiation.

It is clear that radiation is of fundamental importance to atmosphere–ocean interaction. There exists an adequate body of literature on the subject from an introductory treatment by Fleagle and Businger (1980) to specialized monographs by Kondratjev (1969), Liou (1980), and Goody and Yung (1989). Here it will suffice to introduce the basic concepts and focus on the applications to the air–sea interface.

## 3.1 Definitions

Radiation in the atmosphere and ocean comes from all directions simultaneously. The radiation energy per unit time coming from a specific direction and passing through a unit area perpendicular to that direction is called the *radiance, I.* The *irradiance, $F_i$,* or *radiant flux density*, is the radiant energy that passes through a unit horizontal area per unit time coming from all directions above it. Therefore

$$F_i = \int_0^{2\pi} I \cos \theta \, d\omega, \qquad (3.1)$$

where $\theta$ is the zenith angle and $d\omega$ is an infinitesimal solid angle. The $\cos \theta$ reflects the projection of the horizontal unit area into the direction from where $I$ comes. The limits 0 and $2\pi$ of the integral reflect the hemisphere of directions above the unit area.

When the radiance is independent of direction it is called *isotropic*. Equation (3.1) may then be integrated to yield

$$F_i = \pi I. \tag{3.2}$$

The irradiance from below the unit area is also called *exitance* and is denoted by $F_e$. The *net irradiance*, $F_n$, is defined by

$$F_n = F_e + F_i. \tag{3.3}$$

For isotropic radiance, the net irradiance $F_n = 0$. The fluxes are positive when upward and negative when downward.

The interactions between radiation and matter may take various forms. They include *refraction, reflection, scattering, diffraction, absorption,* and *emission.* All these interactions are described by the theory of electromagnetic waves (e.g., Panofsky and Phillips, 1962). The full theory will not be developed here, but a number of basic and useful relations will be introduced to describe the characteristics of the interactions mentioned previously.

Visible and near infrared radiation propagates through the atmosphere without much interference or attenuation. However, even on an exceptionally clear day, the speed of light in the atmosphere, $c_a$, is slightly less than the speed of light in vacuum, $c_0$. The radio $c_0/c_a = m_a$ is called the *index of refraction* for air. This ratio differs by less than 3 ppt from unity and varies little with wavelength. The index of refraction for water, $m_w$, is much larger and varies from 1.332 for red light with a wavelength of $0.656\,\mu$m to 1.344 for violet light of $0.404\,\mu$m. The refractive index also varies with temperature and salinity of sea water. These variations are described by, for example, Jerlov (1976).

The abrupt change of the index of refraction going from air to water or vice versa led to the formulation of *Snell's* law.

$$\sin\theta_a = \frac{m_w}{m_a}\sin\theta_w \simeq 1.34\sin\theta_w, \tag{3.4}$$

where $\theta_a$, $\theta_w$ are the angles between the direction of propagation and the normal to the interface and $m_a, m_w$ are the refractive indices in air and water, respectively. The angle $\theta_w$ cannot exceed a limiting value $\theta_w^\circ =$ arc sin $(1.34)^{-1} \approx 48°$. Light that reaches the interface from the water at a larger angle of incidence cannot pass into the air but is totally reflected, and rays that enter the water from the air are always deflected to an angle $\theta_w < \theta_w^\circ$.

In general, some radiation is reflected and some is transmitted at an interface. For $\theta_w < \theta_w^\circ$ the *reflectance*, that is, the ratio of the reflected radiance, $I'$, to an incident unpolarized radiance, $I$, is given by *Fresnel's law*

$$\frac{I'}{I} = \frac{1}{2} \left[ \frac{\sin^2 (\theta_a - \theta_w)}{\sin^2 (\theta_a + \theta_w)} + \frac{\tan^2 (\theta_a - \theta_w)}{\tan^2 (\theta_a + \theta_w)} \right] \equiv r(\theta), \qquad (3.5)$$

where $\theta$ is the angle of incidence as defined earlier for rays that reach the interface from either side. The reflectance increases with $\theta$ as illustrated by Fig. 3.1.

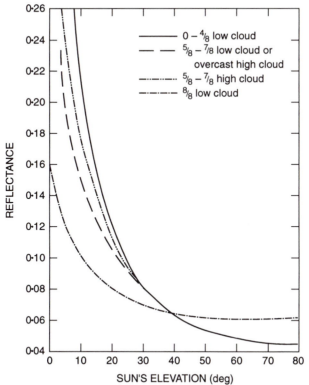

**Fig. 3.1.** Short-wave reflectance of the sea surface. After Deacon and Stevenson (1968).

The net irradiance through a partly reflecting horizontal interface is, using (3.1) and (3.3),

$$F_n = \int_0^{4\pi} [1 - r(\theta)] I \cos \theta \, d\omega. \tag{3.6}$$

If the incident irradiance is diffuse, we can write (3.6) in the form

$$F_n = F_e \left[ 1 - \int_{\pi - \theta_w^\circ}^{\pi} r(\theta) \sin 2\theta \, d\theta \right] + F_i \left[ 1 - \int_0^{\pi/2} r(\theta) \sin 2\theta \, d\theta \right]$$

$$= F_e [1 - A_1(\theta_w^\circ)] + F_i (1 - A_2) \tag{3.7}$$

For visible light at an air–water interface $A_2 \sim 0.07$, which means that about 93 per cent of a diffuse irradiance from above will be transmitted.

Usually not all radiation reaching a surface is reflected and transmitted; some or all of it may be absorbed. For a given wavelength, the fraction of the incoming radiation that is reflected, $r_\lambda$, the *reflectance*, the fraction that is transmitted, $\tau_\lambda$, the *transmittance*, and the fraction that is absorbed, $a_\lambda$, the *absorptance*, must add up to unity, therefore

$$r_\lambda + \tau_\lambda + a_\lambda = 1. \tag{3.8}$$

For a *black body* $a_\lambda = 1$ and $r_\lambda = \tau_\lambda = 0$ for all wavelengths.

A substance that absorbs radiation of a particular wavelength, $\lambda$, also emits radiation of the same wavelength. The rate at which emission takes place is a function of the temperature and the wavelength. *Kirchhoff's law*, a consequence of the second law of thermodynamics, states that the *emittance*, $\epsilon_\lambda$, which is the ratio of the radiance and the maximum possible radiance, is equal to the absorptance, therefore

$$\epsilon_\lambda = a_\lambda. \tag{3.9}$$

This means that a black body emits the maximum possible amount of radiation for a given temperature and wavelength. Black body irradiance,

$F_\lambda^*$, as a function of temperature and wavelength, is given by *Planck's radiation law.*

$$F_\lambda^*(\lambda, T) = 2hc^2\lambda^{-5}(e^{hc/k\lambda T} - 1)^{-1}, \qquad (3.10)$$

where $h$ is Planck's constant, $c$ is the speed of light in vacuum, and $k$ is Boltzmann's constant. The total black body irradiance is found by integrating (3.10) over all wavelengths with the result

$$F^*(T) = \sigma T^4. \qquad (3.11)$$

This is the *Stefan–Boltzmann law*, according to which the irradiance emitted by a black body varies as the fourth power of the absolute temperature. The constant $\sigma$ has the value $5.67 \times 10^{-8} \, \text{W m}^{-2} \, \text{K}^{-4}$ and is called the *Stefan–Boltzmann constant*.

The wavelength of maximum irradiance, $\lambda_m$, for black-body radiation is found by differentiating (3.10) with respect to wavelength and equating the result to zero. This results in *Wien's displacement law.*

$$\lambda_m = \alpha/T, \qquad (3.12)$$

where $\alpha = 2897.8 \, \mu\text{m K}$. Figure 3.2 shows the irradiance of the sun at its

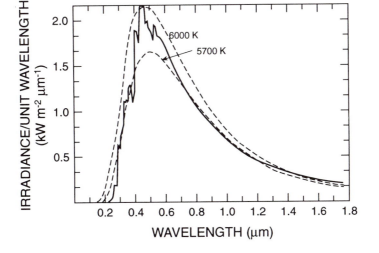

**Fig. 3.2.** Observed solar spectral irradiances corrected to mean solar distance (solid line) and corresponding black-body irradiances for temperatures of 6000 and 5700 K (dashed lines). After Johnson (1954).

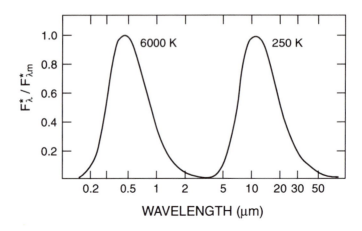

**Fig. 3.3.** Normalized black-body irradiance per unit wavelength calculated from (3.2) and (3.10) for temperatures of 6000 and 250 K. After Fleagle and Businger (1980).

mean distance, as observed by Johnson (1954), and the black body irradiance for a few temperatures close to the solar temperature, illustrating (3.10)–(3.12). Figure 3.3 displays the normalized black body irradiation for 6000 K and for 250 K, characteristic of the *short-wave* incoming radiation and of *long-wave* outgoing radiation, respectively. The figure clearly shows that there is virtually no overlap between short-wave and long-wave radiation. This allows us to separate the discussion of solar radiation from the discussion of terrestrial radiation as is done in Sections 3.2 and 3.3.

Radiation that passes through a medium such as the atmosphere or the ocean usually encounters absorption, emission, and scattering. In general, following a monochromatic beam of radiance, $I_\lambda$, through a medium, the interactions will be proportional to the thickness of the layer, the density, $\rho_c$, of the medium, and the radiance,

$$\frac{dI_\lambda}{dx} = -k_\lambda \rho_c (I_\lambda - I_\lambda^*) - k_{s\lambda} I_\lambda + \Gamma_\lambda, \tag{3.13}$$

where $k_\lambda$ is the *monochromatic absorption coefficient* of the layer, $I_\lambda^*$ the *black-body radiance* $(I_\lambda^* = F_\lambda^*/\pi)$, $k_{s\lambda}$ the *scattering coefficient*, which indicates how much radiation is scattered into other directions per unit radiance, and $\Gamma_\lambda$ the contribution to radiance by scattering from other directions into the direction of the beam.

Equation (3.13), which is known as *Schwarzschild's equation*, usually can only be integrated numerically. However, some simple cases may be treated analytically. Of special interest is the case in which $I_\lambda^*$ and scattering are negligible. In this case (3.13) may be integrated to

$$I_\lambda = I_{\lambda 0} e^{-k_\lambda \rho_c x}, \tag{3.14}$$

which is known as *Beer's law*. This equation has many practical applications, including the measurement of concentrations of trace constituents. The quantity $\rho_c x$ is frequently referred to as *optical thickness, u.*

If scattering is the only process that takes place in (3.13) it can be written in a more specific form

$$\frac{dI_\lambda}{dx} = -k_{s\lambda} \rho_c I_\lambda + \Gamma_\lambda = \int_0^{4\pi} [I_\lambda(\theta', \alpha') - I_\lambda(\theta, \alpha)] S_{\theta,\alpha,\theta',\alpha'}(\lambda) \, d\omega$$

$$= \Gamma_\lambda(\theta, \alpha) - \int_0^{4\pi} I_\lambda(\theta, \alpha) S_{\theta,\alpha,\theta'\alpha'}(\lambda) \, d\omega, \tag{3.15}$$

where $\theta$ and $\alpha$ define the beam in the x-direction, $\theta'$ and $\alpha'$ define another particular direction, and $S$ denotes the scattering function. The last integral measures the amount of radiation scattered out of the beam into all other directions. $I_\lambda(\theta, \alpha)$ can be taken out of the integral and a scattering coefficient, $k_{s\lambda}$, may be defined by

$$k_{s\lambda} = \frac{1}{\rho_c} \int_0^{4\pi} S_{\theta,\alpha,\theta',\alpha'}(\lambda) \, d\omega. \tag{3.16}$$

Analytical expressions for $S$ and $k_s$ have been derived by Gustav Mie in 1908 and presented in great detail by van de Hulst (1957). For spherical particles the Mie series may be written as

$$k_{s\lambda} = \frac{2\pi}{3} \frac{N}{\rho_c} \left(\frac{m^2 - 1}{m^2 + 2}\right)^2 d^2 \left[\left(\frac{\pi d}{\lambda}\right)^4 + \frac{6}{5}\left(\frac{m^2 - 1}{m^2 + 2}\right)\left(\frac{\pi d}{\lambda}\right)^8 + \cdots\right] \tag{3.17}$$

where $N$ is the number of scatterers per unit volume, $d$ is their diameter, and $m$ their refractive index relative to the medium for radiation with wavelength $\lambda$. When $(d/\lambda)^4 \ll 1$, we can neglect all but the first term of the expansion in (3.17), and we then speak of *Rayleigh scattering*. The scattering function for Rayleigh scattering is

$$S = \frac{16\pi}{3}(1 + \cos^2 \varphi)\beta_s, \qquad (3.18)$$

where $\varphi$ is the angle of intersection between the directions $(\theta, \alpha)$ and $(\theta', \alpha')$ and $\beta_s \equiv k_s \rho_c$. Equation (3.18) shows Rayleigh scattering to be twice as large along the direction of the ray as it is in the normal direction.

## 3.2   Solar radiation

The sun is an average star in our galaxy. It is a gaseous sphere with a diameter of $1.42 \times 10^6$ km. Its distance to the earth is $150 \times 10^6$ km, whereas the next closest star is $3 \times 10^5$ times as far away. Each square centimeter of the solar surface emits on the average about 6.2 kW of radiation. Only a minute fraction of the output of the sun is intercepted by the earth; nevertheless, it is sufficient to generate winds and ocean currents and to sustain life on earth. The amount of solar radiation received per unit time and per unit area, perpendicular to the sun's rays at the top of the atmosphere, is called the *solar constant*. The solar constant has been monitored by satellite and is found to be 1370 Wm$^{-2}$ ± 1% (Willson, 1984). The variability appears to be less than 0.1 per cent per year.

Electromagnetic radiation emitted by the sun is not very different from radiation emitted by a black body. This similarity allows us to estimate the temperature of the surface of the sun. The Stefan–Boltzmann law (3.11) together with the observed solar constant indicates a solar surface temperature of 5733 K. Alternatively, Wien's displacement law (3.12), in combination with the wavelength of maximum radiance, 0.4750 $\mu$m, yields a temperature of 6100 K.

### 3.2.1   *The net short-wave irradiance at the sea surface*

The short-wave radiation that reaches the sea surface has passed through the atmosphere. The atmosphere is only partly transparent for short-wave radiation. Ultraviolet radiation with wavelengths shorter than 0.29 $\mu$ is absorbed by ozone. Although the total amount of ozone in the atmosphere is less than 1 part in 10,000, its presence shields life on land and in the upper ocean from lethal high-frequency radiation. The atmosphere is also partly

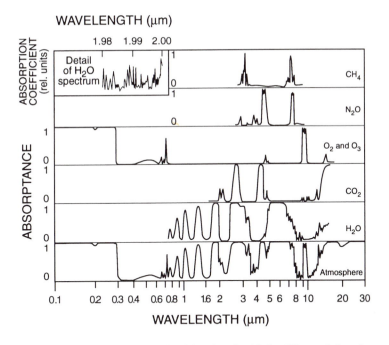

**Fig. 3.4.** Absorption spectra for $H_2O$, $CO_2$, $O_2$, $O_3$, $N_2O$, $CH_4$, and the absorption spectrum of the atmosphere. After Fleagle and Businger (1980).

transparent for infrared short-wave radiation (0.7–4.0 $\mu$m). Oxygen, water vapour, and carbon dioxide are the main absorbing constituents, as can be deduced from Fig. 3.4. The amount of short-wave infrared radiation, which passes through the atmosphere and is absorbed at the sea surface, can lead to significant warming of the surface layer itself.

The atmosphere is quite transparent to visible radiation (0.4–0.68 $\mu$m), which contains nearly 60 per cent of the energy emitted by the sun. On a clear, cloudless day, the visible radiance of direct sunlight is reduced only by scattering. On the other hand, scattering increases the light that comes down to the surface from other directions of the sky. The diameter of air and water molecules is so small compared to the wavelength of light that higher-order terms in the expansion (3.17) are practically negligible, and Rayleigh scattering, inversely proportional to the fourth power of the wavelength, is produced. The sun radiates most intensely at a wavelength of 0.47 $\mu$m in the blue-green part of the spectrum. This light is scattered about four times more effectively than red light with a 0.66 $\mu$m wavelength. The well-known explanation for the blue colour of the sky and of the sea is based on this phenomenon.

The composition of sunlight varies with the sun's elevation. When the sun is about 20° above the horizon, the rays have to pass through three times as much air as when the sun is overhead. The longer the path through the atmosphere, the more blue light is scattered out of the direct solar beam. As a result, the peak of the spectrum of direct sunlight is shifted at low solar angles from blue-green in planetary space to yellow or red at the earth's surface. Part of the blue light that has been scattered out of the direct beam is lost into space; the remainder reaches the surface as sky radiation, which is nearly—though not entirely—diffuse. Its contribution to the irradiance of the sea surface in clear weather increases with the sun's zenith distance from 16 per cent, when the height of the sun is 60°, to about 37 per cent, when it is 10° above the horizon.

The radiance is more nearly omnidirectional when clouds reduce the amount of directly transmitted light. Empirical formulas for the reduction of irradiance are given in Section 3.4.

### 3.2.2   Reflection at the sea surface

Some of the radiation that reaches the sea surface is reflected back into the atmosphere. The reflectance of the surface, defined as the ratio of the reflected to the incident radiance, is not identical with the albedo. The albedo of the sea surface is the ratio of all short-wave radiation leaving the surface to the incident irradiance. It includes light from specular reflection at the surface alone, with upwelling light that passes the interface from below, after scattering within the water.

The reflectance increases with the angle of incidence and can be computed from Fresnel's law (3.5). When the sun is low, more light is reflected and the fraction of radiant energy that penetrates the water is reduced. The same applies to that part of the diffused sky radiation which comes from the regions close to the horizon. Clouds increase the reflectance when the sun is high because their presence then increases the fraction of radiation that comes from near the horizon. When the sun is low, clouds allow more light to penetrate the water because they then increase the light scattered down vertically. All these effects are illustrated in Fig 3.1.

When the sea is rough, the local reflectance becomes a function of time. Radiation that comes from a high elevation is in this case more likely to meet a sloping wave surface, causing a higher fraction of the irradiance to be reflected. When radiation from near the horizon reaches a rough sea, it does not have a grazing incidence as on a calm surface; rather, it meets wave flanks that slope toward the source of the light. The local angle of incidence is therefore smaller, causing less of this horizontal radiation to be reflected and more to penetrate into the sea. The resultant variations of the reflectance were used by Saunders (1968) who worked with infrared

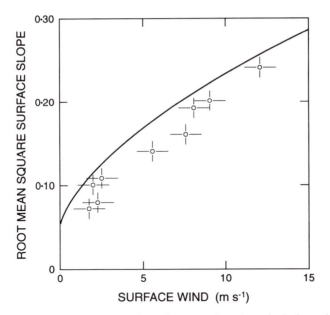

**Fig. 3.5.** Root mean square sea surface slope as a function of windspeed. The solid line represents the relation obtained from glitter pattern observations by Cox and Munk (1954). The individually plotted points are surface slope estimates based on infrared measurements by Saunders (1968).

radiation, as reported shortly, to develop a method that relates variations of the mean square slope of the sea surface at different winds to the observed variations of the reflected radiance. Similar results were achieved earlier by Cox and Munk (1954) who related the slope variance to the distribution of sparkles produced by the specular reflection of the sun on the many facets of a waving surface that have the appropriate orientation toward the observer. Results obtained by both methods are shown in Fig. 3.5. In Section 4.3 we will show the mean square slope to be an important parameter in the characterization of sea states. The Cox and Munk method in particular can be used to specify not only this parameter, but the whole spectrum of wave slopes.

The reflectance of the sea surface is influenced by other factors as well. Bubbles scatter light, and this gives whitecaps a diffused reflectivity which generally is much larger than that of any singly connected water surface. The albedo in sea ice can vary from less than 0.4 for melting, dirty ice to as much as 0.9 for ice which is covered with fresh snow. Maykut and Untersteiner (1971) assume an average albedo of 0.64 for an ice cover, in agreement with Soviet references quoted in their paper. Though small, the amount of light that does penetrate into the ocean through an ice cover is of great biological importance.

### 3.2.3  *Absorption of solar radiation in the ocean*

Most of the solar radiation that does pass the interface is visible and is ultimately absorbed by the ocean. In clear water, blue radiation may penetrate to considerable depths, whereas red radiation is mainly absorbed in the first few meters. Figure 3.6 shows the transmittance at various wavelenths of a one-meter layer of water for several ocean areas, coastal waters, and for Crater Lake. We see a large difference in transmittance between clear and turbid water. In turbid water the transmittance is much smaller. Selective absorption by chlorophyll and other pigments of the small-scale sea flora, by dust and by dissolved so-called yellow substances reduces the amount of violet and blue light transmission, causing a shift of colour from blue to green or brown or even reddish, as indicated by the lowest curve in Fig. 3.6.

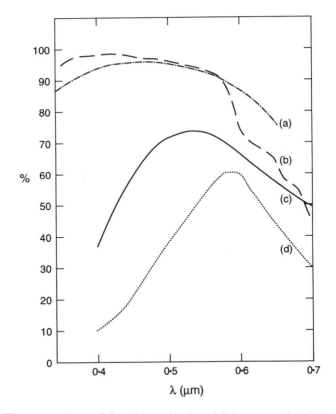

**Fig. 3.6.** The percentage of irradiance in the visible range that is transmitted through 1 m of water: (a) clear Mediterranean water (Morel, 1965); (b) Crater Lake (Smith and Tyler, 1967); (c) coastal water of medium; and (d) low transparency (Jerlov, 1965).

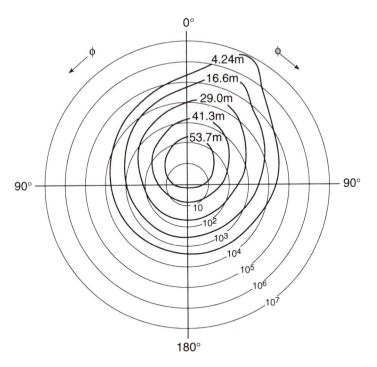

**Fig. 3.7.** The radiance distribution in the vertical plane through the sun at elevation 56.6° as a function of zenith angle and depth, as measured by Tyler and Preisendorfer (1962) in Crater Lake. The intensities in relative units are given by the circles. Depth labels on measured curves are in metres.

As in the atmosphere, scattering in the ocean reduces the directionality of the flux. Any light that passes the surface is refracted so that at the very start the rays form an angle of less than 49° with the vertical in agreement with (3.4). Scattering and the sifting-out by absorption of the more oblique rays causes the radiance to become more nearly symmetrical about the vertical as the depth increases. This effect can be computed by solving (3.13) as an integral equation. It is illustrated in Fig. 3.7, which is based on lake measurements by Tyler and Preisendorfer (1962), and represents the radiance distribution in the vertical plane through the sun, at different depths. A similar three-dimensional display is called a *radiance distribution solid.* In a medium containing only scattering centres, the limiting shape of this figure at great depths is a sphere; in a medium without scattering and with absorption only, it would be a vertical line.

Figure 3.7 shows that some of the light comes up from below (i.e., from a zenith angle of more than 90°). In mountain lakes and in clear ocean areas, this upwelling light may have reached considerable depth before it returns to the surface. All red and yellow light has been filtered out during

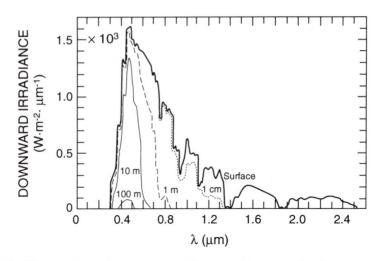

**Fig. 3.8.** The complete solar spectrum of downward irradiance in the sea at various depths. After Jerlov (1976).

this long journey; the remainder gives these water bodies their deep blue or even violet colour—Homer's wine-dark seas. In turbid coastal waters, which contain much suspended matter, the light cannot penetrate deeply. Much of it is scattered back from a limited depth to emanate from the surface at a shallow angle. As a result, such waters are not only less blue but appear to have a lighter hue than clear water when observed from ship or aircraft.

All radiation that is absorbed in the ocean is converted into internal energy. This solar heating is distributed over various depths as we have seen from the transmittance variability in Fig. 3.6. The spectra of downward irradiance in clear ocean water are presented in Fig. 3.8. We see that in the infrared practically all radiation is absorbed in the top 1 m and that a substantial fraction of this is absorbed in the top 1 cm. The strong absorption in this surface layer leads to relatively rapid warming near the surface with some interesting consequences which will be discussed in Section 5.5. The larger amount of radiation in the visible range is distributed over a deeper layer and contributes slowly to the heating of that layer.

The absorption of short-wave radiation may be approximated by a series of exponentials in the form

$$F_{is}(z) = F_{is}(0) \sum_n a_n e^{b_n z}, \tag{3.19}$$

where $F_{is}$ is the short-wave irradiance that penetrates the sea surface; $a_n$

and $b_n$ are coefficients, with the condition that $\Sigma a_n = 1$. Soloviev (1982) suggests three terms with $a_1 = 0.28$; $a_2 = 0.27$; $a_3 = 0.45$, and $b_1 = 71.5$; $b_2 = 2.8$; $b_3 = 0.07 \, \text{m}^{-1}$. Price et al. (1986) use two terms: $a_1 = 0.4$; $a_2 = 0.6$, and $b_1 = 15.0$; $b_2 = 0.5 \, \text{m}^{-1}$. (Note that $z < 0$ in the ocean.)

This type of equation allows a fraction (infrared) of the radiation to be absorbed near the surface and a fraction (blue) to penetrate rather deep into the ocean.

The rate of sensible heating of the oceans by the sun is slow because of the great heat capacity of the oceans. Even in the tropics, where the irradiance may be as high as $1000 \, \text{Wm}^{-2}$, the heating is only 0.036°C per hour if all of this radiation is absorbed in the upper 10 m. The cumulative effects are, of course, very large. The amount of solar energy that is absorbed directly by the oceans is between 1.5 and 3 times as large as that absorbed directly by the entire atmosphere. It is more than three times as large as that absorbed by all the global land surfaces. A relatively small fraction of this absorbed energy is transported over large horizontal distances by ocean currents. The larger part is stored locally or in the immediate vicinity, and is later transmitted to the atmosphere mainly by evaporation and by long-wave radiation.

## 3.3   Terrestrial radiation

### 3.3.1   *Long-wave emission from the sea surface*

Long-wave radiation is absorbed and emitted in the top 1 mm of the sea. For this radiation the sea surface approximates a black body. The exitance, $F_{el}$, from the sea surface to the atmosphere may be approximated by

$$F_{el} = \epsilon_f \sigma T_s^4 - (1 - \epsilon_f)F_{il}, \tag{3.20}$$

where $\epsilon_f$ is the *flux emittance* of the sea surface for all wavelengths of the black-body spectrum. The actual emittance, $\epsilon$, is about 0.98 for wavelengths from 3 to 50 $\mu$m with a slight dependence on temperature and salinity. $T_s$ is the temperature of the sea surface, which may be slightly different form the 'bulk' temperature near the sea surface, i.e., a few cm below the surface (see 5.5).

The flux emittance is related to the actual emittance by

$$\epsilon_f = 2 \int_0^{\pi/2} \epsilon(\theta) \cos\theta \sin\theta \, d\theta, \tag{3.21}$$

assuming that $T_s$ appears to be the same for all directions. This is usually the case but it may not be quite valid when there is a strong temperature gradient from the surface down. The emittance, $\epsilon(\theta)$, is a function of the zenith angle because in this case

$$\epsilon(\theta) = 1 - r(\theta). \tag{3.22}$$

Equation (3.22) has been illustrated in Fig. 3.1 for various conditions of cloudiness. The reflectance, $r$, behaves in similar ways for long-wave and short-wave radiation. Thus, under a clear sky, the long-wave radiance of a calm sea is largest when seen directly from above ($\theta = \pi$). As the zenith angle, $\theta$, decreases, the radiance decreases because $(1 - r(\theta))\sigma T_s^4/\pi$ decreases faster than the reflected radiance, $r(\theta)I_i(\theta)$, increases. This trend continues to a zenith angle of about 95°, when the reflected radiance becomes the dominant term. Both $r(\theta)$ and $I_i(\theta)$ increase rapidly upon approaching the horizon ($\theta = 90°$). The calm sea has, therefore, a minimum radiance about 5° below the horizon. At the horizon itself $\epsilon = 0$ and $r = 1$, and the radiance of sea and sky merge into each other. In Fig. 3.9 this distribution is illustrated by observation of the radiance between

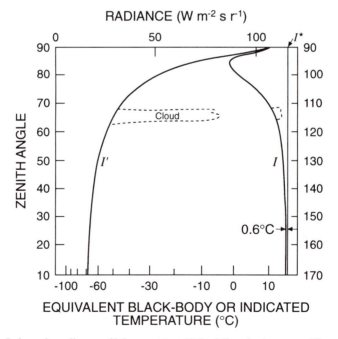

**Fig. 3.9.** Infrared radiance ($8.3\,\mu\mathrm{m} < \lambda < 12.5\,\mu\mathrm{M}$) of the sea ($I$) and of a cloudless sky ($I'$) as a function of zenith angle. After Saunders (1968). $I^*$ is the black-body radiance at the surface temperature of 15°C. The lower scale represents black-body temperatures corresponding to the radiance marked on the upper scale.

8.3 $\mu$m and 12.5 $\mu$m by Saunders (1968). For this wavelength interval, the cloud-less sky is nearly transparent and its radiance, $I'$, is low, as can be seen in Fig. 3.9 from the equivalent black-body temperature. However, close to the horizon the optical path is so long that the atmosphere becomes nearly opaque. Therefore, the long-wave radiance of the sky has a minimum at the zenith and a maximum at the horizon, where it approaches the radiance of a black body at surface air temperature.

When clouds are present, the long-wave radiation from the sky is almost diffuse and the difference between vertical and horizontal radiances diminishes greatly. The long-wave absorption and emission of clouds depends on their liquid-water content. For a liquid-water path of 40 g m$^{-2}$, which corresponds to a cloud thickness of 100–300 m at the top of the boundary layer and somewhat more for higher clouds, the radiance is almost equal to black-body radiance at cloud temperature. Figure 3.10 gives more detail on this issue for stratocumulus clouds in the boundary layer. The result is that, with a cloud cover, the radiance from the sea surface, including the reflected radiance from the clouds, approaches the black-body radiance.

When the sea surface is rough, the reflectance near the horizon is less than unity, as we have seen in Section 3.2, and its radiance differs discontinuously from that of the sky at the horizon. The sea surface

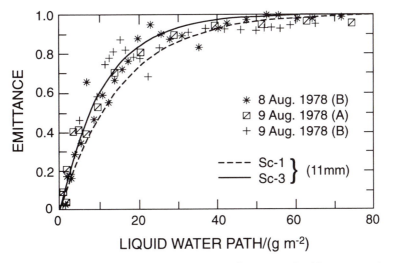

**Fig. 3.10.** Cloud emittance (9.5 $\mu$m $< \lambda <$ 11.5 $\mu$m) measured with a narrow-beam radiometer as a function of the liquid water path. The profiles were obtained from an aircraft during descent (A) and ascent (B) through stratocumulus cloud. Curves are calculated for two cloud droplet spectra corresponding to cloud top (Sc-1) and cloud base (Sc-3). After Schmetz et al. (1983).

radiance in this case is much closer to being uniform than it is for the smooth sea surface and (3.20) becomes an excellent approximation.

When the sea surface is rough and the sky is overcast, the radiance from the surface is equal to the black-body radiance at the sea surface temperature for all practical purposes. Nevertheless, there still may be transfer of radiation from the sea surface to the cloud base. This may have some impact on the cloud layer, especially when the cloud layer is cooler than the sea surface. As we shall see in Fig. 6.20, the cloud base may be heated by several degrees Celsius per day at the base, which enhances the unstable stratification and turbulence within the cloud layer.

### 3.3.2 *Radiative transfer in the lower atmosphere*

The upward and downward long-wave irradiances depend on the vertical distributions of temperature, water vapour, $CO_2$, and other trace gases in the atmosphere. The dependence can be treated numerically. A relatively simple treatment can be found in Fleagle and Businger (1980), while Goody and Yung (1989) give a more complete discussion of the subject.

We shall restrict ourselves to a basic sketch of the problem. We start by asking what the contribution to the irradiance is of a volume element at height $z$ with thickness $dz$ to a reference level $z_r$, as illustrated in Fig. 3.11.

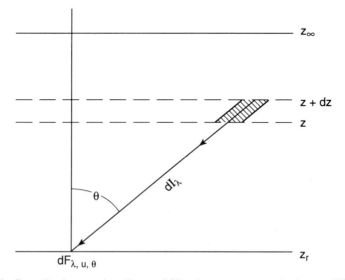

**Fig. 3.11.** Contribution to irradiance $(dF_{\lambda,u,\theta})$ at $z = z_r$ made by a differential element at level $z$ and zenith angle $\theta$.

The monochromatic radiance, $dI_\lambda$, emitted by the element in the directions $\theta$ is given by Kirchhoff's law in the form $k_\lambda I_\lambda^* \sec \theta \, du$, where $u$ is the optical thickness measured in the vertical. The radiance is attenuated according to Beer's law (3.14), so that the portion of the radiance increment $dI_\lambda$, that arrives at the reference level, is

$$dI_\lambda \exp(-k_\lambda u \sec \theta) = I_\lambda^* K_\lambda \sec \theta \, du \exp(-k_\lambda u \sec \theta). \qquad (3.23)$$

We now assume that the atmosphere is uniformly stratified in the horizontal, so that the radiance coming from all elements with the same angle $\theta$ is the same and has the same attenuation. Thus, all volume elements seen at zenith angle $\theta$ contribute an equal increment to the irradiance at the reference level. Keeping in mind that $d\omega = 2\pi \sin \theta \, d\theta$ and using (3.1), the increment to the irradiance $dF_{\lambda, u, \theta}$ from the ring may be expressed by

$$dF_{\lambda, u, \theta} = 2\pi I_\lambda^* k_\lambda \sin \theta \exp(-k_\lambda u \sec \theta) \, d\theta \, du. \qquad (3.24)$$

This expression must now be integrated over all zenith angles from 0 to $\pi/2$, over all layers $du$, and over all wavelengths that contribute to long-wave radiation.

To proceed from here, it is convenient to introduce the spectral *flux transmittance*, $\tau_{f\lambda}$, of a finite layer between the reference level and the emitting layer, as illustrated in Fig. 3.11. Flux transmittance is defined as the proportion of the incidence irradiance that is transmitted through the layer. Thus

$$\tau_{f\lambda} \equiv \frac{F_\lambda}{F_{\lambda 0}} = \frac{F_\lambda}{\pi I_{\lambda 0}}, \qquad (3.25)$$

and the transmitted irradiance, $F_\lambda$, may be calculated by integrating the transmitted radiance, $I_{\lambda 0} \exp(-k_\lambda u \sec \theta)$, over the solid angle, similar to the step from (3.1) to (3.2). This yields

$$\tau_{f\lambda} = 2 \int_0^{\pi/2} \exp(-k_\lambda u \sec \theta) \sin \theta \cos \theta \, d\theta. \qquad (3.26)$$

Upon differentiation we find that

$$\frac{\partial \tau_{f\lambda}}{\partial u} = -2k_\lambda \int_0^{\pi/2} \exp(-k_\lambda u \sec \theta) \sin \theta \, d\theta. \qquad (3.27)$$

After integrating (3.24) over the solid angle, (3.27) may be introduced into it with the result

$$dF_{\lambda,u} = -F_\lambda^* \frac{\partial \tau_{f\lambda}}{\partial u} \, du. \tag{3.28}$$

The flux emittance, $\epsilon_{f\lambda}$, is related to $\pi_{f\lambda}$ by $\epsilon_{f\lambda} \equiv 1 - \tau_{f\lambda}$, and therefore

$$dF_{\lambda,u} = F_\lambda^* \frac{\partial \epsilon_{f\lambda}}{\partial u} \, du = F_\lambda^* \, d\epsilon_{f\lambda}.$$

Upon integration over wavelength and optical depth, the radiation from a cloudless atmosphere reaching the reference level may be expressed formally by

$$-F_i = \int_0^{\epsilon_{f^\infty}} F^* \, d\epsilon_f = \int_0^{\epsilon_{f^\infty}} \sigma T^4 \, d\epsilon_f, \tag{3.29}$$

where $\epsilon_{f^\infty}$ represents the flux emittance from the top of the atmosphere to the reference level.

Determination of $\epsilon_f$ as a function of height is especially useful if $\epsilon_f$ is independent of temperature over the range encountered in the vertical column of air that contributes to the irradiance. This is the case when the area under the black-body envelope, which represents energy absorbed in an atmospheric layer, does not change as the black-body envelope changes with temperature. This condition is approximately fulfilled for water vapour and carbon dioxide.

The upward irradiance consists of a contribution from the atmosphere below the reference level plus a contribution from the ocean surface. This may be written as

$$F_e = \int_0^{\epsilon_{fs}} \sigma T^4 \, d\epsilon_f + (1 - \epsilon_{fs})\sigma T_s^4, \tag{3.30}$$

where the index $s$ refers to the surface. The reflected radiation from the

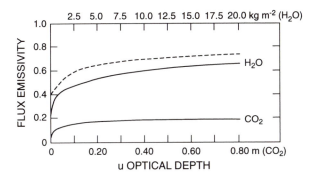

**Fig. 3.12.** Experimental observations of emittance of pure water vapour, of carbon dioxide, and of an atmospheric mixture of $CO_2$ (0.032 per cent) and $H_2O$ (mixing ratio $5 \, g \, kg^{-1}$) as a function of optical depth for temperature of 10°C and pressure of 1013 mb. After Elsasser and Culbertson (1960).

sea surface has been neglected in this equation because it is a minor contribution.

Equations (3.29) and (3.30) can be evaluated when $\epsilon_f$ is known as a function of optical depth. In Fig. 3.12, experimental observations of flux emissivity are given for water vapour and carbon dioxide. When the effects of $H_2O$ and $CO_2$ are combined, care must be taken to avoid doubling the contribution from overlapping absorption bands.

Because the temperature and the absorbing gases are not analytical functions of height, (3.29) and (3.30) may be integrated graphically by dividing the atmosphere in finite layers and by summation over the layers. This allows us to calculate the divergence of the net long-wave radiation and, consequently, whether the layers are heating or cooling, and to what degree. From conservation of energy we find

$$\rho c_p \frac{\partial T}{\partial t} = -\frac{\partial}{\partial z}(F_{il} + F_{el}) = -\frac{\partial}{\partial z} F_{nl}. \tag{3.31}$$

Under clear skies, the divergence of the net long-wave radiation typically cools the air in the atmospheric boundary layer by 1–3 K per day. When clouds are present, the divergence of net long-wave radiation is small in the mixed-layer below the clouds, because the clouds usually approximate black bodies. At cloud-top, however, the cooling may be very large because the cloud droplets may radiate directly to space through the atmospheric window from about 8 to 12 $\mu$m. Measurements reported by

Schmetz et al. (1983) indicate cooling rates up to 60 K or more per day, as is shown in Fig. 6.22.

## 3.4  Empirical formulas for estimating the surface radiation budget

The radiation budget at the sea surface is made up of four components, i.e., short-wave irradiance, $F_{is}$, long-wave irradiance, $F_{il}$, short-wave exitance, $F_{es}$, and long-wave excitance, $F_{el}$. These four components add up to the net radiative flux, as discussed in Section 3.3,

$$F_n = F_{is} + F_{es} + F_{il} + F_{el}. \tag{3.32}$$

The ocean is large, and few direct observations are available. Yet a knowledge of the net radiation at the sea surface is essential for determining the heat budget of the oceans. Therefore, many efforts have been made to obtain empirical relations for the terms in (3.32). A paper by Katsaros (1990) reviews this subject.

### 3.4.1  *Short-wave irradiance*

The total short-wave irradiance at the sea surface from the sun and the sky under cloudless conditions is tabulated in the Smithsonian Tables No. 147–149 as a function of vapour content and zenith angle of the sun. The formula for $F_{is}$ under a clear sky, assuming a transmittance of the atmosphere of 0.7, is given by Seckel and Beaudry (1973).

The determination of the short-wave irradiance becomes much more difficult when clouds are present. A number of parameterizations exist for average daily insolation, when clouds are present, that have been examined critically by Lind and Katsaros (1986) and Dobson and Smith (1985, 1988). In general,

$$F_{is} = F_{is0} f_s(n), \tag{3.33}$$

where $F_{is0}$ is the irradiance of the clear sky, $n$ is the fractional cloud cover in tenths of the sky as found in the synoptic code, and $f_s(n)$ is a function of the cloud cover that has the value 1 for no-cloud cover ($n = 0$). A simple parameterization was proposed by Kimball (1928) in the form

$$f(n) = 1 - 0.71n. \tag{3.34}$$

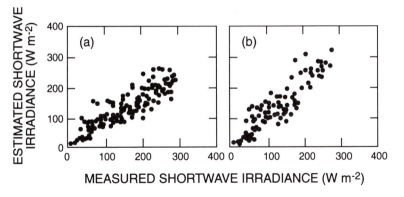

**Fig. 3.13.** Comparison of surface-based short-wave irradiance schemes to daily averaged observations between (a) Kimball (1928) and (b) Lumb (1964). Reprinted by permission of Kluwer Academic Publishers.

Several variations of (3.34) have been proposed, without much improvement. Lumb (1964) proposed a slightly different approach

$$f(n) = A(n) + B(n)\cos\theta, \qquad (3.35)$$

where $A$ and $B$ are constants for each of nine cloudiness categories, and $\theta$ is the zenith angle of the sun. This scheme was developed for midlatitude ship observations.

In Fig. 3.13, we compare (3.34) and (3.35) to observations obtained in a number of experiments. The Lumb parameterization is better than Kimball's and variants thereof. Lumb's scheme has the drawback that it requires more specific cloud information, which usually is not available from ship reports.

More promising for worldwide coverage are satellite observations. Several models have been developed for inferring surface irradiance from satellite-based radiance measurements. There are some inherent difficulties in this approach, such as: inadequate calibration of satellite sensors; difficulty in obtaining ground truth at the surface compatible with the satellite footprint; corrections for various viewing angles; uncertainty of satellite navigation, and so on. Nevertheless, the short-wave irradiance at the surface can be estimated to about 10 per cent accuracy for daily averages from geostationary satellite radiance data (for references, see Katsaros, 1990).

On an hourly basis, the differences between ground-based measurements and satellite estimates are much larger, typically 20 per cent (e.g., Buriez et al., 1986). These can be partly attributed to uncertainties in satellite navigation and to a mismatch between temporal and spatial

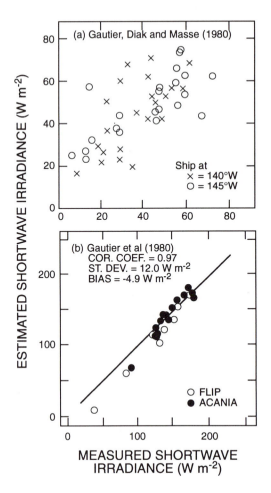

**Fig. 3.14.** Comparison of daily averaged shortwave irradiance estimated with the satellite-based model of Gautier et al. (1980) to pyranometer measurements made at sea: (a) in the STREX experiment at 50°N. After Gautier and Katsaros (1984); (b) in the MILDEX experiment. After Frouin et al. (1988a). Reprinted by permission of Kluwer Academic Publishers.

resolution of the satellite sensor and the surface pyranometer, especially when the cloud cover is variable.

A comparison between satellite estimates and direct measurements is given in Fig. 3.14. The estimates are obtained using the model described by Gautier et al. (1980). The measurements were obtained during STREX and MILDEX. STREX occured in late fall at 50°N, where the sampling from a geostationary satellite is at a large incidence angle. Although these conditions are marginal for the model by Gautier et al., the results are quite acceptable. It is clear from these comparisons that satellite modelling of the short-wave irradiance at the sea surface will be a useful technique.

### 3.4.2  Short-wave exitance

Short-wave exitance at the sea surface can be calculated from model-produced estimates of the short-wave irradiance and a value for the albedo of the sea surface. Therefore a model of the sea surface albedo is needed. In Section 3.2.2 the albedo, $A$, was defined as

$$A \equiv -\frac{F_{es}}{F_{is}}. \tag{3.36}$$

A successful model for the albedo was introduced by Payne (1972). He assumed that the albedo is a function of the flux transmittance, $\tau_f$, of the atmosphere, given by

$$\tau_f = \frac{F_{is}}{F_{is0}} \, \tau_{f0} = f(n) \tau_{f0}, \tag{3.37}$$

where $\tau_{f0}$ is the flux transmittance of the cloud-less atmosphere, $\tau_{f0} \approx 0.7$. Substitution of (3.37) into (3.36) yields

$$F_{es} = -A[f(n)]F_{is0}\tau_{f0}f(n). \tag{3.38}$$

The sea surface albedo is a rather weak function of cloudiness with a value of about 0.05 for $n = 0$ (i.e., clear sky) to about 0.08 for $n = 10$ (i.e., complete overcast). It also varies with the roughness of the sea. For windspeeds from 4 to 12 m s$^{-1}$, the effect is less than $\pm 20$ per cent of $A$. For stronger winds the albedo increases because of breaking waves and the associated foam-cover. Monahan and O'Muircheartaigh (1987) estimate an increase in albedo of about 10 per cent for windspeeds of 15 m s$^{-1}$ and of about 20 per cent for windspeeds of 20 m s$^{-1}$.

By and large, the variations in albedo at the sea surface are modest. By assuming a constant value of 0.07, the maximum error is probably $\pm 30$ per cent. This translates to about 2 per cent of $F_{is}$. Therefore, although the error in the emittance at the sea surface is rather large, it has only a minor impact on the radiation budget because the magnitude of $F_{es}$ is relatively small.

### 3.4.3  Long-wave irradiance and exitance

Parameterization of the long-wave irradiance consists of two parts: (1) an estimate of the flux emittance of the cloudless atmosphere, $\epsilon_f$, using the surface temperature, $T_s$, for the black-body radiation of the sky; and (2) a formulation of the contribution from clouds. This may be written in the form

$$F_{il} = -\epsilon_f \sigma T_s^4 f_l(n), \tag{3.39}$$

where $f_l(n)$ is a function of the cloud cover that has the value 1 for $n = 0$, and $f_l(n) > 1$ for $n > 0$. The flux emittance is a function of the water vapour content of the atmosphere, and in many parameterizations this is explicitly introduced. Swinbank (1963) suggested a very simple scheme by assuming that $\epsilon_f$ is only a function of $T_s$, which implies that the water vapour content and the temperature are correlated. This is a reasonable assumption for a neutral planetary boundary layer, where the air is more or less in equilibrium with the ocean surface. One may also argue that most of the atmosphere is approximately in equilibrium with the ocean underneath. This scheme is clearly in error where large air–sea temperature differences occur, such as in cold-air outbreaks over the Gulf Stream.

A more sophisticated parameterization was developed by Lind and Katsaros (1982). It requires detailed cloud reports, similar to Lumb's (1964) scheme for hourly short-wave irradiance and allows for different radiative temperatures for each cloud deck. In addition to variables reported in the surface code, information about the vertical structure of temperature and humidity is required, which may be read from synoptic weather charts at 850, 700, and 500 m at the location of interest. The cloud emissivities are a function of the liquid water content of the cloud as indicated in Fig. 3.10.

Comparisons of the parameterizations of Swinbank (1963), and of Lind and Katsaros (1982) with observations are given in Fig. 3.15. For daily

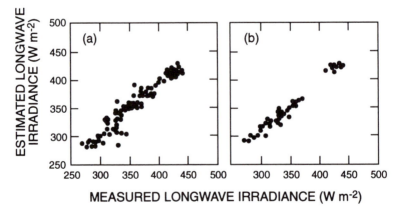

**Fig. 3.15.** Comparison of surface-based longwave irradiance schemes to daily averaged observations for models: (a) by Swinbank (1963); (b) by Lind and Katsaros (1982) for GATE, JASIN and STREX experiments. Reprinted by permission of Kluwer Academic Publishers.

averaged irradiance, the Swinbank scheme was found to have a mean error of $-4 \, \text{W m}^{-2}$ and a root mean square error of $12.5 \, \text{W m}^{-2}$, whereas Lind and Katsaros (1982) obtained errors of $1.2 \, \text{W m}^{-2}$ and $8.6 \, \text{W m}^{-2}$, respectively. The Lind and Katsaros scheme requires more information than is available from commercial ships. They simplified their model to include the less-complete reports, which resulted in a slightly larger error (Lind and Katsaros, 1986).

The long-wave irradiance at the surface is difficult to obtain from satellites. Nevertheless, some promising results have been obtained. Schmetz et al. (1986) developed an interesting model for the estimation of $F_{il}$, employing geostationary data. The model uses short- and long-wave data from METEOSAT-2 in order to derive cloud parameters in conjunction with grid point thermodynamic fields. It calculates the cloud contribution from satellite estimates of cloud amount, cloud base height, and temperature, using a parameterization developed by Martin and Berdahl (1984). Surface humidity is obtained by assuming 85 per cent relative humidity, similar to Swinbank's (1963) scheme. The model is limited to daytime applications and to latitudes lower than 50°.

The model developed by Frouin et al. (1988b) uses sounding data from the TIROS Operational Vertical Sounder (TOVS) for the temperature and humidity profiles and hourly GOES (Geostationary Operational Satellite) cloud data obtained with the Visible and Infrared Spin-Scan Radiometer (VISSR). The cloud-base altitude is deduced from the cloud-top altitude and the liquid water path interpreted from the visible reflectance. Cloud-base temperature can then be obtained from the atmospheric temperature profile from the TOVS data. In Fig. 3.16, a comparison of the model of Frouin et al. (1988b) with surface measurements by Lind and Katsaros (1987) is made. Although improvement of this technique is desirable, the results are impressive.

The surface long-wave exitance, $F_{el}$, is the largest and also the simplest term in the budget equation. As was discussed in Section 3.3.1, a good approximation for $F_{el}$ is given by

$$F_{el} \simeq \sigma T_s^4, \qquad (3.40)$$

where $T_s$ is the sea surface temperature, and $\sigma$ the Stefan Boltzmann constant. Therefore any parameterization that gives $T_s$, provides an estimate of $F_{el}$, which is slightly overestimated, as can be seen from (3.20). The error is very small for overcast conditions and can be up to 1 per cent for clear skies.

It is clearly important to obtain accurate values of the sea surface temperature. As we shall see in Section 5.5, $T_s$ is usually lower (up to

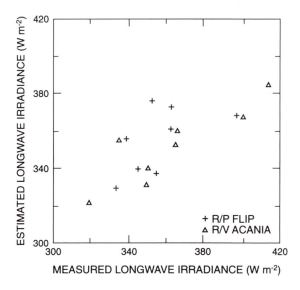

**Fig. 3.16.** Comparison of longwave irradiance estimates calculated with satellite data by Frouin et al. (1988b), using visible satellite data for the cloud amount, with pyrgeometer measurements obtained by Lind and Katsaros (1987). Reprinted by permission of Kluwer Academic Publishers.

0.5°C) than the bucket temperature that is typically obtained on a ship. Satellite estimates of $T_s$ are good to $\pm1$°C, and are improving. This corresponds to an error in the exitance of $\pm7\,\mathrm{W\,m}^{-2}$. For daily and monthly averages, the error may be reduced significantly.

In short the accuracy, with which the parameterization schemes can determine the surface radiation budget, is not quite good enough for long-term monitoring of climate-change. This requires an accuracy of $5-10\,\mathrm{W\,m}^{-2}$. The determinations of the short- and long-wave irradiances especially need to be improved.

# 4

## SURFACE WIND WAVES

Rhythmic and monotonously repetitive, but quite unpredictable in its details, the structure of the sea surface is an epitome of the natural world. Surface waves have been studied actively by mathematicians and physicists since the dawn of modern science. Though the phenomenon seems deceptively simple, it cannot be explained or predicted rigorously by existing theories. Nonlinear interactions between wind, waves, and currents cause theoretical problems as well as make it difficult to obtain comprehensive, interactive data sets.

In response to wind and pressure changes at the air–sea interface, the ocean reacts with waves that occupy some nine spectral decades: from capillary waves, which undulate within a fraction of a second over distances smaller than one centimeter, to planetary waves with periods measured in years and wavelengths of thousands of kilometers. The dynamics of all these waves can be related to the set of equations discussed in Section 4.1. For that reason, a consideration of all wave forms could have been combined in the same chapter, but we found it more convenient to divide the subject into two parts. The present chapter deals exclusively with wind-generated waves at the sea surface. They determine the small-scale configuration of the air–sea interface and that affects the turbulent transfers, which are the topic of the following chapter. On the other hand, information and energy transports from the sea surface into the ocean interior by internal and inertial waves, depend upon the state of the upper layers of the ocean. This made it desirable to discuss these wave forms in Chapter 7, after the consideration of planetary boundary layers in Chapter 6.

Small-amplitude or linear, harmonic surface waves are considered in Section 4.2. Analysis of these waves has been the classic approach to the topic. Linear waves represent an idealized abstraction, but their analysis does provide basic insights into actual wave dynamics. Linear approximations have to be abandoned when one considers the energy and the momentum of wave fields. This is the topic of Section 4.3.

In Section 4.4 we discuss the various sources and sinks of wave energy and momentum. This section is divided into three parts, which deal, respectively, with interactions between surface waves, with energy losses by dissipation and breaking, and with energy inputs from the wind. The topic is of particular interest in a book on atmosphere–ocean interactions.

The sea state (i.e., the actual configuration of the sea surface) has a crucial influence upon the transfer of energy and momentum from the air to the water. It also affects the vertical fluxes of other properties. Sea states can be represented by wave spectra and this is the topic of Section 4.5.

## 4.1  Basic dynamics of harmonic waves in fluids

Comprehensive treatments of waves in fluids can be found in many textbooks and monographs, notably Gill (1982), Lighthill (1978), Phillips (1977), Whitham (1974), or Landau and Lifshitz (1959). The present brief and general introduction to the topic is needed to provide a basis and reference for both the discussions in this chapter as well as for Chapter 7.

Waves are distinguished from turbulence by the dominance of a conservative restoring force, which increases with increasing departure from equilibrium. In surface waves, this force is due to gravity and to surface tension. In other types of fluid waves, pressure, inertial, or Coriolis forces can play an analogous role. The presence of a restoring force establishes a functional relationship between the wave frequency and the wavenumber, which is known as the *dispersion relation*. It is convenient to specify the frequency, $\omega$, of a harmonic wave in radians (i.e., by $2\pi$ divided by the wave period). The wavenumber vector $\mathbf{k}$ and its components $l, m,$ and $n$ are specified similarly by the number $2\pi$ divided by the distance between wave crests, as measured along the horizontal $x, y,$ and the vertical $z$ directions.

Harmonic waves tend to appear in perturbed fluids when the products and squares of the dependent variables are small. Allowing for the Boussinesq approximation and for the relative insignificance of the vertical Coriolis force component, the governing equations are

$$\frac{\partial u}{\partial t} - fv + \frac{\partial}{\partial x}\frac{p'}{\rho_0} = F_x \qquad \text{(a)},$$

$$fu + \frac{\partial v}{\partial t} + \frac{\partial}{\partial y}\frac{p'}{\rho_0} = F_y \qquad \text{(b)},$$

$$(4.1)$$

$$\frac{\partial u}{\partial x} + \frac{\partial v}{\partial y} + \frac{\partial w}{\partial z} = 0 \qquad \text{(c)},$$

$$\left(\frac{\partial^2}{\partial t^2} + N^2\right)w + \frac{\partial}{\partial t}\frac{\partial}{\partial z}\frac{p'}{\rho_0} = \frac{\partial F_z}{\partial t} - F_T \qquad \text{(d)}.$$

These equations are the linearized forms of (1.22) and (1.27), with $N^2$ defined in (1.37). Lower-case symbols are used to denote component velocities because we are now dealing with fluctuating quantities. The forcing terms $F_x$, $F_y$, and $F_z$ here represent external forces and the effects of viscosity and of nonlinear interactions including Reynolds stresses. The last equation (4.1d) was obtained from (1.27) in its inviscid form, after elimination of $g' = -b'$ with the aid of (1.37). The forcing term $F_T$ stands for the effects of horizontal density advection and of non-adiabatic density changes, that were not considered in the derivation of (1.37).

General solutions of the set of equations (4.1) can be expressed in many cases by a sum of terms of the form

$$[u, v, p'] = [U_n^*(x, t), V_n^*(x, t), P_n^*(x, t)]n^{-1}\frac{\partial \chi_n(z)}{\partial z}e^{i(lx+my-\omega t)}$$

$$(4.2)$$

$$w = W_n^*(x, t)\chi_n(z)e^{i(lx+my-\omega t)}$$

The summation must be carried out over all possible values or modes of $n$, an as yet unspecified inverse length, which can be imaginary; $\chi_n$ is a nondimensional function of $n$ and $z$. The starred amplitudes are constant only if the forcing functions are zero or negligible. The solutions (4.2) have then the character of progressive harmonic waves. They will continue to have this character even when some forcing is present, as long as the relative change of the amplitudes remains negligibly small over the distance of a wavelength or the time of a wave period. We shall use the notation $l^2 + m^2 \equiv k^2$ to indicate the magnitude of the horizontal wavenumber.

When the expressions (4.2) are introduced into (4.1), one finds that all left-hand terms in the first three equations contain the factor $\chi_n$, which can be factored out. The $z$-dependence can be removed from (4.1.d) if one stipulates that $\chi_n$ satisfies the subsidiary equation

$$\frac{\partial \chi_n^2}{\partial z^2} + n^2\chi_n = 0, \qquad (4.3)$$

which indicates that $\chi_n$ varies harmonically or hyperbolically, dependent on the sign of $n^2$. Substitution of (4.2) into the homogeneous part of the system (4.1) yields a set of linear, algebraic equations. A nontrivial solution of this set is possible only if the determinant of the coefficients is zero.

$$\begin{vmatrix} -i\omega & -f & 0 & il \\ f & -i\omega & 0 & im \\ il & im & n & 0 \\ 0 & 0 & (N^2-\omega^2) & i\omega n \end{vmatrix} = -i\omega[\omega^2 n^2 - (N^2-\omega^2)(l^2+m^2) - f^2 n^2] = 0 \tag{4.4}$$

It follows from (4.4) that

$$n^2 = k^2 \frac{(N^2-\omega^2)}{(\omega^2-f^2)} \tag{4.5}$$

The corresponding dispersion relation has the form

$$\omega^2 = \frac{N^2 k^2 + f^2 n^2}{k^2 + n^2}. \tag{4.6}$$

The structure of the waves is determined by the sign of $n^2$ in (4.5). In general $\omega^2 \geq f^2$. If the fluid is weakly or neutrally stratified ($N^2 < \omega^2$) the fraction $(N^2-\omega^2)/(\omega^2-f^2)$ and hence $n^2$ are both negative. In this case (4.3) indicates an exponential or hyperbolic variation of the perturbation along the vertical, which can then only propagate horizontally. If the fraction is positive ($n^2 > 0$), (4.3) is satisfied by $\chi_n \propto \exp(\pm inz)$. A perturbation with this type of $z$-dependence can travel upward or downward as a harmonic internal wave with vertical wavenumber $n$.

The wave phase, as indicated by the location of the wave crest, can be observed to propagate in the $x$, $y$, and $z$ directions with *phase speeds*

$$c_x = \omega/l \qquad c_y = \omega/m \qquad c_z = \omega/n. \tag{4.7}$$

These phase speeds are always larger than the relevant components of the absolute phase velocity $\mathbf{c} = \omega/\mathbf{k}$, which are

$$\mathbf{c} \cdot \mathbf{i} = \omega l/k^2 \qquad \mathbf{c} \cdot \mathbf{j} = \omega m/k^2 \qquad \mathbf{c} \cdot \mathbf{n} = \omega n/k^2. \tag{4.7'}$$

To illustrate the difference between (4.7) and (4.7'), consider a surface wave train that propagates toward a beach from a direction that makes an angle $\Psi$ with the shoreline. The component parallel to the shoreline, of the absolute phase velocity $\mathbf{c}$, is $\mathbf{c}\cos\Psi$. The much faster rate at which crests appear to run along the beach is given by $\mathbf{c}/\cos\Psi$. The two become equal only in waves that propagate parallel to the shore.

Wave amplitudes are specified by the maximum departure $A^*$ of fluid particles from a position of equilibrium. In linear waves, the starred quantities in (4.2) are all proportional to this amplitude. The *energy density* (i.e., the energy of a wave field in a unit of volume) is proportional to the mean square amplitude of all the waves in that volume. The transport or flux of energy that is caused by the waves is known as *wave radiation*. True monochromatic radiation at a discreet frequency would require an infinite wave train with uniform amplitudes. The specific energy of such a wave would be everywhere the same. It therefore could not produce any flux of energy nor could it transmit any information. For that very reason, waves without beginning or end have no physical reality. When we turn on a light bulb or drop a pebble into a pond, we do not get monochromatic radiation. Waves in nature come in finite packets that involve a whole set of waves with different wave numbers and frequencies. The rate of energy transmission by these wave packets is the *group velocity* $\mathbf{c}_g$. It usually differs in magnitude and direction from the phase velocities of the constituent harmonic waves that make up the packet.

To illustrate the group velocity concept, we consider just two sinusoidal waves as shown in the upper part of Fig. 4.1. Superposition of these two waves produces the pattern or wave group shown in the lower part of the figure. The amplitude of the deformation produced by this group is large where the two basic waves are in phase and reinforce each other. It is small where the two waves are in opposite phase. If the phase speed of the first

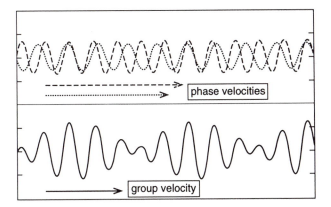

**Fig. 4.1.** Group and phase velocities of two progressive waves.

wave is $\omega_1/k_1$ and that of the second $\omega_2/k_2$, then the combined pattern propagates with a group velocity equal to $(\omega_1 - \omega_2)/(k_1 - k_2)$. In the atmosphere and in the ocean, waves never come in discrete pairs, but in groups that have a continuous spectrum. The group velocity is then specified by the vector.

$$\mathbf{c}_g = \frac{\partial \omega}{\partial \mathbf{k}}. \tag{4.8}$$

The group velocity can have different values in different parts of the spectrum. In other words, different parts of a wave group propagate generally with different velocities. The process causes the energy to the group to be dispersed over increasingly larger volumes or areas. The resulting loss in energy density or *wave dispersion* is proportional to $\partial \mathbf{c}_g/\partial \mathbf{k} = \partial^2 \omega/\partial k^2$.

Waves have not only energy but also momentum as can be observed by anyone walking into a surf. The energy density, $E$, and the momentum vector, $\mathbf{M}$, are both proportional to the mean-square wave amplitude or to the corresponding integral of the wave power spectrum. The two quantities are related by the equation

$$E = \mathbf{c} \cdot \mathbf{M}. \tag{4.9}$$

Both $E$ and $\mathbf{M}$ are advected by the group velocity $\mathbf{c}_g$. The radiation vector

$$\mathbf{I} = E\mathbf{c}_g \tag{4.10}$$

is known as the *Poynting vector* in electromagnetic theory or as the *radiance* in optical terminology. For mechanical waves in fluids, it can be identified with the kinetic energy flux vector defined by (1.32). The transport of wave momentum by the wave, is described by the *radiation stress tensor*

$$\Gamma_{ij} = c_{gi}M_j. \tag{4.11}$$

In fluids, this is simply a particular manifestation of the Reynolds stress tensor (see Section 1.3). The momentum flux converges when a wave is absorbed or slowed down. It then exerts an inertial force equal to $\partial \Gamma_{ij}/\partial x_j$. The radiation pressure, exerted by solar radiation in space or by waves pounding on a sea wall, represents the isotropic part of the radiation stress.

The transfer of energy between different components of a wave spectrum involves turbulent Reynolds stresses, which are included as part of the forcing functions on the right-hand side of (4.1). Alternatively, it requires explicit consideration of nonlinear terms in the Navier–Stokes equations. Energy exchange between wave modes can occur only if the combination of two wave trains results in the generation of a third, freely travelling wave. Denoting the characteristics of the interacting waves by the suffixes 1, 2, 3, one can formalize a resonance condition by the relations

$$k_3 = k_1 \pm k_2 \qquad n_3 = n_1 \pm n_2 \qquad \omega_3 = \omega_1 \pm \omega_2 \qquad (4.12)$$

This is the simplest possible formulation of a resonance process. It will be indicated later that significant energy exchanges between surface waves tend to involve more than three different wave modes.

## 4.2   Small-amplitude waves at the air–sea interface

This section deals with a linear pattern of small-amplitude, long-crested waves, which do not interact with each other. Though never realized exactly, this pattern can provide some insight into the actual ocean wave dynamics. Obviously, it is not realistic. Any casual look at the sea surface suggests the emergence of new wave crests while others vanish. The length of these crests is rarely more than two or three times larger than the normal distance between them. Nonlinearities are always present. They cause interference between wave trains of different frequency and direction, which results in phase and amplitude changes. Non-linearities make the configuration of the surface unpredictable in detail, limiting us to a stochastic point of view. The physics and statistics of nonlinear, finite-amplitude waves goes beyond the scope of this book. We shall quote results of relevant studies only if they bear upon our understanding of the transport of energy and momentum across the air–sea interface. A relatively comprehensive account can be found in Phillips (1977). A review paper by Donelan and Hui (1990), contains more recent references.

The departure $\zeta(\mathbf{x}, t)$ of a water surface from its equilibrium level, such as the deformation produced by a pebble thrown into water, can be represented by the real part of the series

$$\zeta(\mathbf{x}, t) = \Sigma_k \zeta_k(\mathbf{x}, t) = \Sigma_k A_k(\mathbf{x}, t) e^{i(\mathbf{k} \cdot \mathbf{x} - \omega t)}.$$

Although this is a mathematical abstraction, the individual Fourier components can be associated—at some distance from the initial perturbation—with observable, sinusoidal waves that travel outward at different speeds. Interactions between these waves tend to become significant only over time intervals that are long compared to the individual wave periods. Until that happens, the amplitudes $A_k$ can be considered constant. It is this fact that makes the theory of small-amplitude harmonic waves applicable to a large range of observed phenomena. The term 'small-amplitude' indicates waves for which $A_k k \ll 1$.

In the absence of wave–wave interaction, one can consider any of the constituent, harmonic waves in isolation. It is convenient in this case to select an $x$-axis in the direction of the wave propagation, making $l \equiv k$ and $m \equiv 0$. The surface deformation produced by such a wave can be expressed by

$$\zeta = A \cos (kx - \omega t). \tag{4.13}$$

In surface wind waves $(f/\omega)^2 \ll 1$. This allows us to set $f = 0$. The density within air or water can be considered constant. The Brunt Vaissala frequency and the buoyancy have the character of delta functions with $N^2 = 0$ everywhere except at the interface. From (4.5) we then obtain

$$n^2 = -k^2. \tag{4.14}$$

It follows that the amplitude of the perturbation must change exponentially or hyperbolically with depth. The boundary conditions for the vertical velocity at the sea surface and at the sea bottom are

$$w = \frac{\partial \zeta}{\partial t} = \omega A \sin (kx - \omega t) \equiv W^* \sin (kx - \omega t) \qquad (z \approx 0)$$

$$\tag{4.15}$$

$$w = 0 \qquad\qquad\qquad (z = -D)$$

As the vertical acceleration $\partial w / \partial t$ must also be zero at a flat sea bottom, it follows from (4.1d) that $\partial p' / \partial z$ must vanish there as well. At the surface, the perturbation pressure

$$p'_{z=0} \approx g\rho\zeta - \gamma \, \nabla^2 \zeta = (g\rho + \gamma k^2)\zeta \equiv g^*\rho\zeta, \tag{4.16}$$

where $\gamma$ is again the surface tension. The symbol $g^*$ has been introduced here for brevity's sake. It represents the combined restoring acceleration caused by gravity and surface tension. From the boundary condtions (4.15) and from the indicated hyperbolic vertical variation of the perturbation amplitude, it follows that

$$p'(z) = g^*\rho \, \frac{\cosh k(z + D)}{\cosh kD} \, \zeta \tag{4.17}$$

Introduction of this expression into (4.1d) yields

$$\frac{\partial w}{\partial t} = g* \frac{\sinh k(z + D)}{\cosh kD} k\zeta.$$

At the sea surface, $(z \approx 0)$ obtains

$$\left(\frac{\partial w}{\partial t}\right)_{z=0} = -g*k\zeta \tanh kD.$$

From $w_{z=0} = \partial\zeta/\partial t$ it follows that

$$\left(\frac{\partial w}{\partial t}\right)_{z=0} = \frac{\partial^2\zeta}{\partial t^2} = -\omega^2\zeta.$$

Together, the last two equations yield the dispersion relation for surface waves

$$\omega^2 = g*k \tanh kD. \qquad (4.18)$$

The density of air has been neglected in the preceding derivative. The dispersion formula for waves at the interface of two real fluids will be derived in Chapter 7, resulting in the more exact expression

$$\omega^2 = \frac{k(\rho_w - \rho_a)g + \gamma k^2}{\rho_w \coth (kD) + \rho_a}, \qquad (4.19)$$

which is obtained by using the relevant equation (7.16). The atmosphere is considered to have infinite depth in (4.19). As the ratio of air and water densities $\rho_a/\rho_w \ll 1$, the relation (4.19) reduces in practice to (4.18).

Two special cases are of particular interest. If the depth $D$ is very much larger than the wavelength, $kD \gg 1$ and $\tanh kD \approx 1$. The dispersion relation (4.18) reduces then to

$$\omega^2 = g*k \quad \text{(deep-water waves)}. \qquad (4.20)$$

On the other hand, when the waves are long and the water is relatively shallow, $kD \ll 1$ and $\tanh kD \approx kD$. The role of surface tension tends to be negligible in these long waves. The relevant dispersion relation therefore has the form

$$\omega^2 = g*k^2D \quad \text{(shallow-water or long waves)}. \tag{4.21}$$

The corresponding phase and group velocity magnitudes can be easily computed from the preceding equations as

$$c = \frac{\omega}{k} = \frac{g*}{\omega} = \sqrt{\frac{g*}{k}} \quad \text{(deep-water waves)} \tag{4.22}$$

$$c_g = \frac{\partial \omega}{\partial k} = (2\omega)^{-1}\left(g* + \frac{2k^2\gamma}{\rho}\right) = \frac{c}{2} + \frac{k\gamma}{\rho c}$$

and

$$c = \sqrt{(gD)} = c_g \quad \text{(shallow-water waves)} \tag{4.23}$$

To compute the local velocity components, one introduces (4.17) and (4.13) into the homogeneous part of the equations of motion (4.1). With consideration of $\tanh kD \approx 1$, one gets for deep-water waves

$$p'(z) \approx \rho g*Ae^{kz} \cos(kx - \omega t) = \rho c^2 (kA)e^{kz} \cos(kx - \omega t)$$

$$w = c(kA)e^{kz} \sin(kx - \omega t) \tag{4.24}$$

$$u = c(kA)e^{kz} \cos(kx - \omega t)$$

The corresponding expressions for shallow-water waves are

$$p'(z) \approx \rho \zeta g = \rho gA \cos(kx - \omega t)$$

$$w = c\frac{A}{D}k(z + D)\sin(kx - \omega t) \tag{4.25}$$

$$u = c\frac{A}{D}\cos(kx - \omega t)$$

The perturbation caused by deep-water waves decreases exponentially with distance from the interface. It will be shown shortly that this is a first-order approximation. In shallow-water waves, the vertical velocity decreases linearly from the surface to the bottom. The horizontal velocity and the perturbation pressure do not change with depth, and the barometric pressure relation is satisfied.

Another matter of interest is the relative importance of gravity and of surface tension. Gravity becomes negligible compared to surface tension in very short waves where $\gamma k^2/\rho g \gg 1$. Waves of this type are known as capillary waves. On the other hand, surface tension is negligible and the dynamics are dominated by gravity when $\gamma k^2/\rho g \ll 1$. The phase speeds associated with these two wave types are

$$c = \left(\frac{g}{k}\right)^{0.5} = \left(\frac{g\lambda}{2\pi}\right)^{0.5} \qquad \text{gravity waves}$$

(4.26)

$$c = \left(\frac{\gamma k}{\rho}\right)^{0.5} = \left(\frac{2\pi\gamma}{\lambda\rho}\right)^{0.5} \qquad \text{capillary waves}$$

It can be seen that the phase speed of gravity waves increases with the wavelength or with decreasing wavenumber, $k$. The phase speed of capillary waves decreases with the wavelength, $\lambda$. Figure 4.2 is a plot of phase speeds of all harmonic surface waves with wavelengths between 0.2 cm and 10 m. The graph indicates a minimum of about 23 cm s$^{-1}$ in the

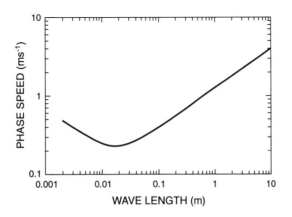

**Fig. 4.2.** Phase speeds of linear surface waves.

intermediate range, where gravity and surface tension are equally important. Harmonic surface waves cannot propagate on deep water at a lower speed. The relevant deep-water group velocities are

$$c_g = \frac{c}{2} \quad \text{gravity waves}$$

$$(4.27)$$

$$c_g = \frac{3c}{2} \quad \text{capillary waves}$$

Capillary waves radiate energy at a rate that exceeds their phase speed.

In the atmosphere above the interface, inviscid linear theory indicates that a harmonic surface wave would cause perturbation velocities, which have the same form as (4.24). The only differences are reversed signs of the exponential and the sine function. This means that the vertical velocity would be continuous at the interface, but that the tangential or horizontal velocities in the two fluids would be 180° out of phase. The resulting discontinuity cannot be maintained in real viscous fluids.

The orbital motion produces a reversing shear and therefore reversing vorticity at the interface. Viscosity diffuses this vorticity into both fluids. The depths $\delta_a'$ and $\delta_w'$, to which this vorticity transport penetrates into the air and into the water before it changes its sign, depends on the wave frequency $\omega$. Dimensional reasoning suggests that

$$\delta_a' \propto \left(\frac{v_a}{\omega}\right)^{0.5} \quad \delta_w' \propto \left(\frac{v_w}{\omega}\right)^{0.5},$$

where $v_w$ and $v_a$ are the kinematic viscosities of water and air. The relative *diffusive capacity* $\mu$, which is the ratio of the two fluid masses affected by viscous diffusion from the interface, is given by

$$\mu = \frac{\rho_a \delta_a'}{\rho_w \delta_w'} = \frac{\rho_a}{\rho_w} \left(\frac{v_a}{v_w}\right)^{0.5} \approx \frac{1}{200}. \quad (4.28)$$

The diffusive capacity does not depend on the particular wave frequency and (4.28) remains valid for laminar motion, even if changes are relatively slow. The affected water mass is much larger than the corresponding mass of air. It can be inferred that the actual velocity of the interface should be much closer to that of the water immediately below, than to that of the air above. We shall return to this topic in Section 5.3.1.

## 4.3   Second-order quantities and approximations

The nondimensional number $kA = 2\pi(A/\lambda)$ specifies the maximum surface slope associated with a particular harmonic wave. An ingenious effort to derive a spectrum of wave slopes from observations of sun glitter on the sea surface has been described by Cox and Munk (1954). It can readily be seen from (4.24) or (4.25) that waves could not persist if either $kA > 1$ or $A/D > 1$ because the local, horizontal water velocity at the wave crest would then become larger than the phase speed, $c$. The wave must break in this case. Alternatively, wave steepness can be related also to the ratio of the maximum vertical acceleration to the restoring force. This ratio can be considered a wave Froude number (1.74). From the preceding expressions, it follows, for both deep and shallow-water waves, that

$$Fr_k = \left(w \frac{\partial w}{\partial z}\right)\bigg/ g^* = (Ak)^2.$$

Disruptive accelerations become larger than the restoring force, when $Fr_k$, and hence $Ak$, exceeds unity. The upper limit $Ak = 1$ is a formalistic abstraction because the whole linear approximation (4.1) is applicable only if $Ak \ll 1$. Higher-order terms in the equations of motion cannot be neglected whenever $Ak$ exceeds a value of about 0.2.

The linear approximations lead to some internal contradictions, even if wave amplitudes remain relatively small. This becomes evident when one compares the time averages of the derived Eulerian and particle velocities at the same point. To determine the path of particles, let $X(x_1, z_1)$ and $Z(x_1, z_1)$ denote the transient position of an individual fluid element, which is in equilibrium at $(x_1, z_1)$ and which, therefore, can be tagged by these numbers. The particle or Lagrangian velocity components are then $dX/dt$ and $dZ/dt$. Since the departure of the particles from their mean position is small in small-amplitude harmonic waves, one might expect the Lagrangian and Eulerian velocities to be equal in a first-order approximation

$$\frac{dX}{dt} \approx u(x_1, z_1), \qquad \frac{dZ}{dt} \approx w(x_1, z_1).$$

When we substitute for the Eulerian velocities $u$ and $w$ from (4.24), integrate with respect to time, square the results, and add them together, we get for each deep-water harmonic wave or Fourier component

$$(X - x_1)^2 + (Z - z_1)^2 = A^2 \exp(2kz_1). \tag{4.29}$$

This is the equation of circular orbits with radii $A \exp(kz_1)$ on which particles move with constant angular velocity $\omega$ around the point $(x_1, z_1)$. For shallow-water waves, these orbits become ellipses that degenerate to straight lines at the boundary $z = -D$.

The contradiction arises because both (4.24) and (4.25) indicate that $\bar{u} = 0$ (i.e., the average of the horizontal velocity at a fixed level is zero). On the other hand, the average particle velocity at the same level could not be zero, if (4.29) is true. Following that equation, water particles should move with constant angular velocity along circles that become smaller with increasing depth. The actual velocity therefore decreases with depth. Particles with an equilibrium position above a fixed level would cross that level faster than those with an equilibrium position below it. As particles from above and below must pass the fixed level in opposite directions, water would be swept past any fixed point, forward and backward, at systematically different rates. This indicates that it is not possible for (4.24) and (4.29) to be correct simultaneously.

In reality, particles do not go around and around in closed circles; rather, they experience a slow mean drift in the direction of the wave propagation. Their forward motion, on top of their orbit below the wave crests, is slightly faster than the backward motion below the troughs. Particles that come from above sweep backward past a fixed point below the wave troughs with exactly the same speed as particles coming from below sweep forward past it under the crests. A fixed-current meter would see particles sweep past it, forward and backward, at the same velocity. The average Eulerian velocity therefore remains zero, although the individual particles experience a mean displacement that depends on the difference of their velocities below the crests and the troughs. Schulejkin (1960) used floating, neutrally buoyant, small lamps in a wave tank to demonstrate the resulting trajectories.

The drift can be associated with the nonlinear terms in the equation of motion. To solve these equations, with various degrees of approximation, one can express the dependent variables as a power series of the small parameter $Ak$. This type of analysis was first described in a classic paper by Stokes (1847). Values of the expansion coefficients up to the eighth power of $Ak$, have been listed by Donelan and Hui (1990). To second-order, the horizontal particle velocity in a Stokes wave is specified by

$$\frac{dX}{dT} = u + c[kA \exp(kz_1)]^2 = u + U_{zs}. \qquad (4.30)$$

The time average of $u$ is zero, and $U_{zs}$ represents the mean drift velocity

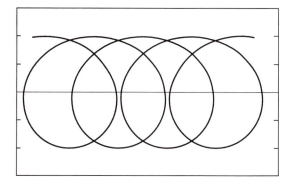

**Fig. 4.3.** Schematic particle trajectory associated with a nonlinear (second-order) wave.

of particles with an equilibrium depth $z_1$. A trajectory derived from (4.30) is shown in Fig. 4.3. A third-order approximation would show a slope of the trajectories in the direction of wave propagation.

When higher-order terms become significant, the shape of a surface wave can not be considered sinusoidal. The crests become sharper and their height above equilibrium level exceeds the depth of the troughs, which become correspondingly wider. A fourth-order approximation of the surface elevation $\zeta$ in a theoretical Stokes wave is shown in Fig. 4.4. The Stokes wave is a better approximation than is a purely sinusoidal form of the shape of waves that are actually observed on the sea surface. The upper limit, at which Stokes waves become unstable and must break, is characterized by a sharp crest with an interior angle of 120°.

Wind waves on the sea surface are generated by the transmission of energy and momentum from the air to the water. Second-order quantities must be considered in all energy studies because energy involves the

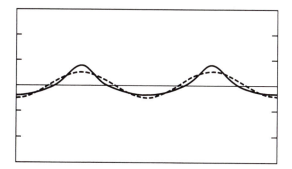

**Fig. 4.4.** Stokes wave profile (4th-order solution).

squares of the perturbation velocities by definition. The wave momentum per unit surface area is equal to the water density, multiplied by the depth integral of the mean Lagrangian particle velocities, summed up over all directional wavenumbers $\mathbf{k}$. It follows from (4.30) that the contribution of any particular wave to the total momentum is a second-order quantity given by

$$\mathbf{M}_k = \frac{1}{2}\rho k(\mathbf{c}A^2)_k, \tag{4.31}$$

Summation or integration over all wavenumbers yields

$$\mathbf{M} = \frac{\rho}{2}\sum [k(\mathbf{c}A^2)_k] \approx \frac{\rho}{2}\int_k\int_\alpha \mathbf{c}_k\Psi(\mathbf{k})k^2\,dk\,d\alpha, \tag{4.32}$$

where $\Psi(\mathbf{k})$ is the mean-square amplitude spectral density in a two-dimensional wavenumber space with polar coordinates $(k, \alpha)$.

The wave momentum can also be observed with Eulerian sensors. Below the level of the surface wave troughs, fixed-current meters can be expected to indicate zero mean horizontal momentum because $\bar{u} = 0$. Conditions are different in the region between the trough and crest levels, which is submerged intermittently. Slugs of water move periodically, with phase velocity $\mathbf{c}$, past any fixed sensor in that region. In Eulerian terms, the averaged contribution of any particular harmonic wave to the total wave momentum is given by

$$\mathbf{M}_k = \int_0^\zeta \rho\mathbf{u}_k\,dz = \overline{\rho[\mathbf{c}\zeta\exp(k\zeta)]_k} \approx \frac{\rho}{2}k(\mathbf{c}A^2)_k, \tag{4.33}$$

which is identical to (4.31). The last term in (4.33) is derived from the preceding term by using the approximation $\exp(k\zeta) \approx 1 + k\zeta$ and substitution of (4.13) for $\zeta$.

Following (4.9), the energy per unit surface area, which is associated with any particular constituent wave, is given simply by

$$E_k = (\mathbf{c}\cdot\mathbf{M})_k = \frac{\rho}{2}k(cA)^2_k. \tag{4.34}$$

The corresponding specific energy associated with waves of frequency $\omega$, regardless of propagation direction, is

$$E_\omega = \frac{\rho}{2}(\omega c A^2)_\omega. \tag{4.35}$$

The specific energy of the complete wave field is obtained again by addition or integration over all wavenumbers or frequencies

$$E = \frac{\rho}{2} \int_\alpha \int_k k^2 c^2 \Psi(k, \alpha)\, dk\, d\alpha = \frac{\rho}{2} \int_\omega k c^2 G(\omega)\, d\omega, \tag{4.36}$$

where $\Psi(k, \alpha)$ and $G(\omega)$ are the mean-square amplitude spectra in two-dimensional wavenumber and frequency space. It should be noted that the last two equations ignore the relatively small amount of wave energy that is located in the air above the surface.

Wave energy conservation for any particular wavenumber $\mathbf{k}$ in the presence of a mean current with velocity $\mathbf{U}$ and a source or sink of energy $S_k$ can be described by the equation

$$\frac{\partial E_k}{\partial t} + (\mathbf{U} + \mathbf{c}_g) \cdot \nabla E_k = S_k. \tag{4.37}$$

If one replaces $E_k$ in (4.37) by the relevant component of the wavenumber spectrum, one gets

$$\frac{\partial \Psi}{\partial t} + (\mathbf{U} + \mathbf{c}_g) \cdot \nabla \Psi = \frac{2}{\rho \omega^2} \frac{d^2 S_k}{dk\, d\alpha}$$

or                                                                                                                (4.38)

$$\frac{\partial G}{\partial t} + (\mathbf{U} + \mathbf{c}_g) \cdot \nabla G = \frac{2k}{\rho \omega^2} \frac{dS_k}{d\omega}$$

Prognostic equations of this form constitute the common basis of analytical surface wave prediction models. Integration of (4.38) over the whole range of spectral wavenumbers would yield changes of the mean-square surface elevation $\zeta^2$ and hence an energy budget for the complete spectrum.

## 4.4   Sources and sinks of surface wave energy

The source/sink term $S_k$ in (4.37) and (4.38) involves three different processes: (1) transfer of energy between different wave modes; (2) dissipation of wave energy; (3) energy input by the wind. The literature dealing with each of these three processes is large. In the framework of the present book we can only give a brief summary of the aspects that seem to us most relevant. Detailed theoretical discussions of the various source/sink processes can be found in Phillips' (1977) monograph.

### 4.4.1   Transfer of energy between waves

Interactions between currents and waves were represented by the second term of (4.37). It is convenient to include these current–wave interactions in the discussion of energy transfers between different wave modes. The advection velocity **U** can cause Doppler effects. When waves propagate from still water into an adverse current, wavelengths contract, phase and group velocities are reduced, and the energy flux converges. The late Bill Richardson (verbal communication) gave a vivid description of the resulting sailing conditions at the edge of the Florida current: "Waves come from one direction, the current from another, and the wind from a yet a different direction $\cdots$ and anything one can do is wrong". Energy cannot be propagated against the current when $c_g \leq U$. The convergence of the wave energy transport must lead to breaking in this case. In reality, wave breaking occurs long before this limit is reached. Phillips (1977) shows that constraints on the shape of the wave energy spectrum tend to enhance breaking of all waves with group velocites $c_g < -U/2$. The longest and fastest propagating waves will be least affected. When an active wind-generated wave field propagates through a current, the shorter waves can experience intense breaking and only a relatively long and gentle swell may emerge on the other side. The resulting changes in surface roughness could affect local, mesoscale atmospheric circulations.

The orbital velocity of long waves can affect short capillary or capillary-gravity waves as an alternating advective current. The resulting Doppler shift may affect remote sensing of the sea surface with micro-waves. Ataktürk and Katsaros (1987) have analyzed this phenomenon in some detail. Doppler shifting tends to cause an increase in the energy density of the short waves on the forward face of the long waves and a corresponding decrease on the rearward face toward the troughs. This can be confirmed by visual observations, which show that long, steep waves are often smooth on their rear slope while their forward face is quite rough. The magnitude of this effect increases with the long-wave amplitude or

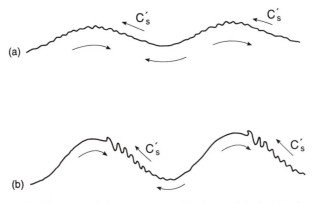

**Fig. 4.5.** Schematic diagrma of shortwave amplitude modulation by longer waves. Arrows below interface represent long-wave orbital velocities, $c_s'$ is the relative phase velocity of the short waves. (a) Non-breaking waves. (b) Preferential breaking on steep forward slope. After Longuet-Higgins (1969).

steepness. Preferential breaking of short waves can occur on the forward slope near the crest of long waves with a sufficiently steep slope. The process is illustrated schematically in Fig. 4.5. The breaking of small short waves near the crest of long waves can produce small, 'spilling' whitecaps there. Theoretical and experimental attempts to estimate the extent of this whitecapping have been made by Banner (1985) and by Cointe (1987). An analysis by Longuet–Higgins (1969) suggested that energy could be transferred locally from very short to very long waves by the indicated process. However, the main outcome of short-wave breaking on long-wave crests seems to be the conversion of wave energy into turbulence kinetic energy.

Nonlinear interactions, which redistribute energy among the components of a spectrum without such an attenuation of wave energy, were first studied by Phillips (1960, 1961) and were later investigated exhaustively by Hasselmann (1962, 1963a,b, 1967). These studies show that the second-order resonance condition (4.12) can lead only to trivial results. Significant resonance and energy exchanges require the interaction of more than two primary wave modes. For a tetrad of primary waves, this can happen if

$$k_1 \pm k_2 \pm k_3 \pm k_4 = 0; \qquad n_1 \pm n_2 \pm n_3 \pm n_4 = 0; \qquad \omega_1 \pm \omega_2 \pm \omega_3 \pm \omega_4 = 0.$$

$$(4.39)$$

The equations of motion cannot be solved with all of these sign combinations, but some solutions do exist. The actual equations that describe the energy transfer between the four wave modes can be found in the quoted literature. They are rather long and their solution is laborious. Methods to shorten the needed computing time have been developed by Komen et al. (1984) and by Hasselmann and Hasselmann (1985).

### 4.4.2   Dissipation and breaking

Wave attenuation appears to be the least well understood component of the source/sink function $S_k$ in (4.37). Wave energy can be dissipated by viscous damping and by wave breaking. Viscosity tends to cause an exponential reduction of the wave energy,

$$\left(\frac{\partial E_k}{\partial t}\right)_\nu = -\sigma(\nu)_k E_k \qquad (4.40)$$

with $\nu$ denoting the kinematic viscosity of sea water. Lambs's expression for the damping coefficient,

$$\sigma(\nu)_k = 4\nu k^2 = 16\pi\lambda^{-2}\nu, \qquad (4.41)$$

shows that viscosity would damp out a capillary wave with a 1-cm wavelength within seconds, whereas a 5-m gravity wave might remain identifiable for days. Table 2.4 indicates that the value of $\nu$ at 0°C is nearly twice that at 20°C. Viscous damping is therefore more effective in high latitudes than in the tropics.

An indirect drainage of energy from somewhat longer waves can be caused by surface tension. This process affects mainly the relatively shorter gravity waves (wavelength 5–30 cm) and occurs mostly when these waves are steep but not yet breaking. Increasing steepness enhances the surface curvature near the wave crests until surface tension becomes important there (see subsection 2.3.2). This permits the local generation of *parasitic capillary waves*, which draw energy from the primary gravity wave until they are damped in turn by viscosity. Since the group velocity of capillaries is larger than their phase velocity, they tend to appear as standing ripples ahead of the crest, on the forward face of the gravity wave. Analytical expressions for the resulting reduction of gravity wave energy can be found in Phillips (1977). The process does not affect the longer gravity waves, because their crests are rarely sharp enough and because their larger orbital velocity prevents the capillaries from maintaining a standing position ahead of the crests.

Viscous dissipation may be the dominant sink of energy for very short waves, but breaking affects all parts of the spectrum. It is probably the main cause of long gravity wave attenuation, although wave–wave interactions must also play some role in keeping the growth of these waves within bounds. Waves must break whenever the slope steepness $Ak \geq 1$. Though this limit is never approached by any individual harmonic wave component, the superposition of many linear waves can result in the local formation of relatively steep slopes and correspondingly sharp crests. Transient, fairly sharp crests, with relatively short separation distances between them, are often seen on the sea surface. Such short and steep seas usually involve some breaking and the appearance of patchy whitecaps.

Shallow-water waves in a surf break in a curling fashion that can entrap a relatively large volume of air. This happens rarely on the open ocean. When deep-water waves break, the crest usually just spills forward. The wave then continues on its way, with a cap of aerated water riding on its forward face. As reported by Donelan (1990) such spilling, once set in motion, can continue for some 10–20 per cent of a wave period, with the forward edge of the whitecap advancing and retreating in a haphazard way. The topic is considered in the papers by Banner (1985) and Cointe (1987), but no full anaytical treatment seems to be available so far.

A different type of breaking process may be due to the existence of a very thin viscous boundary layer immediately below the surface. Breaking can occur when the drift velocity $U_0$ of this thin layer and the orbital velocity $u$ together exceed the phase speed. This can only happen to slowly propagating waves, which have a correspondingly short wavelength ($\lambda < 10 \, cm$). When longer waves or swells move across the sea surface, shorter waves, which are overtaken by them, experience an augmented surface drift near the long wave crests. This can increase the critical speed $u + U_0$ and hence the range of short waves that could be destabilized by this mechanism. The process has been analyzed in detail by Phillips (1977), who refers to it as *microscale breaking*. Its significance for the development of the wave spectrum has not been fully substantiated.

To summarize, waves break when inertial accelerations exceed the restoring force, or when particle velocities at the wave crest outstrip the phase velocity of the wave. These critical conditions can arise from a variety of causes. Various mechanisms have been investigated, but their probability of occurrence has not been fully established and their bearing upon the evolution of wave spectra is therefore uncertain. Wave breaking manifests itself through whitecaps, which can be observed with remote sensing techniques. Additional statistical or monitoring studies that would relate whitecap incidence to wave energy attenuation on the one hand, and to wind, sea state, and other bulk variables on the other hand, could be very useful at this stage.

### 4.4.3   *The generation of waves by the wind*

The part of the source/sink functions $S_k$ that is directly associated with the wind energy input is probably of greatest interest in the present context. We shall therefore discuss it in somewhat greater detail.

Any deformation of the sea surface must result in spreading surface waves. Deformations can be caused directly by local air pressure fluctuations or indirectly by horizontal changes in the local windstress. The initial amplitude of the resulting sea surface perturbation is likely to be exceedingly small, but it can be amplified if the atmospheric disturbance continues to act over some finite time. This happens if the atmospheric disturbance contains components that match the length scale and propagation speed of free waves on the surface. The resulting resonance between wind advection and incipient waves is the basis of Phillips' (1957) wave generation theory.

Wavy patterns that one sees on the surface of a wheat field are footprints of eddies that are being advected downwind in the air above. Something similar happens when 'cat's paws'—patches of capillary or capillary-gravity waves—flit over the surface of the sea. There is a difference: Each wheat stalk is a separate entity, but water is a continuum. The water waves, which are generated by the surface perturbation, propagate energy and momentum. Additional energy can be added to travelling waves that remain in phase with the moving eddy or air pressure anomaly above. If the advection velocity of the atmospheric disturbance is denoted by $\mathbf{U}_a$ and if $\alpha$ represents an angle between $\mathbf{U}_a$ and the direction of wave propagation, waves that travel with a speed

$$c = U_a \cos \alpha \qquad (4.42)$$

would remain in contact with the atmospheric disturbance. These waves may continue to receive energy and momentum from the air. This happens if the atmospheric eddy involves a surface pressure deficit over the forward face of the wave where water rises upward, or a corresponding pressure excess behind the crest, where water sinks down. Alternatively, they could receive additional energy by an excess forward stress in the direction of the orbital motion near the wave tops and by a reduced stress in the wave valleys. Either process can occur only if at least part of the atmospheric disturbance is characterized by wavelengths or wavenumbers that match the free-travelling water wave. In other words, advection frequencies $\mathbf{U}_a \cdot \mathbf{k}$ must be close to some free surface wave frequencies $\omega = (g^*k)^{0.5}$. This is most likely at the leading edge of the atmospheric eddy. The overall size of these eddies is much too large to match the short length of the waves that compose the cat's paws below. Longer waves can be excited by the same process, but their growth tends to be exceedingly slow.

The statistical configuration of the advected air pressure and stress fluctuations can be specified by a time-dependent wavenumber-frequency spectrum $\Pi(\mathbf{k}, \omega, t)$, with dimension mass$^2 \times$ time$^{-3}$. Matching resonance between the turbulent air fluctuations and free surface waves can transfer energy from the air to the water at a rate

$$\frac{d\Psi(\mathbf{k})}{dt} \propto \frac{\Pi(\mathbf{k}, \omega, t)}{(\rho c_k)^2}. \tag{4.43}$$

which can be compared to the first equation (4.38). If the spectrum $\Pi$ is statistically stationary, (4.43) implies linear increases of the various $\Psi(k)$ components with time. The inverse dependence on the square of the phase speed accounts for the very slow growth of the longer, faster moving gravity waves. A detailed and rigorous discussion of this resonance process can be found in Phillips (1977).

Once waves have been generated by essentially random air pressure fluctuations, they perturb the mean air flow above in turn. This induced perturbation travels with the wave. Wind moves somewhat faster over the wave crests, where streamlines are closer together, than over the wave troughs, where the streamlines are more widely spaced. The Bernoulli equation (1.20) indicates that such a flow pattern should cause increased perturbation pressure over the troughs and a small pressure deficit over the crests. This type of perturbation, which causes an upward suction on the crests and a downward push on the troughs, is the subject of the classic Kelvin–Helmholtz theory of inviscid instability.

There are two reasons why Kelvin–Helmholtz instability cannot be the primary cause of wave generation on the sea surface. First, it can develop only provided $\rho_a U^2 > \rho_w c^2$, where $U$ is a hypothetical surface wind velocity. This suggests that waves could grow only if $U > 6.5$ m s$^{-1}$, because $c$ has an absolute minimum of about $0.23$ m s$^{-1}$, as shown in Fig. 4.2. In fact, waves are observed on the sea surface at much lower windspeeds. Second, air pressure perturbations in phase or antiphase with the surface elevation do not transfer momentum to the water. Travelling waves cannot be amplified unless they also receive an appropriate amount of momentum. This can be supplied only by tangential stresses that vary in phase with the surface elevation, or by pressure anomalies that are in phase with the wave slopes and hence in quadrature with the elevation.

Jeffreys (1924, 1925) appears to have been the first fluid dynamicist to consider the effect of out-of-phase pressure variations. He postulated that the airflow separates from the surface as it overtakes the travelling wave crest below. This sepration causes a 'sheltering effect' with a persistent pressure deficit in the lee of the crests and a pressure excess on the windward side. Such a mechanism could draw both energy and momentum from the wind and transfer it to the water below. Jeffrey's hypothesis was widely accepted for 30 years until it was realized that it failed to allow for the nonlinear change of wind with height above the sea surface (see Chapter 5). Air in immediate contact with the surface cannot move much faster than the water in the viscous boundary layer below. This causes the mean airflow just above the water to be slower than the phase propagation of most waves. Boundary layer separation in the air, therefore, is a rather rare event.

The crucial influence of the wind profile curvature upon the supply of momentum to the waves was first recognized by Miles (1957, 1959a,b). Figure 4.6 illustrates Lighthill's (1962) physical interpretation of the Miles theory. It shows streamlines of a hypothetical nonturbulent airflow over a sinusoidal wave, as they would appear to an observer who travels with that wave. Above a level $z_c$, where $U(z_c) - c = 0$, the wind would overtake this observer; below this level it would move in the opposite direction. Air particles close to $z_c$ tend to remain within the vicinity of the observer. Those immediately above travel slowly forward relative to the wave. As they approach the preceding trough they run into a adverse perturbation pressure, which forces them back again across the undulating $z_c$ surface. Then they continue to move slowly backward relative to the wave until

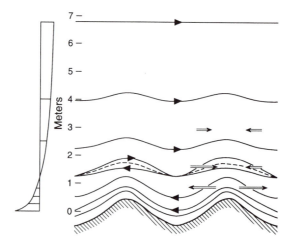

**Fig. 4.6.** Wave amplification by induction. Logarithmic wind profile at left, streamlines relative to the moving wave at right. The double arrows on the right indicate schematically the horizontal component of the 'vortex force' at different locations. The dashed line indicates the matched layer.

they recross the $z_c$ surface and the positive pressure anomaly over the following trough pushes them forward once more. The resulting streamlines in the moving system form closed loops as shown in the figure. The layer where this occurs is designated as the *matched layer*.

Turbulence determines the shape of the mean wind profile. Apart from that, it played no role in Miles' and Lighthill's papers quoted above. This has since been recognized as unrealistic, as is the configuration shown in Fig. 4.6 with streamlines that are in phase at all heights. Phillips (1977) published a similar diagram with the loops displaced over the forward face of the wave. However, the mechanism that produces this shift is not very clear. These differences are almost certainly important for the quantitative application of the theory, but they make it more complicated without affecting the principle on which it is based. The following explanation, therefore, is based on Miles' original assumption of laminar air flow.

The horizontal accelerations in a steady-state, two-dimensional, incompressible flow can be expressed in two ways by the identities

$$\frac{du}{dt} = \frac{u \, \partial u}{\partial x} + \frac{w \, \partial u}{\partial z} \equiv \frac{\partial u^2}{\partial x} + \frac{\partial(uw)}{\partial z} = \frac{1}{2}\frac{\partial(u^2 + w^2)}{\partial x} + \eta w,$$

where $\eta = \partial u/\partial z - \partial w/\partial x$ is here the *perturbation vorticity* in the $y$-direction. When the last identity is averaged horizontally one gets

$$\frac{\partial(\overline{uw})}{\partial z} \equiv \overline{\eta w}, \qquad (4.44)$$

where $\overline{\eta w}$ is the mean of the vortex force defined in (1.17). The mean product $\overline{uw}$ denotes a Reynolds stress, which is in this case equivalent to the vertical transport of specific horizontal momentum. Physical interpretation of (4.44) indicates a balance between the convergence of this transport and the mean of the vortex force at the same level. Vertical integration of this relation yields an expression for the wave-induced stress $\tau_z^w$ at the level $z$

$$\int_z^\infty \overline{\eta w} \, dz = -(\overline{uw})_z = \frac{\tau_z^w}{\rho_a}. \qquad (4.45)$$

Essentially, the Miles theory is based on the argument that the horizontal mean of the vortex force must vanish at all levels where the flow is sinusoidal, but that it does not vanish within the matched layer. During their slow circulation there, air parcels below $z_c$ can acquire the relatively high vorticity of the near-surface wind by turbulent and viscous diffusion.

When they rise again above $z_c$ they lose this excess vorticity and acquire the lower vorticity that characterizes the wind above. The downward-moving air, therefore, has less vorticity than does the upward-moving air. The difference is approximately

$$\Delta\eta \approx \delta_m \frac{\partial\eta}{\partial z} \approx -\delta_m\left(\frac{\partial^2 U}{\partial z^2}\right)_{z_c}, \tag{4.46}$$

where $\delta_m$ is the approximate thickness of the matched layer and $U$ is the mean windspeed. Different values of the vorticity in the rising and sinking branches of the circulation give rise to correspondingly different values of the vortex force. This is indicated schematically in Fig. 4.6. It can be seen that the average vortex force and hence the mean of the horizontal acceleration does not vanish in the matched layer. The argument can be represented in formal terms by

$$\begin{aligned} \overline{\eta w} &= 0 && \text{for } z \neq z_c \pm \frac{\delta_m}{2} \\[2mm] \overline{\eta w} &\approx -\overline{w\delta_m}\frac{\partial^2 U}{\partial z^2} \neq 0 && \text{for } z = z_c \pm \frac{\delta_m}{2}. \end{aligned} \tag{4.47}$$

Vertical integration of these expressions and introduction into (4.45) yields

$$\begin{aligned} \tau_z^w &= 0 && \text{for } z > z_c + \frac{\delta_m}{2} \\[2mm] \tau_z^w &= \tau_0^w \approx \rho_a(\overline{\delta_m\eta w})_{z_c} \approx -\rho_a\overline{w\delta_m^2}\left(\frac{\partial^2 U}{\partial z^2}\right)_{z_c} && \text{for } z < z_c - \frac{\delta_m}{2} \end{aligned} \tag{4.48}$$

Miles' original (1957) analysis implied that

$$\delta_m^2 \propto \frac{\lambda}{2}\left[\frac{w}{\partial U/\partial z}\right]_{z_c} = \frac{\pi}{k}\left[\frac{w}{\partial U/\partial z}\right]_{z_c}. \tag{4.49}$$

Combination of (4.48) and (4.49) results in

$$\frac{dM(k)}{dt} = \tau_0^w \approx -\frac{\rho_a\pi}{k}\left(\overline{w^2}\frac{\partial^2 U/\partial z^2}{\partial U/\partial z}\right)_{z_c}. \tag{4.50}$$

The first equality, (4.50), simply indicates that the wave-induced downward momentum flux from the air is transformed necessarily into an increase of wave momentum. One can obtain the corresponding energy flux to the wave by multiplying both sides of this equality with the phase speed $c$ as indicated by (4.34). Wave energy and wave momentum are both proportional to the square of the wave amplitude $A$. On the other hand, the wave induced vertical velocity $w$ is also likely to vary with the wave amplitude. It follows from the last equality in (4.50) that

$$\frac{d(A^2)}{dt} \propto A^2 \quad \text{or} \quad \frac{d\Psi}{dt} = \sigma(U, \mathbf{k})\Psi(\mathbf{k}), \tag{4.51}$$

where $\sigma(U, \mathbf{k})$ is a wind-induced growth factor, which measures an exponential amplitude growth of waves with wavenumber $k$. This exponential growth has been confirmed by many observations (see, e.g., Hsiao and Shemdin, 1983). Al-Zanaidi and Hui (1984) have developed a simple approximate equation for the growth factor $\sigma$ in the form

$$\frac{\sigma}{\omega} = r \frac{\rho_a}{\rho_w} \left(\frac{U_\lambda}{c} \cos\alpha - 1\right)^2, \tag{4.52}$$

where $r$ is an empirical proportionality factor with values $r = 0.04$ for smooth or transitional flow, and $r = 0.06$ for rough flow; $U_\lambda$ is the windspeed at a height of one wavelength above the surface; and $\alpha$ is the angle between wave direction and wind direction. Being dependent on $U_\lambda/c$ and not on the wind profile curvature, (4.52) is perhaps more consistent with Jeffrey's (1924) sheltering hypothesis than with the Miles mechanism. The factor $r$ plays then the role of the sheltering coefficient. Donelan and Hui found (4.52) to be in reasonably good agreement with observations, although it seems to underestimate some field data. The equation accounts only for the momentum input by the wind and not for wave–wave interactions. It is obviously not applicable to waves that travel faster than the wind at the level $z = \lambda$.

It is preferable to relate $\sigma$ to the friction velocity $u_*$ rather than to $U_\lambda$. This can be accomplished with the use of (5.17) in the form

$$U_\lambda = \frac{u_*}{k_*} \ln\frac{\lambda}{z_0}$$

or

$$\left(\frac{u_*}{U_\lambda}\right)^2 = \left(\frac{\ln\dfrac{\lambda}{z_0}}{k_*}\right)^2 = C_D',$$

where $k_*$ is the von Karman constant and $C_D'$ is the drag coefficient related to the height $z = \lambda$. This is slightly different from definition (5.53). Assuming that $C_D'$ is approximately independent of windspeed, Plant (1982) suggested the simple empirical relation

$$\frac{\sigma}{\omega} = 0.04 \left( \frac{u_*}{c} \right)^2 \cos \alpha, \qquad (4.53)$$

which is valid in the range $0.08 < u_*/c < 3$. Figure 4.7 shows a comparison of (4.52) and (4.53) with observations for $\cos \alpha = 1$.

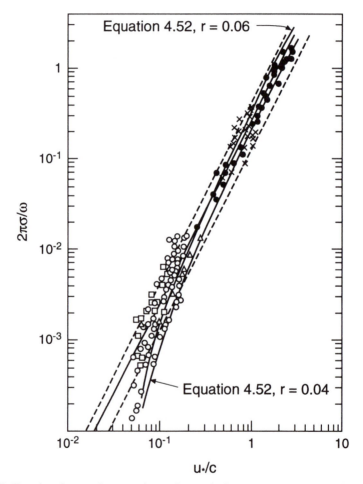

**Fig. 4.7.** Fractional rate of energy input from wind to wave per wave period versus friction velocity per unit phase velocity. Straight solid and dashed lines (4.53), Plant (1982); △ Shemdin & Hsu (1967); ● Larson and Wright (1975); ○ Snyder et al. (1981), fixed sensors; □ Snyder et al. (1981), wave-following sensor; solid curves (4.52). After Al-Zanaidi and Hui (1984).

The preceding discussion represents a highly simplified and idealized version of the real problem. It did not consider the possible existence of wave orbital motions above the interface, nor did it make allowances for the effect of turbulence upon the shape of the streamlines, or for the presence of a multitude of waves with a corresponding spectrum of matched layers. The actual mechanism that transmits momentum, from the level $z_c$, where it is withdrawn from the wind, to the surface below, had not been specified. This mechanism must lead to a surface stress or pressure distribution that can amplify the wave, but its exact form remains somewhat controversial. Davis (1972) examined the effect of different turbulence parameterizations and found that the deduced surface pressure distribution was very sensitive to both the turbulence closure assumptions as well as to details of the near-surface wind profile.

A rigorous and much more detailed treatment of some of these problems was given by Phillips (1977). Measurements since then (Snyder et al., 1981) suggest that Miles' original, inviscid theory underestimates the actual growth rate of waves by a factor of 2–3. Refinements and extensions of the theory, by Riley et al. (1982) and by Janssen and Komen (1985), indicate that this growth rate might be enhanced under certain circumstances, but that the enhancement is too small to eliminate the discrepancy.

## 4.5  The evolution and parameterization of surface wave spectra

For air–sea interaction studies and for many practical applications we are more concerned with the statistical state of the sea surface than with the dynamics of particular waves or wave components. To our knowledge, the first frequency spectrum of sea surface elevations was obtained by Barber (1946). Other detailed descriptions of surface wave spectra and their formation can be found in publications by Mitsuyasu et al. (1975), Kahma (1981), and Donelan et al. (1985). Analytical prognostic models, based on equations of the type (4.37), are affected by uncertainties associated with the source/sink function $S$, especially that part of it that is caused by wave-breaking. Most of the operational wave prediction models are therefore empirical, or have at least an important empirical component. A review of this topic can be found in the (1985) report of the (Sea Wave Modelling Project) (SWAMP) group.

An obvious first-order parameterization of sea states is the mean wave energy or the equivalent surface variance

$$\overline{\zeta^2} = \int G(\omega) \, d\omega = \iint \Psi(k, \alpha) \, dk \, d\alpha. \tag{4.54}$$

A simpler parameter, the *significant wave height*, $H_{1/3}$, was introduced by Sverdrup and Munk in 1947 and is still used for routine ship observations today. It represents the mean height of the highest third of all the waves. The corresponding significant frequency $\omega_{1/3}$ is usually close to the frequency $\omega_p$, which characterizes the peak of the frequency spectrum $G(\omega)$. This is the reason why $H_{1/3}$ tends to approximate the wave height reported by visual observers. An analysis by Longuet–Higgins (1952) suggested

$$\bar{H} = 1.77(\overline{\zeta^2})^{0.5}$$

$$H_{1/3} = 2.83(\overline{\zeta^2})^{0.5} \tag{4.55}$$

$$H_{1/10} = 3.60(\overline{\zeta^2})^{0.5}$$

The parameterization of spectra is based on their apparent general similarity. The shape of a dimensionally consistent similarity spectrum can be expected to depend on $g$ or $g^*$, the windspeed $U$, the fetch $F$, and the wind duration $T$. This suggests a general expression of the form

$$G(\omega) = \beta g^2 \omega^{-5} f[(g/\omega U), (U^2/gF), (U/gT)]. \tag{4.56}$$

When a wind blows sufficiently long and far over the ocean $F, T \rightarrow \infty$. The spectral energy of the waves is assumed, therefore, to become asymptotically independent of the wind fetch and duration. The spectrum of such a *fully developed sea* would have to be of the form

$$G(\omega) = \beta g^2 \omega^{-5} f(g/\omega U). \tag{4.57}$$

Pierson and Moskowitz (1964) used the formula

$$f(g/\omega U) = \exp\left[-\alpha(g/\omega U)^n\right], \tag{4.58}$$

where $U \approx U_{20}$ is the observed wind velocity at the 19.5 m level and $\alpha = 0.74$. The exponent $n$ was tentatively chosen to have values 2, 3, or 4. The resulting difference is relatively small, but $n = 4$ seemed to give the best fit to the existing observations.

Dimensional reasoning suggests that in a fully developed sea, the surface variance $\overline{\zeta^2}$ and the frequency $\omega_p$ of waves with maximum energy, which occupy the spectral peak, can be represented by expressions of the form

$$\overline{\zeta^2} \propto g^{-2} U^4 \qquad \omega_p \propto g U^{-1} \tag{4.59}$$

Both expressions are compatible with (4.57) and (4.58) and with observations. It is often assumed that in a fully developed sea the phase speed $c_p$, which corresponds to the spectral peak frequency $\omega_p$, approaches the windspeed $U_{10}$ at the 10 m level. However, Donelan and Hui (1990) suggest that a 20 per cent excess of $c_p$ over $U_{10}$ would provide a more appropriate criterion. Waves that propagate with such a velocity could not be raised by the Miles induction process; rather, they would have to receive their energy through wave–wave interaction.

The evolution towards a fully developed sea is asymptotic. It may, therefore, never be realized exactly on the real ocean, even if the requisite equilibrium were possible theoretically. Komen et al. (1984) considered the theoretical arguments for and against the existence of a universal fully developed spectrum. It may not be possible to base realistic parameterizations of such spectra on wind data alone. Hasselmann et al. (1973) used peak energy and a measure of the spectral width to categorize their spectra. Donelan et al. (1985) represented spectra from Lake Ontario by a complicated equation that contains $\omega_p$, $U_{10}$ and the peak width as parameters.

In general, sea states must depend also on wind persistence and fetch. The fetch, $F$, is usually introduced in the nondimensional forms

$$F_* = \frac{Fg}{u_*^2} \quad \text{or} \quad F_{10} = \frac{Fg}{U_{10}^2},$$

where $u_*$ is the friction velocity and $F_*$ tends to be a relatively large number. Phillips (1977) derived the following empirical relations for the peak frequency and the surface elevation variance in a set of observed, fetch-limited spectra:

$$\omega_p = 2.2\frac{g}{u_*}F_*^{-1/4} = 2.2g^{3/4}u_*^{-1/2}F^{-1/4}$$

$$\zeta^2 = 1.6\times10^{-4}\frac{u_*^4}{g^2}F_* = 1.6\times10^{-4}g^{-1}u_*^2F$$

(4.60)

Corresponding expressions, derived by Kahma (1981), have the form

$$\omega_p = 20gU_{10}^{-1}F_{10}^{-1/3} = 20g^{2/3}U_{10}^{-1/3}F^{-1/3}$$

$$\zeta^2 = 3.6\times10^{-7}(U_{10}^4/g^2)F_{10} = 3.6\times10^{-7}g^{-1}U_{10}^2F$$

(4.61)

These equations can be valid only over a limited fetch range. An indefinite

linear increase of the spectral energy with increasing fetch is hardly
possible. Ultimately, $\omega_p$ and $\zeta^2$ must become independent of the fetch as
$F \rightarrow \infty$.

Both empirical relations (4.60) and (4.61) have been shown by their
respective authors to agree very well with different sets of observations.
The two expressions for $\zeta^2$ become identical if one assumes that
$u_*^2 = 2.2 \times 10^{-3} U_{10}^2$, but the expressions for $\omega_p$ have a dimensionally
different structure. This may be less significant than it appears, when one
considers that they both involve products of quantities that have very
different magnitudes. Though they both seem to conform rather well with
observations on a log–log plot, that does not mean that they provide much
more than good order of magnitude estimates. We have a slight preference
for Kahma's specification of $\omega_p$ because it is based on more recent data. It
is also almost identical with a formula derived by Hasselman et al. (1973)
from the results of a large field program, which had been organized
specifically for the study of spectra in offshore winds at different distances
from the coast. Some of the observational results obtained during that
program, are reproduced in Fig. 4.8. The figure illustrates the energy

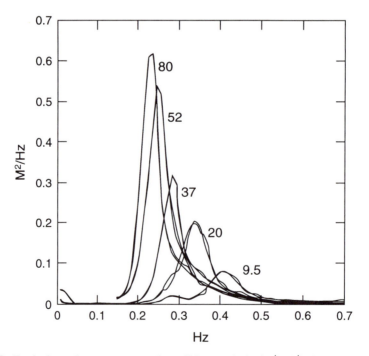

**Fig. 4.8.** Evolution of wave spectra for offshore winds ($11^h$–$12^h$, Sept. 15, 1968).
The number next to the different curves indicate the fetch in kilometres. After
Hasselmann et al. (1973).

increase and progressive move of the spectral peak toward lower frequencies and correspondingly longer wavelengths as the fetch increases
from 9.5 to 80 km.

An interesting feature of Fig. 4.8 is the decrease with distance in the
energy density of waves that formed the peak closer inshore. For example,
at a distance of 20 km, waves characterized by frequencies of about
2.2 rad s$^{-1}$ and a corresponding wavelength of about 14 m have three times
the energy of the same waves 60 km further downwind, although the total
energy is much higher there. An energy excess seems to characterize the
wave components that form temporarily the spectral peak. This excess is
removed later as the sea state matures and the spectral peak becomes
occupied by longer waves with a lower frequency. The same 'overshoot
effect' had been observed earlier by Barnett and Sutherland (1968). It
suggests that the peak wave components are most prone to be attenuated
by wave–wave interactions or wave dispersal. These attenuating processes
will always tend to broaden the spectral shape.

Waves cannot grow indefinitely without their slopes becoming too
steep. There must be a frequency range within which further wave growth
is limited by breaking and other attenuating effects. Phillips (1958, 1977)
argued that the energy of waves in this *equilibrium* or *saturation range* is
entirely determined by these limiting processes, regardless of windspeed. It
follows that the frequency spectrum within that range could be specified by

$$G(\omega) = \beta g^2 \omega^{-5} \qquad \left(\omega_p \ll \omega \ll \frac{2g}{u_*}\right). \tag{4.62}$$

If $\alpha$ denotes the direction of the wavenumber vector $k$, the corresponding
wavenumber saturation spectrum is described by

$$\Psi(k, \alpha) = k^{-4} f(\alpha) \qquad \left(k_p \ll k \ll \frac{2g}{u_*^2}\right). \tag{4.63}$$

The expressions in brackets indicates the saturation range. Its upper limit
decreases with increasing wind or friction velocity.

Expressions (4.62) and (4.63) are well-known and have been applied by
many investigators. Phillips cites a substantial number of observations with
which they are in agreement. From the same data he deduced also a mean
value of $\beta = 1.23 \times 10^{-2}$ for the proportionality constant in (4.62). However, some more recent observations seem to contradict the basic concept

of a wind-independent saturation range. In particular Kahma (1981) found that an expression of the form

$$G(\omega) = g U_{10} \omega^{-4} \tag{4.64}$$

provided a better fit for his data than (4.62). Donelan et al. (1985) indicate that within the $(1.5\omega_p < \omega < 3.0\omega_p)$ range, their spectra are best characterized by an implicitly wind-dependent $\omega^{-4}$ power law. They also maintain that the spectrum at higher frequencies could not be simply a function of the 'local' $\omega$, due to Doppler shifting of the short wave frequencies by the orbital velocities of longer waves.

Earlier on, it had been argued by Toba (1973) that the mean energy of all components in a wind-generated wave spectrum must remain a function of the local wind velocity. This is certainly true for the very short, high-frequency waves, for which remote sensing by microwave reflectivity indicates a strong windspeed sensitivity. The reasons for this phenomenon are not yet well understood. Waves that contribute to microwave reflectivity are attenuated generally within seconds. It is conceivable that they are regenerated at a faster rate over a larger fraction of surface area, when the winds are strong. Alternatively, the above mentioned Doppler shift may have some bearing on the issue.

The preceding discussion suggests that the theory of surface wave spectra is still somewhat controversial. There is no truly universal spectrum. Sea states may approximate some idealized form, but they depend on so many undetermined variables—atmospheric conditions, past history, topography, and so on—that this form is never realized exactly. Existing models and parameterizations have provided useful explanations and predictions of actual sea states, but there is potential for further improvement.

# 5

# TURBULENT TRANSFER
# NEAR THE INTERFACE

The atmosphere and the ocean are in intimate contact at their interface, where momentum, water substance, heat, and trace constituents are exchanged. This exchange is often modest when a light breeze strokes the surface; sometimes the processes are violent, when gale force winds sweep up ocean spray into the atmosphere and when braking waves engulf air into the ocean. It may even appear that the transition between ocean and atmosphere becomes gradual and indistinct. The transition from ocean to atmosphere is usually an abrupt transition of one fluid to another. The interface may then be considered a continuous material surface. On both sides of the interface the fluids are usually in turbulent motion and properties are transported readily, but upon approaching the interface turbulence is largely suppressed so that on both sides of the interface a very thin layer exists where the molecular diffusion coefficients play a major role in the transport. The interface is consequently a significant barrier to the transport from ocean to atmosphere and vice versa, with little or no turbulent transport of scalar quantities across it.

The quantitative determination of the thickness of the molecular sublayers and the strength of the gradients and shear layers within them are discussed in Section 5.1. We also examine the transition from the molecular sublayers to the well-mixed turbulent layers that exist beyond them, and the structure of these turbulent layers on either side of the interface. In Section 5.2 we discuss the effect of stratification on the structure of these surface layers. Some of the nonstationary interactions between the wind and the sea surface are described in Section 5.3. Sections 5.4 and 5.5 deal with practical applications: a formulation of gas transfer across the interface and of the sea surface temperature. Several observational techniques are discussed in Section 5.6.

## 5.1   The structure of the interface and adjacent layers

### 5.1.1   *The profiles in the molecular sublayers*

In order to formulate the transition from the interface to the relatively well-mixed turbulent surface layer in a quantitative form, it is easiest to work with the heat-flux and the related temperature profile because much experimental work has been done in this area.

Considerable success has been obtained with a simple model which allows small parcels of air or water adjacent to the interface to be replaced intermittently by air or water from the turbulent layers away from the interface. This so-called surface renewal model has been introduced many years ago in chemical engineering applications (Higbie, 1935; Danckwerts, 1951). Liu and Businger (1975) applied the model to the air–sea interface.

Although the sea surface itself is not easily replaced, the fluid from below or above can come very close to it. Consider a fluid element from the mixed layer that has been moved adjacent to the sea surface. It initially had a uniform temperature equal to the bulk temperature. As it is exposed to the surface, which has a different temperature, thermal conduction takes place and, assuming that the horizontal gradients are negligible in comparison to the vertical gradients, the temperature $T(z, t)$ is governed by the heat conduction (1.40), which may be written in the form

$$\frac{\partial T}{\partial t} = \kappa \frac{\partial^2 T}{\partial z^2}, \tag{5.1}$$

where $\kappa$ is the thermal diffusivity. The classic error-function integral solution of this equation for constant initial condition, $T(z, 0) = T_b$, and constant boundary condition, $T(0, t) = T_0$, is given by (see, e.g., Carslaw and Jaeger, 1959)

$$T(z, t) = (T_0 - T_b)\, \text{erfc}\left(\frac{z}{2(\kappa t)^{0.5}}\right) + T_b, \tag{5.2}$$

where erfc denotes the complementary error function integral, the index 0 refers to the surface and $b$ (bulk) to the well-mixed interior. The corresponding heat flux density at the interface is

$$Q(0, t) = -k\left(\frac{\partial T}{\partial z}\right)_{z=0} = -k\frac{T_0 - T_b}{(\pi \kappa t)^{0.5}}, \tag{5.3}$$

where $k$ is the thermal conductivity.

We now define a distribution function $\phi(t)$, which represents the fractional area of the surface containing fluid elements that have been in contact with the interface for a time $t$. The average temperature and heat flux are then given by

$$T(z) = \int_0^\infty \phi(t)T(z, t)\, dt \tag{5.4}$$

and

$$Q(0) = \int_0^\infty \phi(t)Q(0, t)\, dt. \tag{5.5}$$

If the turbulence in the interior of the fluid provides the mechanism that replaces the surface elements, we might argue that each element has the same chance to be replaced. This may be expressed by

$$\frac{d\phi(t)}{dt} = -\frac{1}{t_*}\phi(t), \tag{5.6}$$

where $t_*$ is the characteristic residence time of fluid parcels at the surface. Therefore,

$$\phi(t) = \phi(0)\exp\left(-\frac{t}{t_*}\right). \tag{5.7}$$

If we require $\int_0^\infty \phi(t)\, dt = 1$, we find that

$$\phi(0) = \frac{1}{t_*}.$$

Therefore (5.7) may be written as

$$\phi(t) = \frac{1}{t_*}\exp\left(-\frac{t}{t_*}\right). \tag{5.8}$$

Substituting (5.2), (5.3), and (5.8) into (5.4) and (5.5) yields the temperature profile in the molecular sublayer in the form

$$\frac{T - T_b}{T_0 - T_b} = \exp\left[-z/(\kappa t_*)^{0.5}\right] \tag{5.9}$$

and the heat flux

$$Q(0) = -k(T_0 - T_b)/(\kappa t_*)^{0.5} = \frac{-\kappa}{z_\theta}c_p\rho(T_0 - T_b), \tag{5.10}$$

where $(\kappa t_*)^{0.5} = z_\Theta$ is the length-scale for the temperature profile, $\rho$ is the density, and $c_p$ the specific heat at constant pressure. This result agrees with the observations of Khundzhua and Andreyev (1974).

In a similar manner we can derive a profile for the velocity and a profile for humidity (only in the atmosphere). The only difference is that thermal diffusivity is replaced by kinematic viscosity, $\nu$, for the velocity profile and by diffusivity, $\delta$, for the humidity profile. Therefore,

$$\frac{U - U_b}{U_0 - U_b} = \exp\left[-z/(\nu t_*)^{0.5}\right] \tag{5.11}$$

and

$$\frac{q - q_b}{q_0 - q_b} = \exp\left[-z/(\delta t_*)^{0.5}\right], \tag{5.12}$$

where $q$ is specific humidity. Similarly $z_u \equiv (\nu t_*)^{0.5}$ and $z_q \equiv (\delta t_*)^{0.5}$ are length-scales for the velocity profile and humidity profile, respectively. Equation (5.12) may also be used for the salinity profile in sea water by replacing specific humidity with salinity and by using the appropriate diffusion coefficient.

### 5.1.2   The matching of surface layers to molecular sublayers

Far away from the interface in terms of the length-scales $z_\Theta$, $z_u$, $z_q$, turbulence is fully developed (the distance may not be more than a few mm), and the profiles behave quite differently. Yet we stay close enough to the interface that the fluxes are essentially constant with height or depth (the surface layer). In this layer the molecular viscosity and molecular diffusivities are no longer important in the fluxes. They are replaced by eddy viscosity, $K_m$, and eddy diffusivities, $K_h$, $K_e$ [see (1.48)]. The fluxes of momentum, $-\tau$, sensible heat, $Q$, and vapour, $E$, may be written as

$$-\tau = -\rho K_m \frac{\partial U}{\partial z}, \tag{5.13}$$

$$Q = -c_p \rho K_h \left(\frac{\partial T}{\partial z} + \Gamma\right) = -c_p \rho K_h \frac{\partial \Theta}{\partial z}, \tag{5.14}$$

and

$$E = -\rho K_e \frac{\partial q}{\partial z}, \tag{5.15}$$

where $\Gamma = g/c_p$, the adiabatic lapse rate, and $\Theta$ is the potential temperature. We shall assume that at the sea surface $T = \Theta$. The flux of momentum, from the atmosphere to the ocean, is equal and opposite to the surface stress, $\tau$. It is convenient to introduce the magnitude of the stress in terms of a scaling velocity $u_*$, the *friction velocity*, so that $\tau \equiv \rho u_*^2$.

Close to the interface, the velocity can only depend on $u_*$ and on the height above the surface. Similarity considerations therefore suggest

$$\frac{\partial U}{\partial z} = \frac{u_*}{k(z + z_0)}, \tag{5.16}$$

where the constant of proportionality, $k$, is the von Karman constant $\simeq 0.4$, and $z_0$ is a roughness length, which is added to account for the fact that $\partial U / \partial z$ is still finite when $z = 0$. This equation may be integrated to

$$\frac{U - U_0}{u_*} = \frac{1}{k} \ln \frac{z + z_0}{z_0}, \tag{5.17}$$

which is the well-known *logarithmic wind profile*.

In the case of a laminar sublayer we can calculate $z_0(\text{smooth})$ by requiring that $\partial U / \partial z = u_*^2 / v$ for $z \to 0$. This yields

$$z_0(\text{smooth}) = \frac{v}{k u_*}.$$

If we substitute this in (5.17) we obtain

$$\frac{U - U_0}{u_*} = \frac{1}{k} \ln \left( \frac{k u_* z}{v} + 1 \right).$$

This profile matches a linear profile near the interface to a logarithmic profile at some distance above or below it. However, this matching occurs without a transition layer. The transition layer is neither linear nor logarithmic and must be accounted for as well. Experimental results obtained by Nikuradse (1933) suggest that a good approximation for flow over a smooth surface is given by

$$\frac{U - U_0}{u_*} = \frac{1}{k} \ln \frac{u_* z}{v} + 5.5, \tag{5.18}$$

which is valid only for $\zeta^* \equiv u_* z / v \gg 1$ because it does not include the transition layer. It reflects the effect of the transition layer on the logarithmic profile beyond it. If we set $U - U_0 = 0$, we find for $z_0(\text{smooth})$, the roughness length for a smooth surface

$$z_0(\text{smooth}) = 0.11 \frac{v}{u_*}. \tag{5.19}$$

We are now in a position to match (5.11) to (5.18). From the preceding discussion it is clear that $v/u_*$ is an appropriate length-scale for the logarithmic profile as well as for the transition profile. This suggests that $z_u u_*/v \equiv \xi_u$ is a constant.

With the boundary condition that $U - U_0 = u_* \zeta^*$ for $\zeta^*/\xi_u \ll 1$, (5.11) may be written as

$$\frac{U - U_0}{u_*} = \xi_u \left[ 1 - \exp \left( -\frac{\zeta^*}{\xi_u} \right) \right]. \tag{5.20}$$

We require that (5.20) and (5.18) match for both $(U - U_0)/u_*$ and $u_*^{-1} \partial U/\partial \zeta^*$. This is accomplished by equating the right-hand sides of (5.18) and (5.20) and their derivatives, yielding two equations, for the two unknowns $\zeta_u^*$ and $\xi_u$,

$$\xi_u \left[ 1 - \exp \left( -\frac{\zeta_u^*}{\xi_u} \right) \right] = \frac{1}{k} \ln \zeta_u^* + 5.5$$

and

$$\exp \left( -\frac{\zeta_u^*}{\xi_u} \right) = (k\zeta_u^*)^{-1}.$$

The solution is $\xi_u = 16$ and $\zeta_u^* = 47$. In Fig. 5.1 the profiles (5.18) and

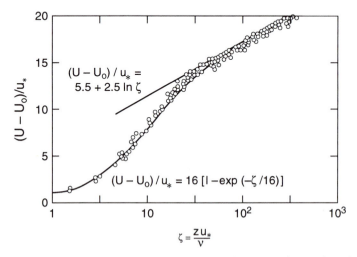

**Fig. 5.1.** Velocity profile from the molecular sublayer to the surface layer as described by (5.18) and (5.20) (solid line) and laboratory measurements by Reichardt (1940) (open circles). After Liu et al. (1979).

(5.20) are compared with laboratory data by Reichardt (1940). The fit is satisfactory.

In order to get a feeling for the dimensions of the molecular transition layer we assume a friction velocity $u_* = 0.2\,\mathrm{m\,s^{-1}}$; this corresponds to a moderate windspeed of 5 to $6\,\mathrm{m\,s^{-1}}$ at $10\,\mathrm{m}$ height. Furthermore, the kinematic viscosity of air at $15°C$ is $v_a \simeq 1.5 \times 10^{-5}\,\mathrm{m^2\,s^{-1}}$. With this information we find that the height of the atmospheric transition layer, $z_u$, is $3.5 \times 10^{-3}\,\mathrm{m}$, or $3.5\,\mathrm{mm}$, and the windspeed at that height is about $3\,\mathrm{m\,s^{-1}}$. Above this height the logarithmic profile extends to several meters above the surface.

In the ocean the dimensions are different because at the surface $\tau_a = \tau_w = \rho_w u_{*w}^2$. So if $u_{*a} = 0.2\,\mathrm{m\,s^{-1}}$, $u_{*w} = (\rho_a/\rho_w)^{0.5}\, u_{*a} \simeq 0.035\, u_{*a} = 0.007\,\mathrm{m\,s^{-1}}$. With the kinematic viscosity for $20°C$ water $1.05 \times 10^{-6}\,\mathrm{m^2\,s^{-1}}$ this gives a depth of transition of about $7\,\mathrm{mm}$ below the surface and the difference in surface drift and drift at this level of about $0.11\,\mathrm{m\,s^{-1}}$. We see that even under modest wind conditions the molecular sublayers are very thin and the shears very large.

A similar matching can be achieved for the temperature profile. Figure 5.2 shows the match with laboratory observations when we take for the molecular and transition layer

$$\frac{\Theta - \Theta_0}{\theta_*} = \xi_\theta \left[ 1 - \exp\left( -Pr\,\frac{\zeta^*}{\xi_\theta} \right) \right], \qquad (5.21\mathrm{a})$$

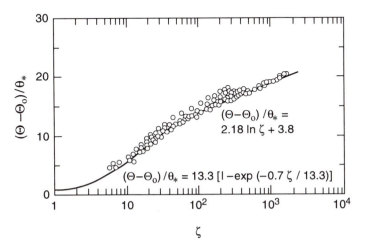

Fig. 5.2. Temperature profile from the molecular sublayer to the surface layer as described by (5.21) and logarithmic relation (solid line) and laboratory measurements by Deissler and Eian (1952). After Liu et al. (1979).

and beyond it, for the turbulent layer,

$$\frac{\Theta - \Theta_0}{\theta_*} = 2.18 \ln \zeta^* + 3.8, \qquad (5.21b)$$

where

$$\xi_\theta = \frac{T_b - T_0}{\theta_*} = +\frac{z_\theta u_*}{\kappa}; \qquad \theta_* \equiv -\frac{\overline{w\theta}}{u_*} = -\frac{\overline{wT'}}{u_*},$$

a temperature scale; and $Pr$ the Prandtl number, defined in (1.79). This combination yields $\xi_\theta = 13$ and $\zeta_\theta^* = 45$. We would have expected $\zeta_\theta^* = \zeta_u^* = 47$. The difference may stem from the different experiments. It may also be related to the fact that in the transition layer the Prandtl number should be replaced by a turbulent Prandtl number, which is close to 1. The ratio $\xi_\theta/\xi_u = Pr^{-0.5}$.

Humidity observations are not available so close to the interface. Therefore, we cannot check the molecular sublayer and its match with the logarithmic turbulent layer for the humidity profile. However, if we assume that scalar quantities behave similarly, it is possible to formulate a model for the transfer of humidity as well as trace gases across the interface. This will be further explored in Section 5.4.

### 5.1.3  Transition from smooth to rough flow

As the wind increases over the sea surface, the surface changes from smooth to rough. The air flow tends to be attached to the surface, but when the waves become very steep, or break, flow separation occurs. When this happens, the stress increases dramatically and the surface is effectively rougher than in other areas. Consequently, with increasing wind, the roughness increases with the flow separation. We are interested in the quantitative specification of the transition from smooth to rough flow.

The concept of a roughness length was introduced in (5.6) and an interpretation of this length for smooth flow in (5.19). The roughness length, $z_0$, was originally conceived for a surface with roughness elements larger than the thickness of the molecular sublayer. It is a parameterization of the surface characteristics that is reflected in the logarithmic profile. In the classic study by Nikuradse (1933) the transition from smooth to rough was examined in pipe flow. He introduced a *roughness Reynolds number* $Rr \equiv u_* z_0 / v$ and found the flow to be smooth for

$$Rr < 0.13$$

and rough for

$$Rr > 2.5$$

with a transition regime in between. From (5.19) we 'actually' find that for smooth flow $Rr = 0.11$.

The roughness of the sea surface is in large measure determined by the local wind stress. To account for this in a dimensionally correct way Charnock (1955) suggested

$$z_0(\text{rough}) = au_*^2/g, \tag{5.22}$$

where $a$ is a constant of proportionality and $g$ is the acceleration due to gravity. Experimental work suggests that $0.011 < a < 0.018$, (Wu, 1980; Garratt, 1977; Smith and Banke, 1975; Walmsley, 1988, and others).

If we combine Nikuradse's criterion for the onset of rough flow with Charnock's relation, using $a = 0.015$, we obtain

$$\frac{u_*^3}{vg} = 167.$$

For $v = 1.5 \times 10^{-5}$ m² s⁻¹ and $g = 9.8$ m s⁻² we find that fully rough flow begins at $u_* = 0.29$ m s⁻¹ and $z_0 = 1.3 \times 10^{-4}$ m. If we assume that Charnock's relation (5.22), may be extended to the limit of smooth flow, $Rr = 0.13$, we find $u_*^3/vg = 8.7$ and hence $u_* = 0.11$ m s⁻¹. If, however, we take $Rr = 0.11$, then $u_*^3/vg = 7.3$ and $u_* = 0.102$ m s⁻¹.

Although it is not clear that the preceding limits were obtained correctly, a transition from smooth to rough flow for $0.11$ m s⁻¹ $< u_* <$ $0.29$ m s⁻¹ seems reasonable. Wu (1980) has used slightly different limits for the transition range, that is,

$$0.17 < Rr < 2.33.$$

Furthermore, he used $a = 0.0185$, which gives $9.2 < u_*^3/vg < 126$ and $0.11$ m s⁻¹ $< u_* < 0.265$ m s⁻¹, a quite similar result.

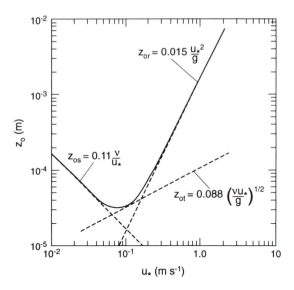

**Fig. 5.3.** The roughness length as a function of $u_*$, from smooth to rough flow (solid line). Asymptotic relations (straight dashed lines).

In Fig. 5.3 the roughness length is plotted as a function of $u_*$. It seems reasonable to assume that the transition from smooth to rough is gradual rather than with a first order discontinuity. The simplest transition is given by Smith (1981)

$$z_0 = z_0(\text{smooth}) + z_0(\text{rough}), \tag{5.23}$$

which encompasses both regimes. This relation is probably good enough for practical applications, but the real world is likely to be more complex. On dimensional grounds one might expect that a transitional regime exists where a transition roughness length $z_0(\text{trans}) \propto (vu_*/g)^{0.5}$. Careful observations are needed to verify this expectation. A line is plotted in Fig. 5.3 that is determined by $z_0(\text{trans}) = 0.088(vu_*/g)^{0.5}$, which is tangent to the curve given by (5.23). If a transitional regime exists, one would expect the coefficient to be larger than 0.088 and (5.23) should be replaced by

$$z_0 = z_0(\text{smooth}) + z_0(\text{trans}) + z_0(\text{rough}). \tag{5.24}$$

In the previous discussion the physics is incomplete because the roughness of the water surface depends also in great measure on the presence of small water waves, and these in turn depend on the surface tension and kinematic viscosity of the water. In view of these dependencies, Hasse (1986) and others have criticized Charnock's relation (5.22) as too simplistic. A more serious shortcoming of the discussion leading to (5.24) is the fact that the intermittent nature of the turbulent wind has not been taken into account. It may well be that in the transition from smooth to rough there is a pattern of smooth and rough patches without a well-defined transition region. In Section 5.3.2 the roughness of the sea surface will be further considered.

We may speculate that a surface renewal model will continue to be valid close to the interface, going from smooth to rough flow. Assuming that $t_* \propto z_0/u_*$, which for smooth flow means that $t_* \propto v/u_*^2$, [see (5.19)], the transitional regime may be written as

$$t_* \propto \left(\frac{v}{gu_*}\right)^{0.5} \tag{5.25}$$

and the rough regime yields

$$t_* \propto \frac{u_*}{g}. \tag{5.26}$$

These equations imply a minimum of $t_*$ for some intermediate value of $u_*$.

Although the transition layer increases in thickness with increasing $u_*$ there is as yet no way to verify experimentally that profiles of the form (5.20) are valid near the surface because the wave height also increases with windspeed. It is very difficult to determine wind profiles at heights lower than the height of the dominant waves.

In summary, for the fully rough flow, $z_0$ is determined by (5.22), and the kinematic viscosity no longer plays a role. The logarithmic profile is valid only for $z/z_0 \gg 1$, which means that in (5.17) $z_0$ in the numerator may be neglected. Therefore, by substitution of (5.22) into (5.17), the logarithmic profile for rough flow may be written in the form

$$\frac{U - U_0}{u_*} = \frac{1}{k} \ln \frac{zg}{u_*^2} + B, \tag{5.27}$$

where $B = k^{-1} \ln a$. $B = 10.5$ for $a = 0.105$. This equation gives a reasonable description of the wind profile over water under neutral conditions for windspeeds $U > 8\,\mathrm{m\,s^{-1}}$.

The profile in the transitional regime, assuming $z_0(\text{trans}) = 0.1(vu_*/g)^{0.5}$, may have the form

$$\frac{U - U_0}{u_*} = \frac{1}{k} \ln \left(\frac{g}{vu_*}\right)^{0.5} z + 5.8.$$

This profile would apply for windspeeds $3 < U < 8\,\mathrm{m\,s^{-1}}$.

In Sections 5.1.2 and 5.1.3 the emphasis has been on profiles in the atmospheric surface layers, see (5.20) and (5.27). As soon as waves play a role the situation in the ocean is more complex. The orbital velocities of the waves are usually much larger than the mean drift velocities from the surface down. It is consequently very difficult to measure mean profiles below the interface. The thickness of the surface layer in the ocean is a factor $(\rho_a/\rho_w)^{0.5}$ smaller than the thickness of the surface layer in the atmosphere (see Section 6.1) that causes the waves to interact with turbulence through a substantial portion of the ocean surface layer. Beside these geometrical limitations, wave-turbulence interactions, as discussed by Kitaigorodskii and Lumley (1983) and Kitaigorodskii et al. (1983), may be a significant producer of TKE. The shear production of TKE, which in the ocean surface layer is given by $u_{*w}^3/kz$ and which is about four orders of magnitude smaller than the shear production in the atmospheric surface layer, might be considerably smaller than the turbulence production by wave interaction and by breaking waves. Recent measurements, reported by Agrawal et al. (1992), show that with strong winds the actual dissipation near the surface is often an order of magnitude larger than the mean shear production of TKE. Therefore the basic mechanism that leads to the logarithmic profile may not be present in the ocean surface layer.

## 5.2   The effect of stratification

So far we have dealt with the neutral case that means that no heat flux or, more precisely, no buoyancy flux exists in the surface layer. A buoyancy flux is commonly present, and the neutral case must be considered a rather special case that happens once in a while, usually when a transition occurs from stable to unstable or vice versa. The buoyancy flux has a dramatic effect on the structure of the surface layer, which we may understand to a certain degree by first considering the turbulence kinetic energy equation. For steady state and uniform horizontal conditions (1.46) may be expressed by

$$-\overline{uw}\frac{\partial U}{\partial z} + \overline{wb'} - \frac{\partial}{\partial z}\left(\frac{\overline{wp}}{\rho} + \overline{wu_i^2}\right) - \epsilon = 0, \qquad (5.28)$$

where $\epsilon$ is the dissipation of TKE and $b'$ is the buoyancy as defined by (1.26). In the atmosphere $b' \equiv (g/\Theta_v)\theta_v$, where $\theta_v$ is the virtual potential temperature fluctuation. The first term of (5.28) represents the shear production; the second term the buoyant production; the third term contains a pressure work term and a turbulent transport term. It was shown by Wyngaard and Coté (1971) that under neutral conditions (5.28) may be approximated by

$$-\overline{uw}\,\frac{\partial U}{\partial z} = \epsilon. \tag{5.29}$$

Because $-\overline{uw} = \tau/\bar{\rho} = u_*^2$, (5.29) may be written, using (5.16), in the form

$$\frac{u_*^3}{kz} = \epsilon. \tag{5.30}$$

The buoyancy term can be either positive or negative (i.e., a productive or consumptive term, respectively). Because the buoyancy flux is more or less constant in the surface layer and the shear production term decreases with height, there is usually a height where the two terms are equal. Obukhov (1946, 1971) realized that this height would be a good scaling height. He equated the neutral form of the shear production term to the buoyant production term and obtained the *Obukhov length* in the form

$$L \equiv -\frac{u_*^3}{k\overline{b'w}}. \tag{5.31}$$

The minus sign was introduced to indicate that the atmosphere is statically unstable when the heat flux is positive. Monin and Obukhov (1954) assumed that the structure of the atmospheric surface layer would be similar if the height were measured in terms of $L$ as a scaling length; that is, if the various quantities were presented in dimensionless form as a function of the dimensionless height,

$$\zeta \equiv z/L. \tag{5.32}$$

The ratio of the buoyant production to the shear production under near-neutral conditions, as indicated by (5.28) and (5.30), is expressed by $-z/L$. Much earlier Richardson (1920) looked at these terms in the gradient form using (5.13) and (5.14) because eddy correlations were difficult or impossible to obtain at that time. Thus he introduced the *flux Richardson* number in the form

$$Ri_f = \frac{g}{\Theta}\frac{K_h\partial\Theta_v/\partial z}{K_m(\partial U/\partial z)^2} = -\frac{\overline{b'w}}{u_*^2\,\partial U/\partial z}$$

By assuming that $K_m = K_h$ the ratio may be expressed by the *Richardson number* in the form

$$Ri \equiv \frac{g}{\Theta} \frac{\partial\Theta/\partial z}{(\partial U/\partial z)^2} = \left(N \Big/ \frac{\partial U}{\partial z}\right)^2 = -\frac{g}{\bar{\rho}} \frac{\partial\bar{\rho}/\partial z}{(\partial U/\partial z)^2}. \tag{5.33}$$

This number, like $\zeta$, can be used as a stability parameter. Although (5.32) and (5.33) both represent the ratio of buoyant production to shear production of TKE, they are not necessarily equal to one another because they were derived under different assumptions. In fact, if no assumptions are made, other than (5.13) and (5.14), we find the identity

$$\zeta = \frac{K_h}{K_m} \varphi_m Ri, \tag{5.34}$$

with

$$\varphi_m \equiv \frac{kz}{u_*} \frac{\partial U}{\partial z} \tag{5.35}$$

as a dimensionless wind shear. Furthermore, the gradients of temperature and humidity may be written in dimensionless form, $\varphi_\theta$ and $\varphi_q$, similar to $\varphi_m$,

$$\varphi_\theta \equiv \frac{kz}{\theta_*} \frac{\partial\Theta}{\partial z}, \tag{5.36}$$

where $\theta_*$ is a temperature-scale, defined following (5.21). Similarly,

$$\varphi_q \equiv \frac{kz}{q_*} \frac{\partial\bar{q}}{\partial z}, \tag{5.37}$$

where $q_* = -\overline{wq'}/u_*$ is a humidity-scale.

These definitions and (5.13–5.15) lead to the following additional identities

$$\frac{\varphi_m}{\varphi_\theta} = \frac{K_h}{K_m} \tag{5.38}$$

and

$$\frac{\varphi_\theta}{\varphi_q} = \frac{K_e}{K_h}. \tag{5.39}$$

The *Monin–Obukhov similarity* argument now requires that $K_h/K_m$, $\varphi_m$, $\varphi_\theta$, $\varphi_q$, and $Ri$ are all functions only of $\zeta$. The explicit form of these functions must be derived from experimental data.

Most of this type of data has been obtained over land because it is more difficult to obtain fluxes and profiles over water, especially over the open ocean. A relatively complete data set was obtained in 1968 in Kansas by the Boundary Layer Group of the Air Force Cambridge Research Laboratories. Figures 5.4 and 5.5 show some results of this experiment. Figure 5.4 also shows results from two experiments in New South Wales (Australia). Good agreement exists between the two data sets; however, later results from the International Turbulence Comparison Experiment,

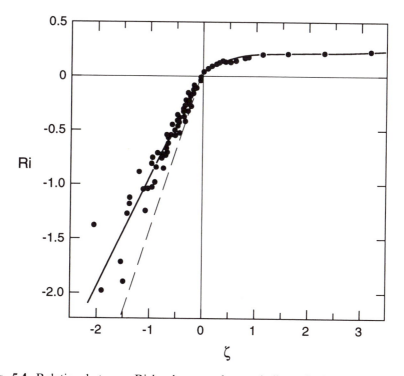

**Fig. 5.4.** Relation between Richardson number and dimensionless height $\zeta$. For unstable stratification ($\zeta < 0$) the solid line represents $Ri = \zeta$ and also the results from Dyer and Hicks (1970). The dashed line represents results from Dyer and Bradley (1982). For stable stratification the line represents $Ri = \zeta(1 + 4.7\zeta)^{-1}$. The dots represent experimental results from the Kansas experiment. After Businger et al. (1971).

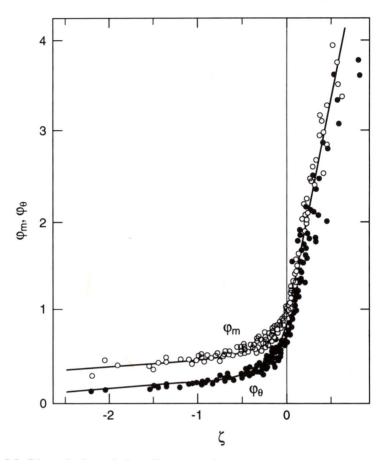

**Fig. 5.5.** Dimensionless wind gradient $\varphi_m$ and temperature gradient $\varphi_\theta$ as function of dimensionless height. For $\zeta < 0$ the solid lines are $\varphi_m = (1 - \alpha\zeta)^{-1/4}$ and $\varphi_\theta = (1 - \alpha\zeta)^{-1/2}$. For $\zeta > 0$ the solid line is $\varphi_m = \varphi_\theta = 1 + \beta\zeta$. After Fleagle and Businger (1980).

done in Australia in 1976, show a somewhat different relationship. Businger (1988) reviewed the trend of $Ri = f(\zeta)$ over the years. More definitive results are needed.

Although possibly not quite correct, it is still attractive to use the simple relation

$$Ri \simeq \zeta \qquad (5.40)$$

for the unstable range. The stable range may be approximated by

$$Ri = \frac{\zeta}{1 + \beta\zeta}, \qquad (5.41)$$

where $\beta$ is a constant ($\sim 5$).

Beside (5.34) and (5.40) or (5.41) one more relationship is needed to define the profiles. For the unstable range this was found by using a simple generalization of Prandtl's mixing length model (Fleagle and Businger, 1980) to arrive at

$$\varphi_m = (1 - \alpha Ri)^{-1/4}. \tag{5.42}$$

This together with (5.40) leads to

$$\varphi_m = (1 - \alpha\zeta)^{-1/4}, \tag{5.43}$$

and with (5.34) and (5.38) to

$$\varphi_\theta = (1 - \alpha\zeta)^{-1/2}, \tag{5.44}$$

where $\alpha$ is a constant ($\sim$16). Equations (5.43) and (5.44) were suggested independently by Businger and Dyer (Businger, 1966; Dyer and Hicks, 1970; see also Sarachik, 1978, and Businger, 1988).

For the stable range, the assumption that closes the set is that $K_h \approx K_m$, which is suggested by the Kansas observations among others. This together with (5.34) and (5.41) leads to the simple form

$$\varphi_m = \varphi_\theta = 1 + \beta\zeta. \tag{5.45}$$

In Fig. 5.5 the functions (5.43), (5.44), and (5.45) are compared with the Kansas data.

Both the unstable and stable expressions for $\varphi_m$ and $\varphi_\theta$ may be readily integrated into dimensionless profiles (Paulson, 1970)

$$\frac{U}{u_*} = \frac{1}{k}\left(\ln\frac{z}{z_0} - \psi_1\right) \tag{5.46}$$

and

$$\frac{\Theta - \Theta_0}{\theta_*} = \frac{1}{k}\left(\ln\frac{z}{z_0} - \psi_2\right), \tag{5.46a}$$

where for the unstable range

$$\psi_1 = 2 \ln \left[ (1 + \varphi_m^{-1})/2 \right] + \ln \left[ (1 + \varphi_m^{-2})/2 \right] - 2 \tan^{-1} \varphi_m^{-1} + \frac{\pi}{2} \qquad (5.47)$$

$$\psi_2 = 2 \ln \left[ (1 + \varphi_m^{-2})/2 \right] \qquad (5.47a)$$

and for the stable range

$$\psi_1 = \psi_2 = -\beta \zeta. \qquad (5.48)$$

No good independent measurements exist of the humidity fluxes together with humidity profiles. However, for rough conditions the data over land indicate that $\varphi_\theta \simeq \varphi_q$ and that the dimensionless humidity profiles therefore have the same shape as the dimensionless potential temperature profiles.

It is clear from the foregoing discussion that the results presented are derived from observations over land. A number of indirect results suggest that the derived equations are also valid over water. Yet, basic differences between land and sea exist with respect to flux-profile relations. This will be discussed to some extent in Section 5.3.

## 5.3   Dynamic interactions between wind and sea surface

So far, the only parameter reflecting the interaction of the wind and the sea surface has been the roughness length, $z_0$. It has been tacitly assumed that this parameter reflects the integrated effects of wave formation, wave breaking, and dissipation on the transfer of momentum to the surface. In this section we shall expand on the interactive nature between the waves and the wind as it was introduced in Chapter 4, while focussing on the transfer of momentum from the atmosphere to the ocean.

### 5.3.1   Surface drift

In the previous sections the surface drift has been introduced in the form of $\mathbf{U}_0$. This quantity is of major importance as a boundary condition for the mixed layer of the ocean. It is usually negligibly small in comparison to the windspeed at 10-m height.

The surface drift is a consequence of momentum transfer from the atmosphere to the ocean. In the absence of waves all momentum is transferred directly into the ocean current. In this case the laminar Ekman solution to be discussed in Section 6.1 indicates a surface velocity

$$\mathbf{U}_0 \simeq 0.005 \mathbf{U}_g, \tag{5.49}$$

where $\mathbf{U}_g$ is the geostrophic velocity as specified in (1.29) and the factor 0.005 is the ratio of diffusive capacities of air and sea water, as defined in (4.28). As indicated in Section 6.1 the surface drift velocity has the same direction as the surface geostrophic wind, when there is no geostrophic current.

When waves are present, the process is more complicated because some of the momentum from the atmosphere is transferred to the waves. The wave momentum is partly dissipated into the surface current and partly radiated away from the area of wave generation. Also, as discussed in Section 4.3, a Stokes drift at the surface, $U_{0s}$, is associated with gravity waves. If the wave spectrum is known we may get an approximate value of $U_{0s}$ by setting $z_1 = 0$ in (4.30), followed by integration over all wavenumbers. Kenyon (1969) used the Pierson–Moskowitz spectrum, (4.58), to compute the mean Stokes drift velocity, $U_{0s}$, at the surface as a function of the empirical exponent $n$. He found that

$$\frac{U_{0s}}{U_{20}} \begin{cases} = 3.6\% & \text{for } n = 2 \\ = 2.1\% & \text{for } n = 3 \\ = 1.6\% & \text{for } n = 1 \end{cases} \tag{5.50}$$

Observations of drifting oil patches and plastic wrappers indicate drift velocities of 3 per cent of the wind velocity and suggest that $n = 2$ or $n = 3$ might provide realistic estimates of $U_{0s}$.

The drift of the interface between two turbulent geostrophic fluids has been considered by Kraus (1977a). It will be shown in Section 6.1 that the distances from the interface, within which surface friction effects remain significant, are of order $D_{Ea} \propto u_*/f$ and $D_{Ew} \propto u_{*w}/f = (\rho_a/\rho_w)_{0.5}\, u_*/f$, respectively. Kraus associated this with an interface drift velocity

$$\mathbf{U}_0 = \frac{\rho_a}{\rho_w} \left[ \frac{D_{Ea}}{D_{Ew}} \right] \mathbf{U}_g = \left( \frac{\rho_a}{\rho_w} \right)^{0.5} \mathbf{U}_g \approx 0.035 \mathbf{U}_g, \tag{5.51}$$

which is relatively close to Kenyon's (1969) estimate. Using direct measurements in a wind-wave tank, Wu (1975) connected $U_0$ with the friction velocity in the air, $u_*$, by the empirical relation

$$U_0 = 0.55 u_*. \tag{5.52}$$

Values of $U_0$ derived from (5.52) are larger than those computed from (5.51) for laminar boundary layers in both fluids. They tend to be smaller by a factor of 0.3–0.5 than those computed from (5.51) for turbulent conditions. Wu (1983) related $U_0$ directly to the windspeed $U_z$, at a level $z$, which was determined as a function of the fetch. For long fetch he found that $U_0 \rightarrow 0.031 U_{20}$, which is slightly smaller than (5.51). Differences in contamination apparently do not affect $U_0$ significantly. Although slick-covered surfaces are driven downwind by capillary-wave absorption, the reduced surface friction compensates for this effect.

There is an unresolved difficulty in the previous discussion. Kenyon and Wu stipulated a surface drift in the direction of the surface wind. The Coriolis force did not enter into their arguments. On the other hand, Ekman theory predicts an interface velocity in the direction of the geostrophic wind or, more precisely, in the direction of the resultant of the geostrophic wind and current vectors. The contradiction between these views does not disappear if the effect of the Coriolis force on the Stokes drift is taken into account (Huang, 1979). Additonal studies are required to resolve this dilemma.

The interface drift velocity, $U_0$, is almost invariably very much smaller than the near surface windspeed. In the atmosphere, the mean windspeed increases rapidly with height. Its value, even quite close to the surface, tends to be significantly larger than the orbital velocities associated with harmonic surface waves. In the sea below the interface, the mean translation velocity is usually smaller than $U_0$. Its value is generally of the same order as or smaller than the orbital wave motion. This seems to be the reason why orbital velocities can be observed in water and why they are very difficult to measure in air.

### 5.3.2 Wind–wave interactions

In (4.37) and (4.38) we introduced a source term for the growth of waves due to wind. This term has been discussed in some detail in Section 4.4.3. We also found that the present theory is still inadequate in describing the actual growth of the waves. Waves usually grow faster than theory predicts.

An element missing from the discussion is the effect of waves on the wind. As we have seen in Section 5.1 and Fig. 5.3, the surface roughness changes with windspeed, and at windspeeds above $8 \, \mathrm{m \, s^{-1}}$ it increases with the windspeed. It is simply not sufficient to assume a given wind profile to derive a wave field. The effect of the wave field on the wind profile also needs to be considered. This is an area of active research where many questions have not yet been answered. A summary has been presented by Dobson et al. (1993).

In this context it is convenient to introduce the *drag coefficient, $C_D$*, given by

$$C_D \equiv \frac{u_*^2}{(U_{10} - U_0)^2},$$ (5.53)

where $U_{10}$ is the windspeed at a 10-m height. In most cases $U_0 \ll U_{10}$ and may be neglected. This coefficient will be discussed further in Section 5.6. In Fig. 5.6 $C_D$ is given as a function of $U_{10}$, using the roughness length from Fig. 5.3 and from (5.17) of the profile description. If $C_D$ is known, the windspeed at a 10-m height may be directly related to the surface stress.

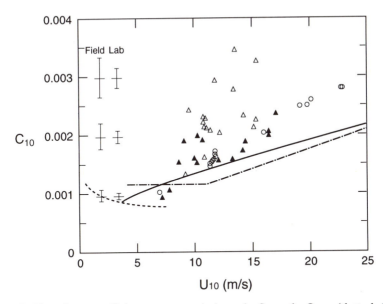

**Fig. 5.6.** The drag coefficient versus windspeed. Smooth flow (dotted line), Charnock's relation with $\mathbf{a} = 0.014$ (solid line) and Large and Pond (1981) (dash dot line). Very young laboratory waves ($6.5 \leq U_{10}/c_0 \leq 15.4$, open circles); young field waves ($3.5 \leq U_{10}/c_p \leq 4.6$, open triangles); mature field waves ($0.8 \leq U_{10}/c_p \leq 2.0$, closed triangles). The vertical bars are a measure of the sampling error, being $\pm 1$ standard deviation about the mean for windspeeds of 10.6 and $6.5 \, \mathrm{m \, s^{-1}}$ (field and laboratory, respectively). After Donelan (1990).

Also, if direct measurements of the stress are made, we can see what $C_D$ should be under the given circumstances. Such measurements, included in Fig. 5.6, show considerable scatter, usually above the curve of steady conditions given by the solid line.

It appears that the drag coefficient may be a good indicator of wind–wave interactions. We shall briefly review the effects of fetch, static stability, secondary flows, seastate, and nonsteady conditions.

In Section 4.4.3 the growth rates as a function of the wavenumber or the wave frequency have been discussed resulting in (4.53) and Fig. 4.7. The figure clearly shows that $\sigma/\omega$ increases rapidly with $u_*/c$. In fact, $\sigma/\omega \propto (u_*/c)^2$, as suggested by Plant (1982), fits most of the data quite well. It is not immediately obvious how this information can be used under limited fetch conditions. However, it is clear that for a short fetch the spectrum is limited to short waves and $u_*/c$ is relatively large. Consequently, all waves in the spectrum have large growth rates. Also, as we have seen in Section 4.5 and Fig. 4.8, the wave energy at the spectral peak for a short fetch is higher than the energy at the same frequency for a longer fetch when the spectral peak has moved to a lower frequency. These facts suggest that the energy input from the wind to the waves is larger for a short fetch than for a long fetch. We therefore expect $u_*$ and $C_D$ to be larger for short fetches than they are for long fetches under the same wind conditions. In Fig. 5.6 some of the points well above the solid line are obtained for short fetches. However, when the fetch is very short, such as in laboratory experiments, all wave amplitudes are very small and, although $\sigma/\omega$ is very large, the product of the growth rate and amplitude is less than with somewhat larger fetches. There appears to be a maximum energy input from the wind to the waves for fetches in the range of 1–10 km. This conclusion is suggested by the measurements presented by Donelan (1990), see Fig. 5.6. Observations by Geernaert et al. (1987) do not include fetches short enough to show the maximum. They suggest an empirical relation between $C_D$ and $c_p/u_*$ of the form

$$C_D = A\left(\frac{c_p}{u_*}\right)^B,$$

where $c_p$ is the phase velocity of the waves at the peak of the spectrum, and $A = 0.0148$ and $B = -0.738$. Theoretical arguments modelling the behavior of the drag coefficient $C_D$ as a function of the wave age $c_p/u_*$ are given by Nordeng (1991) and Janssen (1989).

The ratio $c_p/u_*$ is often used to specify the wave age. It is not necessarily a function of the fetch only. A sudden increase in windspeed is also possible in the open ocean. In this case the wave spectrum will initially resemble the fetch-limited spectrum with a small value of $c_p/u_*$. The drag coefficient would initially increase with time until a maximum is reached and then gradually decrease again to the equilibrium value corresponding to the new windspeed.

We should note here that, when the windspeed suddenly decreases and the dominant waves are moving faster than the wind, the drag coefficient may become negative and momentum is transferred from the waves to the atmosphere. An interesting example is given by Holland et al. (1981), which has been reproduced in Fig. 5.7. However, because waves are not as efficient in generating wind as wind is in generating waves, observations as presented in Fig. 5.7 occur only when the wind is very light. After their formation, long waves tend to coast along until they eventually reach a shore where they dissipate in the surf.

The effect of stability also needs to be considered in relation to the

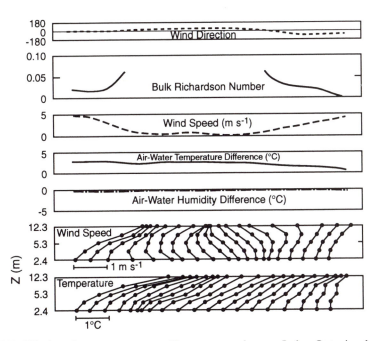

**Fig. 5.7.** Wind and temperature profiles measured over Lake Ontario showing formation and decay of a wave-driven wind. The profiles are running averages over 30 min plotted at 10-min intervals. The humidity difference is expressed in buoyancy equivalent degrees Celsius. The time series of wind and air–water differences are obtained from measurements at the top level and the surface water temperature. After Holland et al. (1981). Note that in the profiles the dots representing the measurements at the centre (5.3 m) are equally spaced and correspond to the time series of wind and temperature of the sequence shown.

non-equilibrium of the sea surface with respect to the wind. There is observational evidence (Keller et al., 1985) that under very unstable conditions $u_*$ is much larger than what would correspond to (5.46), an equation based on observations over land. The evidence has been obtained with a scatterometer and we must assume that the scatterometer cross-section is a measure of $u_*$ rather than of the mean wind $U$. There is growing evidence (see Section 5.6.5) that this assumption is correct. Yet it would be more conclusive if more direct measurements of $u_*$ existed under very unstable conditions.

An argument in favor of a stronger effect of stability over water than over land is that strong convective conditions in the atmospheric boundary layer are associated with large energetic eddies and secondary flows, such as helical rolls (see Section 6.2). Under influence of these eddies the surface winds are highly variable. As a result, the surface never comes to equilibrium with the wind. The stress increases therefore, not only because of the effect of stability per se [i.e. (5.46)], but also because of non-equilibrium between wind and wave field. Therefore, the drag coefficient will be larger than the one resulting from (5.46), and this is what the scatterometer indicated.

Because of the responsive character of the sea surface, the convective elements in the boundary layer may be modified. This would be especially true for helical rolls. Over uniform land terrain, helical rolls encounter a constant surface roughness. Over water, however, the rolls modify the sea surface, which in turn must modify the rolls. There will be a much larger than average stress in the region of the rolls' downdraft than anywhere else. We might expect that in this region heat and vapour transfer are also enhanced. This relatively localized input of heat and water vapour may in turn modify the character of the helical rolls. Some calculations in this area may be useful.

There are many other reasons why the sea surface is not in equilibrium with the wind input. Nikolayeva and Tsimring (1986) considered the effect of large-scale gustiness of the wind on wave growth. Their model calculations show a substantial enhancement of energy transfer from wind to waves due to gustiness. Besides gustiness, variability of the wind direction also modifies the seastate and consequently $u_*$ in a complex manner. In this case the two-dimensional character of the wave spectrum plays an important role.

It is clear from the previous discussion that the drag coefficient $C_D$ at a selected height is a function of a number of parameters beside the windspeed. These parameters include:

1. The effect of wave age, $c_p/u_*$.

2. The effect of stability, $z/L$ or $h/L$ or $\sigma/L$, where $h$ is the height of the boundary layer and $\sigma$ the significant wave height.

3. The effect of gustiness, $\sigma_u/U$, where $\sigma_u$ is the square root of the variance of the windspeed.

4. The effect of fetch, $F/z_0$, where the distance to the shore, $F$, is normalized by the roughness length.

5. The effect of seastate as characterized by the two-dimensional wave spectrum. The angles $\alpha_n$ between the wind and various waves, indicating swell, need to be considered.

$C_D$ may be expressed formally as

$$C_D = C_{Dn}K\left(\frac{c_p}{u_*}, \frac{h}{L}, \frac{\sigma_u}{U}, \frac{F}{z_0}, \alpha_n \cdots \right), \qquad (5.54)$$

where $C_{Dn}$ is the neutral-equilibrium drag coefficient given by the solid line in Fig. 5.6, and $K$ is the complex function which incorporates all effects mentioned above. The parameters in parentheses on the right-hand side of (5.54) are not all independent of each other (e.g., the wave age is a function of fetch and gustiness as well as change in wind direction). A careful study of the interrelationships between these parameters is needed, both theoretically and experimentally. It clearly must include a major, systematic, and well-organized observation program of seastate and the structure of the atmospheric boundary layer with the aim of determining the semi-empirical relations suggested by (5.54). Relations similar to (5.54) may be formulated for the bulk transfer coefficients for sensible heat flux and vapour flux.

The seastate affects the wind profile as well as the drag coefficient. The wind profile in the vicinity of the waves must be modified by the transfer of momentum directly to the waves. This is usually referred to as the *form drag* of the waves and relates to the induction process formulated by Miles, see Section 4.4.3. Calculations by Plant (pers. comm.) indicate that the profiles are only modified within one wave height from the surface. The effect of the waves on the wind profile is therefore very difficult to measure and most of the observations tend to confirm the existence of a logarithmic profile under near-neutral conditions because the lowest measurement level is well above the height where the effect of the form drag may be felt.

Phillips (1977) estimated that the transfer of momentum from wind to water is shared in approximately equal portions between induced wave drags and tangential, viscous shears. However, the very short waves that support most of $\tau_w$ tend to break within a short distance. Their momentum, as well as that part of the wind momentum that sustains the tangential shears, is therefore transformed locally into mean current momentum. This process is continuously renewed. The fraction of wind momentum that goes directly or indirectly into long waves, which can radiate it away over significant distances, is probably less than 20 per cent in most cases. This suggests that more than 80 per cent of the sea surface stress is used to accelerate currents locally. Very little is known about changes of the momentum flux along the vertical below the surface, or about the character of the associated mean current profile. Observations of slow mean currents within the wave zone are notoriously difficult and the analysis of the problem is fraught with uncertainties. We suggest that this is an important topic for further research.

### 5.4   Transport of trace gases across the interface

The transfer of trace gases across the interface has important climatological implications. For instance, knowledge of the $CO_2$ cycle is essential for climate modellers concerned with global warming. Yet this knowledge is quite incomplete because the transfer of $CO_2$ across the air–ocean interface is very difficult to observe and few direct measurements exist. The same is true for other important trace gases such as $CH_4$ and $O_3$. The development of better and less costly sensors is needed and is ongoing. We may see significant breakthroughs in this area in the next 10 years.

In addition to direct measurements, indirect techniques can be used to arrive at the fluxes at the interface. It is possible to measure the concentration of a trace species in the ocean and the atmosphere simultaneously and to use a model to derive the fluxes across the interface with this information.

#### 5.4.1   *Application of the surface renewal model*

The surface renewal model provides the thickness of the molecular sublayers and transition layers on both sides of the interface. In Section 5.1.2 we found

$$\xi_u \equiv \frac{z_u u_*}{v} = 16$$

for the velocity profile and the equivalent thickness for the temperature profile $\xi_\theta = 13$. Because we are dealing with scalar fluxes we shall assume that also $\xi_c = 13$.

The flux equation for a trace gas $n$ may be written in a form analogous to (5.10), that is,

$$G_n = \delta_{na} \frac{C_0 - C_a}{z_a} = \delta_{nw} \frac{C_w - C_0}{z_w}, \tag{5.55}$$

where $z_a$ and $z_w$ are the thicknesses the surface layers would have if the molecular layers would extend to the well-mixed turbulent layers, and where $C_a$ and $C_w$ refer to the bulk concentrations in air and water, respectively. From these equations we can eliminate the unknown concentration at the interface, $C_0$, to arrive at

$$G_n = \frac{\delta_{nw} \delta_{na}}{\delta_{na} z_w + \delta_{nw} z_a} (C_a - C_w). \tag{5.56}$$

The fraction on the right-hand side of this equation is called the *transfer velocity*, also known by chemical oceanographers as the *piston velocity*. If $\delta_{nw} \ll \delta_{na}$, which is the case for most trace gases, then (5.56) may be approximated by

$$G_n \simeq \frac{\delta_{nw}}{z_w} (C_a - C_w) = v_n (C_a - C_w), \tag{5.57}$$

where $v_n$ is the transfer velocity for constituent $C$. If $\delta_{nw}$ is known and we have measurements of $C_a$, $C_w$, and $u_*$, then the flux can be calculated using $z_w = z_u Sc_w^{-1/2} = 16 v_w u_{*w}^{-1} Sc_w^{-1/2}$, where $Sc$ is the Schmidt number, as defined in (1.80). Therefore, (5.57) may be rewritten in the form

$$G_n \simeq 0.023 u_{*w} Sc_w^{-1/2} (C_a - C_w) \tag{5.58}$$

or

$$v_n = 0.023 u_{*w} Sc^{-1/2}.$$

We see that in this case the transfer velocity, and hence the flux, is proportional to $Sc^{-1/2}$ and to the friction velocity, $u_*$. We also found a

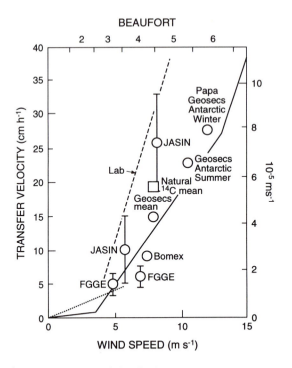

**Fig. 5.8.** Oceanic measurements of the liquid phase gas transfer velocity plotted as a function of windspeed measured at a height of 10 m. The dotted line is calculated from Eq. (5.58) representing the surface renewal model for light winds. The dashed line represents the results for intermediate windspeeds from laboratory wind tunnel studies. All data are converted to $S_c = 600$, corresponding to $CO_2$ at 20°C. The solid line represents the functional relation proposed by Liss and Merlivat (1986). After Liss and Merlivat (1986). Reprinted by permission of Kluwer Academic Publishers.

quantitative estimate of the constant of proportionality and therefore the transfer velocity, see Fig. 5.8. The coefficient 0.023 is somewhat larger than the 0.017 Ledwell (1984) reported using a similar model. With decreasing temperature the kinematic viscosity increases and the diffusivity decreases. As a result the Schmidt number increases and the transfer velocity decreases. Therefore, one might expect to find slower transfer rates in colder regions. From a climatological point of view, however, this change with temperature is overcompensated by higher windspeeds during winter at higher latitudes.

Equation (5.58) is only valid for light wind conditions and it does not apply to strong winds, when the sea surface may be considered rough. The stress, and therefore $u_*$, is enhanced greatly when flow separation occurs at the interface. Similarly the transfer of trace gases is enhanced strongly in breaking waves; (5.58) is likely to underestimate the flux.

### 5.4.2 The stagnant water film model

Broecker and Peng (1982) discuss an even simpler model, the *stagnant water film model*. In this model the tranfer rates are controlled by a thin stagnant water film which separates two well-mixed reservoirs with constant concentrations in the air above and the water below. The flux, that passes through the stagnant film of thickness, $z_f$, may be written in a form similar to (5.57)

$$G_n = \frac{\delta_n}{z_f}(C_w - C_a). \tag{5.59}$$

The transfer velocity $\delta_n/z_f$ is different from the one in (5.57) because $z_f$ is different from $z_w$. The thickness $z_f$ is assumed to be proportional to the thickness of the laminar sublayer, therefore

$$z_f \propto \frac{\nu}{u_*}. \tag{5.60}$$

Substituting this into (5.59) yields

$$G_n \propto u_* Sc^{-1}(C_w - C_a). \tag{5.61}$$

The constant of proportionality must be obtained through experiment. We see that for this model the transfer velocity is proportional to $u_*$ and inversely proportional to $Sc$ in contrast to (5.58), where the transfer velocity is inversely proportional to the square root of $Sc$. In Table 5.1 Schmidt numbers are given for several gases.

**Table 5.1.** Schmidt Numbers of Trace Constituents in Sea Water of 0°C and 24°C*

| Constituent | | 0°C | 24°C |
|---|---|---|---|
| Nitrogen | $N_2$ | 1625 | 430 |
| Oxygen | $O_2$ | 1490 | 390 |
| Argon | A | 2235 | 600 |
| Helium | He | 890 | 225 |
| Neon | Ne | 1275 | 320 |
| Krypton | Kr | 2550 | 640 |
| Xenon | Xe | 2550 | 640 |
| Radon | Rn | 2550 | 640 |
| Carbon Dioxide | $CO_2$ | 1790 | 475 |
| Nitrous Oxide | $N_2O$ | 1790 | 475 |
| Sodium Chloride | NaCl | 2415 | 712 (at 20°C) |

* Based on Tables 2.4 and 2.5.

### 5.4.3   *Experimental methods and results*

The layer thickness, $z_f$, and transfer velocity, $v_n$, have been derived experimentally by two types of radioactive tracer studies, which have been described in detail by Broecker and Peng (1982). One is based on the relative abundance of the radioactive carbon isotope $^{14}C$ in the atmosphere and the oceans. These atoms are created in the atmosphere by cosmic rays through the collision of energetic neutrons with nitrogen atoms. In a steady state the net number of $^{14}C$ atoms being transferred downward through the interface must equal the number being lost within the ocean by radioactive decay during the same period. Evaluation of this second number, together with (5.59) yields a stagnant film thickness of $4 \times 10^{-5}$ m as a global climatological estimate.

The second method is based on measurements of radioactive radon, $^{222}Rn$, in the surface air and in the subsurface mixed layer. In contrast to $^{14}C$, which is generated in the atmosphere, $^{222}Rn$ is the result of radioactive decay of radium, which takes place in the earth and reaches the ocean through bottom sediments. Measurements of the fraction of $^{222}Rn$ that escapes from the ocean mixed layer into the atmosphere before undergoing radioactive decay permits local determination of the thickness $z_f$. This method indicates that $z_f$ decreases with increasing windspeed from a maximum of about $8 \times 10^{-5}$ m in the equatorial zone to $2 \times 10^{-5}$ m in the southern ocean. Apparently no measurements have been carried out during severe storms or during periods of no wind.

The radon deficiency method is the most frequently used field method for determining the transfer velocity. It is superior to the $^{14}C$-method because it allows for the determination of local values of $v_n$ rather than a global average. The assumptions inherent in this method have been formulated by Liss (1983). The technique has been used extensively during the GEOSECS cruises in the Atlantic and Pacific Oceans (Peng et al., 1979) and in JASIN and FGGE (Roether and Kromer, 1984). The radon deficiency method gives results that, on the average, are in reasonable agreement with the $^{14}C$ method. The $^{14}C$ method yields an average transfer velocity of about $5.5 \times 10^{-5}$ m s$^{-1}$, whereas the $^{222}Rn$ measurements yield an average of about $4.4 \times 10^{-5}$ m s$^{-1}$. This result may reflect a bias in the $^{222}Rn$ measurements because the technique is usually not carried out under high wind conditions when the transfer velocities are large.

A summary of experimental results is presented in Fig. 5.8. Most of these have been obtained using the radon method. The averaging time of several days makes it difficult to relate this method to controlling environmental parameters such as windspeed, sea state, and local concentrations of the trace gas under consideration. There is a need to use more direct techniques, such as the eddy correlation technique, to determine the local transfer velocity. These techniques are technically more difficult and require development of better sensors than are currently available. Furthermore, it is necessary to establish that other trace gases behave similarly to radon.

Smith et al. (1991) used the eddy correlation technique for determination of the flux of $CO_2$. They used a newly developed sensitive and fast-response $CO_2$ sensor and were successful in obtaining some fluxes of $CO_2$. However, the measurements were taken at the shore of Bedford Basin rather than over the open ocean. The results are therefore only of local interest and cannot yet be used to test the radon method. Further refinement of the instrumentation and a stable platform in the open ocean are needed to pursue this endeavor.

## 5.5   The Sea Surface Temperature (SST) and the energy budget

Although it is difficult to measure the interface temperature directly, it has been known for many years that the SST is often a few tenths of a degree cooler than the so-called bucket or bulk temperature of the ocean. Bulk temperature is measured routinely from ships with thermometers in an engine water intake, or by taking a bucket of water from the ocean near the surface and measuring its temperature. This temperature represents the ocean temperature some distance below the surface, where the water is usually well-mixed. Woodcock (1940) may have been the first to measure the difference between the SST and the bulk temperature. His results were confirmed by Woodcock and Stommel (1947) and by many other investigators in later years. Katsaros (1980a,b) reviewed this subject. The first measurements of the surface temperature were obtained with small mercury thermometers or thermocouples. Although relatively small, they were large in comparison to the molecular sublayer and would not quite measure the SST. They have now been replaced by the remote infrared thermometer that measures from 8 to 12 $\mu$m in the atmospheric window and samples a layer of the order of 100-$\mu$m thickness.

To appreciate the significance of the temperature difference $\Delta T = T_b - T_0$, where $b$ refers to bulk and 0 to surface, it is necessary to consider the energy budget at the surface. This may be written as

$$Q_w = F_n + Q_a + LE, \qquad (5.62)$$

where $Q_w$ is the heat flux in the ocean, $Q_a$ the sensible heat flux in the atmosphere, and $LE$ the latent heat flux. The net radiation flux, $F_n$, is given by (3.3). All fluxes are by definition positive when upward.

In general the fluxes are positive at night because the solar radiation absorbed in the ocean warms the ocean and this heat can only escape through the interface. Therefore, averaged over the global oceans, $Q_w$ must be equal to the average short-wave absorption in the ocean, which requires that, on the average, $\Delta T > 0$. Local conditions may vary greatly from the average and all the terms on the right-hand side of (5.62) need to be considered in order to evaluate $Q_w$ and $\Delta T$.

A value of $\Delta T$ may be obtained at night, when the short-wave radiation is zero. By taking for $Q_w = 200 \text{ Wm}^{-2}$, which is approximately the daily solar radiation absorbed in the ocean at midlatitudes, $\Delta T$ may be calculated using (5.21) as

$$\Delta T = -\frac{13 Q_w}{c_p \rho u_*}. \tag{5.63}$$

The coefficient 13 is the result of matching the exponential profile to the logarithmic profile, see Fig. 5.2. If we now assume light wind conditions with $u_{*a} = 0.1 \text{ m s}^{-1}$, which means that $u_{*w} = 0.0035 \text{ m s}^{-1}$, we find that $\Delta T \simeq 0.2°C$. This is not a large temperature difference, but because it is systematic it needs to be considered in climatological studies as well as in applications of the bulk transfer relations for $Q_a$ and $LE$. The coefficient $\xi_\theta$ has also been obtained from field and laboratory studies. Paulson and Parker (1972) determined a range from 4 to 17 and found that the presence or absence of waves has a significant effect. The smallest values are obtained in cases with wind-driven waves.

When there is no wind (i.e., $u_* \to 0$), (5.63) is no longer valid because $u_*$ is in the denominator in the right-hand side of the equation. In this case we must consider conditions of free convection. Similarity arguments for free convection lead to

$$Nu = A \, Ra^{1/3}, \tag{5.64}$$

where the Nusselt number $Nu \equiv Q_w / (k \Delta T / D)$, $Ra$ is the Rayleigh number, (1.81), and $D$ is the vertical length-scale (e.g., the depth of the convective layer). The constant $A$ was found to be 0.156 (Katsaros, 1977). If we assume that (5.64) is also applicable over the open ocean, we find that, for $Q_w = 200 \text{ W m}^{-2}$, $\Delta T \simeq 0.85°C$. For moderate and strong winds $\Delta T$

becomes much smaller because $u_*$ increases more than linearly with the windspeed and $\xi$, decreases as previously mentioned.

When short-wave absorption near the surface is no longer negligible in comparison to the other terms in (5.62), the situation is more complex. A substantial fraction, about 25 per cent, of short-wave radiation is absorbed in the upper 5 cm of the ocean as indicated by (3.19). The long-wave radiation, which is emitted and absorbed at the sea surface, comes from a layer less than 1-mm thick. Therefore, the short-wave radiation tends to warm up a layer several centimeters thick and the long-wave radiation tends to cool the surface.

Turbulent mixing is usually sufficiently strong for the heat from short-wave absorption to be distributed over the ocean mixed layer. Part of this heat is transported downward and increases the temperature of the mixed layer, and part of it is transported upward and contributes to $Q_w(0)$. $\Delta T$ consequently decreases and the SST increases.

Sometimes, when the wind is very light, the turbulent mixing may be suppressed by the heating near the surface. This happens when the local flux Richardson number, $Ri_f$, reaches a critical value (i.e., $Ri_f \geq 0.2$), where

$$Ri_f \equiv -\frac{\overline{b'w}}{u_*^2 \dfrac{\partial U}{\partial z}}$$

Once there is a layer where the turbulence is suppressed, the upper layer becomes effectively uncoupled from the deeper layers and its temperature may rise relatively rapidly. From the temperature maximum to the surface, the stratification remains superadiabatic. Because the layer is so shallow, however, convection will be relatively light or even fully suppressed depending on whether or not the Rayleigh number for this layer has reached its critical value.

The actual SST in this case will also be uncoupled from the temperature of the well-mixed layer (e.g., at a 20-cm depth) and $\Delta T$ may become positive, although $(\partial T)/\partial z)_0$ at the interface must remain negative. This condition may explain the observation of anomalously high sea surface temperatures during the daytime over the tropical oceans.

An increase in SST is accompanied by an increase in $Q_w$ because all terms on the right-hand side of (5.62) increase. This represents a negative feedback. Although the SST may be uncoupled from the well-mixed layer below, it remains rather uniform horizontally due to the feedback mechanism.

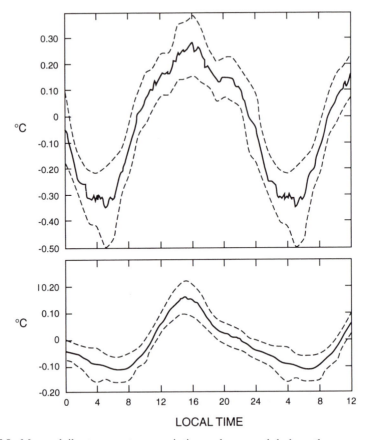

**Fig. 5.9.** Mean daily temperature variations above and below the sea surface. Upper curves represent air temperatures $(z = 8.9\,\text{m})$ and lower curves sea temperatures $(z = -0.2\,\text{m})$. Standard deviations are marked by dashed lines. Measurements were obtained from Sept. 21 to Oct. 9, 1965, at location 0°N, 30°W. After Hoeber (1969).

In this context it is of interest to look at some observations of daily variations in air and sea temperature on the equator, made by Hoeber (1969), given in Fig. 5.9. The daily variation in temperature is considerably larger in the atmosphere than it is in the ocean. Although the SST is not measured, the data indirectly suggest the strong influence of the short-wave radiation on the SST, and the temperature structure near the surface. The figure also suggests that the average temperature is constant over the period of observation. This would mean that an equilibrium is reached between $Q_w$ and the fraction of short-wave radiation that is absorbed in the mixed layer.

Evaporation and precipitation have a significant effect on the SST and

are important in the formation of water masses. Oceanic surface areas where the preciptation is larger than the evaporation tend to have a salinity that is below the ocean average and vice versa for areas where the evaporation is larger than the precipitation. In coastal areas, the run-off of rivers needs to be taken into account as well.

Under conditions of light wind, rainfall has a stabilizing effect on the water near the surface. The lower salinity, and therefore lower density, of the surface water may suppress turbulence in the same way that higher temperatures do. In this case the flux Richardson number is related to the density fluctuations due to salinity rather than to temperature, but the argument is otherwise the same as given earlier. The stable layer formed by rainfall or by river run-off is sooner or later eroded by wind-generated turbulence, and the low density surface water is mixed into the ocean mixed layer.

If the calm conditions persist after the rain has stopped and the sky has cleared, anomalously high surface temperatures may be observed during the day, as discussed earlier, and anomalously low temperatures during the night. The low night-time temperatures are the result of surface cooling while convection to deeper layers is suppressed because the low surface salinity maintains a stable stratification.

In subtropical regions where evaporation is much larger than precipitation, the salinity tends to increase in the immediate vicinity of the surface. The cooling at the surface combined with the increase in density due to the higher salinity creates an unstable stratification at the surface and enhances convective stirring of the mixed layer. However, as pointed out earlier, strong solar irradiance at the surface and light wind conditions may lead to a stable layer just below the surface and turbulence may be suppressed. The result is not only a higher surface temperature, but also enhanced evaporation with an increase in salinity. This process has a daily as well as an annual cycle. During summer the salinity may increase in a shallow boundary layer over large ocean areas. In fall and winter this saline water is mixed through a greater depth.

## 5.6  Methods to observe the fluxes in the atmospheric surface layer

It is very difficult to measure fluxes in the oceanic layer, which includes the wave motions, and very few measurements of the fluxes of heat and momentum in this region have been made. An interesting technique to measure the heat flux $Q_w$ remotely was developed by McAlister and McLeish (1970). They measured the emission of the water surface in two

wavelength bands in the infrared. These bands were chosen in such a way that the effective depths from which the emission comes are 25 and 75 $\mu$m, respectively. These measurements allowed them to derive the temperature difference over the 50-$\mu$m interval and consequently to calculate the heat flux with the first equation (5.3), assuming that the measurements are within the molecular sublayer. The technique works reasonably well in light winds, but even here corrections for incidence angle and slope of the sea surface complicate interpretation of the data. As far as we know the technique has not been used in recent years. The momentum flux in the upper ocean is possibly even more difficult to measure than the heat flux, as has been discussed in Section 5.1.3. In this section we will restrict ourselves to a discussion of flux measurements in the atmospheric surface layer.

### 5.6.1  The eddy correlation method

The eddy correlation method is the only direct method available for obtaining the fluxes. For this method to be valid, several conditions must be fulfilled. The covariance measured at a point, which gives the complete budget of transport up and down at that point, must be representative of the ensemble average of the flux of the area of interest. This requires that conditions are horizontally uniform and stationary (see Section 1.4). In addition, the flux must be essentially constant with height from the surface to the level of measurement.

These conditions are usually met over water when the fetch is large enough, $F > 1$ km, and the height of observation $z \leq 10$ m. If the height of observation is greater than 10 m, a correction may be applied for the change of flux with height.

The flux $G_n$ of quantity $C$, including water vapour, may be expressed by

$$G_n = \overline{wc} + WC. \tag{5.65}$$

For momentum, sensible heat, and vapour fluxes the second term on the right-hand side of (5.65) is usually negligible. For trace gases this is a different matter. Webb et al. (1980), pointed out that the fluxes of heat and water vapour may introduce large errors in the eddy correlation of trace constituents. These errors are not random and can be corrected for when the fluxes of sensible heat and water vapour are known.

To understand these corrections in a qualitative sense, we will consider an inert species, like argon, with an average mass flux of zero. We assume

that there is a sensible heat flux with associated density fluctuations. These fluctuations will also be reflected in $C$, the concentration of argon, because the mixture is uniform (i.e., the mixing ratio of argon does not fluctuate). Thus

$$\frac{c}{C} \approx -\frac{\theta}{\Theta} \quad \text{and} \quad \overline{wc} \propto -\overline{w\theta} \neq 0.$$

A similar effect is a consequence of the vapour flux. The partial pressure of water vapour is to be subtracted from the total pressure in order to obtain the partial pressures of dry air and trace constituents. Fluctuations in water vapour density, $\rho_v$, are therefore associated with fluctuations in $c$; consequently, the water vapour flux induces an apparent flux of $c$.

The result of the analysis by Webb et al. (1980) is that the mean vertical velocity, $W$, is given by

$$W = 1.61 \frac{\overline{w\rho_v'}}{\rho_a} + (1 + 1.61\bar{q}) \frac{\overline{w\theta}}{\Theta}. \tag{5.66}$$

The coefficient $1.61 = m_a/m_w$ is the ratio of the molecular mass of air over the molecular mass of water vapour; $\rho_a$ is the density of dry air; and $q$ the specific humidity (see Section 2.2.1). The specific humidity should be replaced by the mixing ratio, but the two are the same to a first-order approximation and $1.61\bar{q} \ll 1$.

After introducing (5.66) into (5.65) we find

$$G_n = \overline{wc} + 1.61 \frac{C}{\rho_a} \overline{w\rho_v'} + (1 + 1.61\bar{q})C \frac{\overline{w\theta}}{\Theta}. \tag{5.67}$$

These corrections can easily be made provided the fluxes of water vapour and sensible heat are measured.

The frequency range required to obtain the fluxes is given by the variance spectrum of the vertical velocity, $w$. It is dependent upon windspeed, height, and stability and has been discussed in detail by Kaimal et al. (1972). Most of the cospectrum is usually confined to the range $10^{-3} < \omega z/U(z) < 10$, where $\omega z/U \equiv n$ is a nondimensional frequency, which is equal to the actual frequency $\omega$ at a height of 10 m.

Kaimal (1975) has formulated criteria for the response time and spacing of the sensors based on the AFCRL's Kansas experiment. It turns out that the cospectra can be approximated by universal functions which peak at $n = 0.03$. The cospectral energy drops off on both sides of the peak. If we select the $-10\,dB$ point (one tenth below the peak value), then we find $n = 10^{-3}$ on the low-frequency side and $n = 1.5$ and 3, for momentum and heat flux respectively, on the high-frequency side. The bandwidth for other scalar fluxes are the same as for the heat flux. The sensor must therefore respond to wavelengths as small as $z/3$ in order to measure these fluxes adequately.

The separation of the sensors should be less than $z/3$ by a factor of $2\pi$, in order to avoid spectral attenuation from spatial averaging. For sensors with a separation distance, $d$, the minimum operating height would be

$$z_{min} = 6\pi d. \tag{5.68}$$

Thus, if the separation of the sensors is 20 cm, the height of observation should be above 4 m. This strict requirement is valid for the momentum flux. The separation criterion may be relaxed for scalar fluxes, as was shown by Kristensen and Fitzjarrald (1984), to

$$z_{min} \approx 5d \tag{5.69}$$

without serious loss of covariance. This is useful because it often is desirable to make the measurements at as low a level as possible.

### 5.6.2   The eddy accumulation method and the conditional sampling method

An interesting variant of the eddy correlation method is the *eddy accumulation method*. This method bypasses the need for a fast sensor for the trace constituents considered. Many trace constituents cannot be measured fast enough to allow application of the eddy correlation method. Instead of determining the concentration quickly, air is sampled conditionally first and the concentration is measured later. The vertical velocity is used to determine the conditional sampling by opening one valve when the vertical velocity is positive and another valve when it is negative. The amount of air being sampled is kept proportional to the magnitude of the vertical velocity. Therefore, the positive and negative reservoirs contain air, respectively given by

$$\overline{w^+(C+c)} \quad \text{and} \quad \overline{w^-(C+c)}.$$

When these two quantities are added, we obtain the flux

$$\overline{w^+(C+c)} + \overline{w^-(C+c)} = (\overline{w^+ + w^-})C + \overline{w^+c} + \overline{w^-c} = \overline{w^+c} + \overline{w^-c} = \overline{wc},$$

because $\overline{w^+ + w^-} = 0$.

This method, first proposed by Desjardins (1972), is attractive on paper but very difficult to realize in the field. A major problem is that most vertical velocity sensors indicate a vertical velocity bias, $W$, which is not equal to zero. $WC$ very quickly becomes a large term, larger in magnitude than $\overline{wc}$. It is also very difficult to open the valves exactly proportional to $w$. Hicks and McMillen (1984) give an analysis of these difficulties.

A simplification of the eddy accumulation method, the *conditional sampling method*, has been proposed by Businger and Oncley (1990). The simplification consists of requiring only that the conditional sampling is determined by updrafts and downdrafts and that the air is sampled at a constant flow rate. The measurements to be made are the variance of the vertical velocity, the average concentration $\overline{c^+}$ during updrafts, $w > 0$, and $\overline{c^-}$, during downdrafts, $w < 0$. In the surface layer the flux is then obtained by assuming that

$$\overline{wc} = b\sigma_w(\overline{c^+} - \overline{c^-}), \tag{5.70}$$

where $b$ is a coefficient, to be determined experimentally, and $\sigma_w \equiv (\overline{w^2})^{0.5}$. The disadvantage of this simplification is the introduction of an empirical coefficient $b$. The advantage is that the method is technically easy to realize.

The conditional sampling method has been simulated with existing data for eddy correlation to determine the value of $b$. The result is that $b \approx 0.6$, independent of stability. It also has been simulated with large eddy simulation models with the result that, for bottom-up convection, $b \approx 0.6$ throughout most of the convective boundary layer. Direct experimental evidence was obtained by Oncley et al. (1993), suggesting that this method is somewhat less prone to errors than the eddy correlation method and that the simulated value of $b$ agrees with the directly measured value.

The fact that $b$ appears to be constant, independent of stability, makes this method attractive for small-scale experiments of fluxes of trace gases because there is no need to measure the momentum flux and heat flux simultaneously.

### 5.6.3   *The gradient method*

This method is the most common one for determining fluxes because it does not need the advanced technology required for the eddy correlation method. The method is based on the assumption that the flux can be obtained by multiplying the vertical mean gradient by an eddy transfer coefficient, see (1.48). In Section 5.2 we discussed the Monin–Obukhov similarity and the semi-empirical profile relationships that provide the basis for the gradient method.

Empirical evidence over land indicates that there is similarity between the gradients of scalar quantities such as potential temperature, $\theta$, specific humidity, $q$, and concentration of trace constituents, $C$. This means that

$$K_h = K_e = K_c, \tag{5.71}$$

where $K_c$ is the eddy diffusivity for $C$ in air, defined by $\overline{wc} = -K_c \partial C / \partial z$.

Over water this similarity may break down under high wind conditions $(U > 20 \text{ m s}^{-1})$ when sea spray is carried into the lower marine boundary layer. The spray is a source for water vapour and a sink for sensible heat. The wind profile and the scalar profiles are, on the other hand, dissimilar, as we have seen in Section 5.2.

In practice, the gradient is determined by making measurements at two heights. The geometric mean height is then used as the height where the tangent to the profile is equal to the gradient, because the profiles are usually close to logarithmic with height. For example, if the concentration, $C$, is measured at 2 m and 8 m, the height, where $dC/dz = (C_8 - C_2)/(z_8 - z_2)$, is $z = 4$ m.

The main drawback of the gradient method is that it requires very accurate measurements of the mean scalar quantity or mean velocity in order to obtain sufficiently accurate gradients, as these may be very small. A good way to avoid instrumental error is the use of the same instruments at two measurement heights. Most instruments have greater relative accuracy than absolute accuracy.

The method works best when the wind is light and the stratification stable, and worst when the wind is strong and the stratification unstable. Businger and Delany (1990) discuss the accuracy requirements for sensors under various atmospheric conditions for the methods discussed so far in this section.

### 5.6.4   The dissipation and inertial dissipation methods

The assumption that turbulence is isotropic in the high-frequency end of the wavenumber domain appears to be valid in the mixed layer of the ocean and the marine boundary layer at some distance to the interface (~10 m in the atmosphere and ~30 cm in the ocean). This allows us to use (1.68) to obtain the dissipation of TKE, $\epsilon$, using a fast response anemometer with a frequency response up to 5 kHz. Expressions similar to (1.68) may be formulated for the dissipation of scalar quantities (see, e.g., Fairall and Larssen, 1986).

Once $\epsilon$ is known (5.30) may be used to obtain the momentum flux for near-neutral conditions. This is called the *dissipation method*. Wyngaard and Coté (1971) showed that the pressure term and the transport term nearly cancel each other for a sizeable range of unstable conditions. A reasonable approximation of the TKE equation (5.28) may therefore be written, using (5.31), (5.32), and (5.35), in the form

$$\varphi_m - \zeta = \varphi_\epsilon, \tag{5.72}$$

where $\varphi_\epsilon \equiv (kz/u_*^3)\epsilon$. This equation requires that, in addition to $\epsilon$, the heat flux also must be known in order to determine the momentum flux. A simple iterative process is then necessary to obtain the momentum flux and the stability simultaneously.

It is technically not easy to measure the dissipation range of the spectrum because it requires a very high data sampling rate and a large storage capacity for the data points, while the signal is very weak. Furthermore, the instrument is usually a hot wire anemometer, which is nonlinear and fragile. The dissipation method is most useful for determining the Kolmogorov constant for the inertial subrange, as discussed in Section 1.4.2.

With knowledge of the Kolmogorov constant, $\alpha$, and the dissipation rate, $\epsilon$, we can relate the momentum flux to the inertial subrange of the variance spectrum, using (1.84) and Taylor's hypothesis (1.65)

$$u_*^2 = \alpha_1^{-1}\left(\frac{kz}{U}\right)^{2/3} \omega^{5/3} G(\omega). \tag{5.73}$$

This method is called the *inertial dissipation method*. It has the advantage that this part of the spectrum is much easier to measure because it is a stronger signal at a lower frequency (10–20 Hz). There are a number of wind sensors available that can resolve the inertial subrange. For diabatic conditions it is necessary to measure the heat flux and to use (5.72) combined with (1.84) and (1.65) in order to obtain the momentum flux. The main attraction of the method is that it can be used on ships because ship motions occur at lower frequencies and the directionality is not very critical. Until recently ship motions made it very difficult to measure the momentum flux with the eddy correlation method.

The inertial dissipation method has been used successfully in many experiments. Fairall et al. (1990) describe a system that was used over the North Sea in strong winds. They also used the inertial dissipation method for temperature and humidity (i.e., for the sensible heat flux and the vapour flux), assuming similarity between scalar quantities.

### 5.6.5   *Fluxes obtained with remote sensing techniques*

Surface winds and stress vectors have been successfully measured with so-called scatterometers. The scatterometer is a radar operating at 14.6 GHz and directed to the sea surface at various angles of incidence. At the surface, short gravity waves and capillary waves scatter the incident radiation back. This signal, called the radar backscatter cross-section, $\sigma_0$, is related to the near-surface wind velocity (Jones et al., 1982; Moore and Fung, 1979; Schroeder et al., 1982; Brown et al., 1982; and Donelan and Pierson, 1987) with the empirical relation given by

$$\sigma_0 = G(\theta, \varphi) + H(\theta, \varphi) \log U_n, \qquad (5.74)$$

where $\sigma_0$ is given in dB, $U_n$ is the equivalent neutral wind at 19.5 m, $\theta$ is the incidence angle between the radar beam and nadir, $\varphi$ is the aspect angle, the difference between the wind direction and the radar azimuth. The empirical coefficients $G$ and $H$ are tabulated for horizontal and vertical polarizations every 2° in $\theta$ and every 10° in $\varphi$. This empirical relation indicates a windspeed accuracy of $\pm 2 \text{ m s}^{-1}$ and direction accuracy of $\pm 20°$. These results were obtained with the Seasat-A Scatterometer System (SASS) (Schroeder et al., 1982).

It is assumed that the radar return is due to Bragg scattering from short waves tilted by long waves. It is further assumed that these short waves are correlated with the windspeed and wind direction. Despite the uncertainties in the physics of the relationship between radar signals and wind, sufficient verification of (5.74) has emerged to give confidence in the results. The data obtained with SASS has found a wide variety of uses (e.g., Brown, 1986).

In many applications, however, the surface stress, $\tau$, is more relevant than $U_n$. It also seems, from a physical point of view, more likely that $\sigma_0$ is more directly related to $\tau$ than it is to $U_n$. Liu and Large (1981) and Li et al. (1989) have explored this possibility. The difficulty is that direct measurements of $\tau$ or $u_*$ are much more difficult to obtain than those of $U_n$. A much smaller data base is therefore available for construction of an empirical relationship between $\tau$ and $\sigma_0$. Liu and Large (1981) used a set of $u_*$ values derived from direct observations and they found that the correlation between $\sigma_0$ and $u_*$ was as good as the correlation between $\sigma_0$ and $U_n$. In the later study by Li et al. (1989), it was found that $\sigma_0$ correlates much better with $u_*$ than with $U_n$, although these results were not yet considered definitive. More simultaneous measurements of $u_*$ and $\sigma_0$ are needed.

It is clear that the scatterometer is potentially a very powerful tool for determining the stress distribution over the oceans on a global scale.

The fluxes of sensible heat and water vapour from the sea surface to the atmosphere are more difficult to obtain with remote sensors. The scatterometer in conjunction with a Scanning Multichannel Microwave Radiometer (SMMR) has been used to obtain estimates of these fluxes (Liu, 1984 and 1989). However, more work is needed before these fluxes are as reliable as the stress determinations with the scatterometer.

### 5.6.6 *The ageostrophic transport or momentum budget method*

The surface stress, $\tau_0 = \rho u_*^2$, can be obtained from (1.44) by integrating this equation from the surface to the top of the boundary later, where $\tau = 0$. If we further assume steady state and horizontal uniformity, we find with simple vector algebra

$$\tau_0 \times \mathbf{n} = f \int_0^h \rho(\mathbf{U} - \mathbf{U}_g)\, dz, \qquad (5.75)$$

where $h$ is the top of the boundary layer, defined by the level where $\mathbf{U} = \mathbf{U}_g$. To evaluate the right-hand side of (5.75), it is necessary to know the wind profile $\mathbf{U}(z)$ throughout the boundary layer, assuming that $\mathbf{U}_g$ is constant with height.

For this method to be successful, the requirements of steady state and horizontal uniformity must be fulfilled over an extensive area limiting its applicability. However, by applying it under extreme conditions such as in hurricanes, where direct measurements are very difficult, we may obtain information of the stress for very high windspeeds. Kondo (1975) and

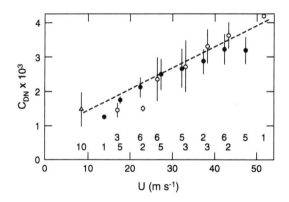

**Fig. 5.10.** Mean values of the neutral drag coefficient as a function of windspeed at a 10-m height based on hurricane studies (○), wind flux experiments (●), and vorticity/mass budget analysis (△). Vertical bars refer to standard deviation of individual data for each mean, with the number of datapoints used in each $5\,\mathrm{m\,s^{-1}}$ interval shown above the abscissa. The dashed line refers to Charnock's relation with $n = 0.014$, the same as in Fig. 5.6. After Garratt (1977).

Garratt (1977) review the results of this method and some variations of it. Figure 5.10 shows that the stress estimates with this type of approach agree quite well with $C_D$, based on (5.22) with $a = 0.014$.

### 5.6.7  *Bulk parameterizations*

Once the flux-gradient relations have been established, it is a simple step to formulate the fluxes with measurements at one height, say 10 m, and knowledge of the sea surface temperature. *Bulk transfer coefficients* are introduced to relate the mean observations at one height to the fluxes. For the momentum flux this was already done in Section 5.3, where the drag coefficient $C_D$ was introduced by (5.53). Here $U_0$ will be neglected so that (5.53) will be rewritten in the form

$$u_*^2 = C_D U^2. \tag{5.76}$$

Similarly we have for the sensible heat flux, $Q$, and the vapour flux, $E$,

$$Q = -c_p \rho C_Q U(\Theta - \Theta_0) \tag{5.77}$$

$$E = -\rho C_E U(\bar{q} - q_0), \tag{5.78}$$

where $C_Q$ is the bulk transfer coefficient for sensible heat and $C_E$ the coefficient for water vapour. As long as there is similarity between the

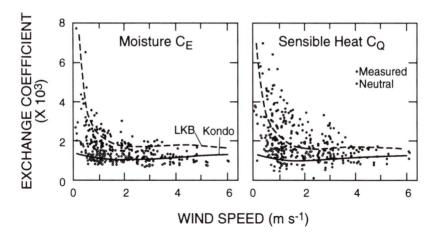

**Fig. 5.11.** Eddy correlation measurements of exchange coefficients for: (a) water vapour $C_E$; (b) sensible heat $C_Q$. Models are from Liu et al. (1979) (dashed curve) for $C_E$ and $C_Q$, and Kondo (1975) (solid line) for $C_{EN}$ and $C_{QN}$. After Bradley et al. (1991).

scalar profiles $C_Q = C_E$. This may be generalized for the fluxes of trace constituents.

Because this method is very convenient, a great deal of effort has gone in determining the bulk transfer coefficients $C_D$, $C_Q$, and $C_E$. Results related to $C_D$ have been discussed in Section 5.3 and in Fig. 5.10. $C_Q$ and $C_D$ have been described by Kondo (1975), Liu et al. (1979), Geernaert et al. (1987), Smith (1988), Bradley et al. (1991), and Katsaros and DeCosmo (1990).

Under near-neutral conditions both $C_Q$ and $C_E$ appear to be independent of windspeed for the range from 5 to 20 m s$^{-1}$ and to have a value of about $1.2 \times 10^{-3}$. There is some indication that $C_E$ increases for windspeeds greater than 20 m s$^{-1}$, due to increasing evaporation from sea spray droplets. Simultaneously $C_Q$ decreases because evaporative cooling of the droplets absorbs much of the upward heat flux and tends to stabilize the atmospheric surface layer.

Under unstable conditions with light winds, the transfer coefficients appear to be large and decreasing with windspeed as shown in Fig. 5.11 (Bradley et al., 1991). There is a great deal of scatter in the observations, which suggests that $C_Q$ and $C_E$, like $C_D$, are complex functions related to fetch, sea state, secondary flows, and so on, as suggested by (5.54).

# 6

# THE PLANETARY BOUNDARY LAYER

The term *planetary boundary layer* (PBL) is often used as a synonym for the Ekman layer (i.e., for the region in which the convergence of the vertical flux of momentum is of the same order as the Coriolis force and the pressure gradient). The definition favored in this chapter is somewhat broader and includes the regions on both sides of the interface in which the vertical fluxes, not only of momentum but also of heat, moisture, and salinity, determine the vertical distribution of these properties. Such a definition may suggest as many different boundary layers as there are transported properties. This may be the case, but the various fluxes are coupled with each other to such an extent that it is usually possible to define a single layer in which interface effects remain significant.

In Section 6.1 we shall deal first with the classic Ekman treatment of the steady-state motion field above and below the boundary of two incompressible, rotating laminar fluids. This will be followed by a discussion of transients and of integral horizontal transports.

Section 6.2 deals with coherent structures—longitudinal rolls, thermal plumes, convection cells, and so on—that are common phenomena, particularly in the gravitationally unstable PBL. This is followed in Section 6.3 by a discussion of various parameterization schemes and models that have been used either to represent vertical fluxes or vertical profiles of conservative properties.

Mixed-layer models, which are the topic of Section 6.4, are distinguished from these parameterization schemes by use of vertical integrals of the conservative properties. The resulting gain in simplicity is offset to some extent by a loss of detail. In Section 6.5 we shall discuss the merits and drawbacks of the different approaches in the two preceding sections.

## 6.1   The Ekman boundary layer

During his arctic expeditions on the ship *Fram,* late in the nineteenth century, Nansen noticed that the pack ice drifted at an angle of about 20–40° to the right of the surface wind. He interpreted this correctly as being due to the deflecting force of the earth's rotation, and inferred further that the water below, which is dragged along by the ice, must be driven even further to the right. After Nansen's return in 1901, the problem was referred to Ekman, who at that time was a student of V. Bjerknes in Stockholm. Bjerknes tells that Ekman produced his famous mathematical solution that same afternoon.

### 6.1.1 _ *The stationary laminar Ekman layer*

Ekman (1905) analyzed the problem as one of viscous laminar flow. This is an idealization that is hardly, if ever, realized in the PBLs. The results remain applicable, however, to some laboratory experiments, and they can be used for an explanation of the basic dynamic principles. These are not changed by turbulence or changes in the vertical distribution of the transport of horizontal momentum, even if the actual flow pattern becomes very different in the presence of these processes.

Ekman's approach was essentially based on the argument that the equation of horizontal motion, (1.22b), involves two driving forces: the vertical stress gradient and the horizontal pressure gradient. The accelerations caused by these two forces can be considered separately if products and squares of the velocity components are negligibly small. This permits us to represent the boundary layer velocity field in the form $\mathbf{U} = \mathbf{U}_p + \mathbf{U}_E$, where $\mathbf{U}_p$ denotes the pressure-driven part of the velocity and $\mathbf{U}_E$ is the part associated with the stress gradient. $\mathbf{U}_p$ satisfies the equation

$$\left(\frac{\partial}{\partial t} + f\mathbf{n} \times \right)\mathbf{U}_p + \frac{1}{\rho}\nabla p = \frac{\partial \mathbf{U}_p}{\partial t} + f\mathbf{n} \times (\mathbf{U}_p - \mathbf{U}_g) = 0, \qquad (6.1)$$

where $\mathbf{U}_g$ is the geostrophic velocity vector as defined in (1.28). During steady and balanced flow, $\mathbf{U}_g = \mathbf{U}_p = $ constant.

The vertical gradient of the horizontal stress component is equal to the convergence of the downward flux of horizontal momentum. The part $\mathbf{U}_E$ of the velocity, which is driven by this force, is called the *Ekman velocity.* It can be derived from

$$\left(\frac{\partial}{\partial t} + f\mathbf{n} \times \right)\mathbf{U}_E = \frac{1}{\rho}\frac{\partial \boldsymbol{\tau}}{\partial z}. \qquad (6.2)$$

In laminar flow,

$$\frac{\partial \tau}{\partial z} = \nu \frac{\partial^2 \mathbf{U}_E}{\partial z^2} \tag{6.3}$$

We will use subscripts $a$ and $w$ to characterize atmospheric and oceanic properties. The boundary conditions are

$$\mathbf{U}_E = 0 \qquad\qquad\qquad \text{for } z = \pm\infty$$

$$\mathbf{U}_0 = (\mathbf{U}_E + \mathbf{U}_g)_a = (\mathbf{U}_E + \mathbf{U}_g)_w \quad \text{for } z = 0 \tag{6.4}$$

In the viscous, steady flow that was considered in Ekman's original (1905) paper, (6.3) is applicable and $\partial/\partial t = 0$. Pressure gradients in the water were assumed to be equal to those in the air. The solution of (6.2) can then be written in the form

$$\mathbf{U}_{Ea} = \mathbf{U}_{ga}\left[ 1 - (1 - \mu)\exp\frac{(i - 1)z}{D_{Ea}} \right]$$

$$\mathbf{U}_{Ew} = \mathbf{U}_{ga}\mu \exp\frac{(1 - i)z}{D_{Ew}} \tag{6.5}$$

This is the classic form of the *double Ekman spiral*. The diffusion capacity, $\mu$, was introduced in (4.28). A relevant plot is shown in Fig. 6.1. The velocity components parallel and normal to $\mathbf{U}_{ga}$ are specified by the real and the imaginary parts of (6.4). The quantities

$$D_{Ea} = \left[\frac{2\nu_a}{f}\right]^{0.5} \qquad D_{Ew} = \left[\frac{2\nu_w}{f}\right]^{0.5} \tag{6.6}$$

are the stationary viscous *Ekman lengths*. They provide an indication of the distance up to which the two fluids are affected by the presence of the interface. The interface velocity $\mathbf{U}_0$ has the same direction as the geostrophic wind. Setting $z = 0$ in either equation (6.5) yields (5.49)

$$\mathbf{U}_0 = \mu \mathbf{U}_{ga} \approx 0.005 \mathbf{U}_{ga}.$$

The preceding analysis would remain applicable if the molecular viscosity, $\nu$, were replaced by some constant eddy viscosity, $K_m$. In reality, $K_m$ is never constant. As shown in 5.2, it varies vertically both with distance from the boundary as well as with the hydrostatic stability. Its

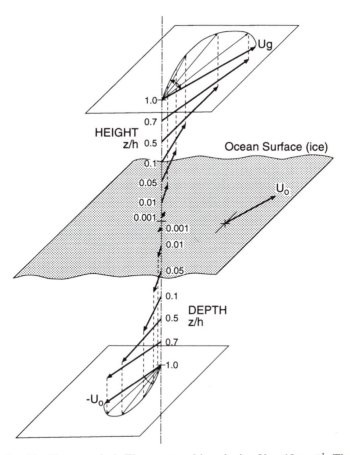

**Fig. 6.1.** Double Ekman spiral. The geostrophic velocity $U_g \simeq 10\,\mathrm{m\,s}^{-1}$. The frame of reference moves with the drift velocity at the ocean surface, $U_0 \simeq 0.3\,\mathrm{m\,s}^{-1}$. The surface drift and the Ekman spiral in the ocean are magnified about 30 times to make them comparable to the velocities in the atmosphere. After Brown (1990).

value is lower in inversions or thermoclines than it is in the adjacent, less stably stratified, layers. The resulting effects are illustrated schematically in Fig. 6.2, which shows hodograms of two laminar Ekman layers over a rigid surface. The smooth curve characterizes the velocity distribution with height if the viscosity is constant. The second curve represents the velocity distribution with the viscosity above the level $z = D_E$ reduced by a factor of four. The figure shows that the velocity turning is then much reduced below the viscosity discontinuity. Most of the shear, particularly the directional shear, is concentrated in the outer region, and the effect of the surface stress does not reach as deep into the interior as it does in the homogeneous case. The atmospheric and oceanic motion fields are not laminar, but the same consideration applies. The scale of turbulence and

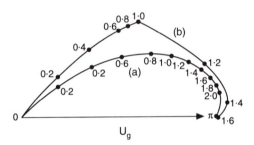

**Fig. 6.2.** Hodograms of velocities in a laminar Ekman layer. The numbers represent non-dimensional heights $z/D_E$. (a) $v = v_c$ (constant). (b) $v = v_c$ for $z/D_E \leq 1$, and $v = 0.25 \, v_c$ for $z/D_E > 1$.

the associated hypothetical eddy viscosity tend to be much larger within the mixed layers than they are in the inversion above or in the thermocline below. This permits us, for many purposes, to assume the same effective depth for Ekman flow and convective mixing.

### 6.1.2  Transient Ekman layers and Ekman transports

Transients of $\rho^{-1} \nabla p$ and hence $\mathbf{U}_p$ can be specified by the Fourier integral

$$\mathbf{U}_p = \int_0^\infty U'(\omega) e^{-i\omega t} \, d\omega. \tag{6.7}$$

The real and the imaginary parts of the expression on the right-hand side of (6.7) correspond to the two components of $\mathbf{U}_g$. The length $U'(\omega)$ is generally a complex quantity. Introduction of (6.7) into (6.1) yields solutions that have the same form as the equation that describes the motion of an ideal forced pendulum with inherent frequency $f$,

$$\mathbf{U}_p = Be^{ift} + \int_0^\infty \frac{f}{f + \omega} U'(\omega) e^{i\omega t} \, d\omega, \tag{6.8}$$

where $B$ is a function of the initial conditions. The first term on the right-hand side of (6.8) represents an *inertial oscillation* with frequency $f$. The corresponding period, $12\text{h}/\sin \phi$, equals half the period of a Foucault Pendulum and is called the *half-pendulum day*.

The solution of (6.2) with the transient boundary condition $\mathbf{U}_{E(z=0)} = -\mathbf{U}_p$ involves vertically propagating, circularly polarized waves that are attenuated with distance from the interface. The Ekman length becomes in this case a function not only of $v$ and $f$, but also of $\omega$

$$D_{E\omega a} = \left[ \frac{2v_a}{f - \omega} \right]^{0.5} \qquad D_{E\omega w} = \left[ \frac{2v_w}{f + \omega} \right]^{0.5}. \tag{6.9}$$

The Ekman length is decreased in the atmosphere and increased in the water below if $\omega < 0$. This corresponds to an anticyclonic backing of $\mathbf{U}_p$ with time. It may occur when a pressure ridge travels with the mean zonal wind or a trough propagates against it. The opposite happens when $\omega > 0$. The interior atmospheric velocity then veers cyclonically. Velocity oscillations without change in either direction can be represented by the superposition of two modes with equal frequencies of opposite signs.

On the spherical earth, $f$ varies with latitude. If the wind field is stationary ($\omega = 0$), then one can expect a singularity of the Ekman layer ($D_E = \infty$) at the equator, where $f = 0$. However, travelling cyclonic troughs ($\omega < 0$) can be affected by a boundary layer singularity that is at a latitude where $f \pm \omega = 0$. This may have a bearing on the off-equator location of the *intertropical convergence zone*.

Laminar flows and true Ekman spirals are not observed in nature. However, the integral friction-driven, horizontal mass transport in the boundary layer is not affected by internal variations of turbulence or viscosity along the vertical. This is easily proved by the vertical integration of (6.2) from the sea surface to the top of the atmosphere ($z = \infty$) and to the bottom of the ocean ($z = -D$)

$$\left(\frac{\partial}{\partial t} + f\mathbf{n} \times\right)\int_0^\infty \rho_a \mathbf{U}_{Ea}\, dz = -\boldsymbol{\tau}_0 = -\left(\frac{\partial}{\partial t} + f\mathbf{n} \times\right)\int_0^{-D} \rho_w \mathbf{U}_{Ew}\, dz. \quad (6.10)$$

The integrals

$$\int_0^\infty \rho_a \mathbf{U}_{Ea}\, dz \equiv \mathbf{M}_{Ea} \quad \text{and} \quad \int_{-D}^0 \rho_w \mathbf{U}_{Ew}\, dz \equiv \mathbf{M}_{Ew} \quad (6.11)$$

represents the *Ekman mass transports*. Locally, atmospheric and oceanic Ekman mass transports are equal and opposite.

A solution of (6.10) for an idealized atmospheric boundary layer was obtained in a classic paper by E. Gold (1908). He stipulated that $\boldsymbol{\tau}_0 = 0$ for $t < 0$ and that it changes at $t = 0$ to a finite, constant, and horizontally uniform value. With these particular forcing conditions, he found that, for $t > 0$,

$$\mathbf{M}_a = \frac{1}{f}\mathbf{n} \times \boldsymbol{\tau}_0(e^{-ift} - 1), \quad (6.12)$$

which represents a steady drift at right angle to the surface stress plus an anticyclonic inertial oscillation. Current meter records of the upper ocean often show rather striking evidence of these oscillations, as illustrated, for example, by Fig. 7.3. This phenomenon seems to be associated almost invariably with sudden and pronounced wind changes.

### 6.1.3  The depth of the turbulent Ekman layer; surface-waves effects

The classic Ekman solution deals with laminar flow in an unstratified fluid. In the marine and oceanic PBLs the Reynolds number is generally much larger than $10^3$. Eddy exchange coefficients exceed the molecular coefficients by several orders of magnitude. The Ekman depths become correspondingly larger. In general, the eddy viscosity $K_m$ varies with distance from the interface and (6.5) and (6.9) are not applicable under these circumstances. Some estimates for the vertical variations of $K_m$ through the PBL will be discussed in subsection 6.3.3, but these estimates are rather uncertain, particularly for the ocean. In general, the convergence of the turbulent, vertical flux of current or wind momentum is not known exactly. This makes it difficult to predict the shape of the vertical $\mathbf{U}_E$ profile, or the depth to which surface friction affects the interior of a geophysical fluid. In the case of neutral stratification, one can assume on dimensional grounds that

$$D_E(\text{turbulent}) \propto \frac{u_*}{f}. \qquad (6.13)$$

The proportionality factor of order is often assumed to be of order 0.25. A value of $u_* \approx 0.3\,\mathrm{m\,s^{-1}}$, would then suggest an atmospheric Ekman depth of about 700 m in midlatitudes. The corresponding oceanic Ekman depth would be about 30 times smaller; however, the proportionality factor cannot be constant. Its value is affected inevitably by internal energy dissipation, vertical buoyancy transports, radiation and other physical processes. The mixed-layer depth provides often a better indication than (6.13) of the depth to which surface friction affects the interior ocean or atmosphere. More detailed arguments concerning this topic can be found at the end of subsection 6.3.3.

In the ocean, additional length scales may be associated with the sea state. When waves break, wave momentum is transferred into current momentum down to a depth that depends on the wavelength, the amplitude and also on the particular type of breaking mechanism. Even in the absence of breaking waves, the action of the Coriolis force on Stokes drift currents (see Sections 4.3 and 5.3) can contribute to the total Ekman transport. None of these processes involves viscosity; they are due to the nonlinear dynamics of inviscid wave fields.

For any particular Fourier wave component, the e-folding depth of the Stokes drift is proportional to the wavelength, as indicated by (4.30). This depth scale is generally different from (6.5) and (6.13) and from the local mixed-layer depth. The resulting effect on Ekman transports has been analyzed by Huang (1979) in a paper that also contains useful field and laboratory data on surface drift currents. The existence of several possible length scales makes it difficult to develop universal expressions for velocity profiles in the upper ocean. Better estimates may require a much better quantitative understanding of the ways in which wind stress, buoyancy flux, waves, and turbulence interact.

## 6.2  Coherent structures in the planetary boundary layer

Unstable disturbances often grow into coherent structures that fill the whole depth of the PBL. The resulting secondary flows tend to modify in turn the horizontally averaged density and velocity profiles. The present section deals with the observed character, theory, and modelling of some of these relatively large-scale perturbations. Reviews by Leibovich (1983) and by Etling and Brown (1993) provide more references to the literature.

Coherent PBL structures can assume a variety of forms. Convective plumes, with or without rotation, are common in the atmospheric boundary layer, particularly over heated land surfaces. Studies related to these coherent structures can be found in Kaimal and Businger (1970) and Wilczak (1984). We are not aware of reports that deal specifically with plumes in the layer immediately below the ocean surface, although they are known to occur under conditions of deep oceanic convection. Winds are usually present over the ocean and the resulting shear tends to organize convective perturbations into adjacent, counterrotating, roll vortices. These constitute the most common and certainly the most studied, finite-amplitude, marine PBL perturbations. Elsewhere, their presence has been invoked to explain the alignment of sand dunes (Hanna, 1969) and the formation of volcanic island or sea-mount chains (Richter, 1973).

### 6.2.1  *Observations of oceanic longitudinal rolls*

The surface signature of these perturbations are wind-rows—parallel bands of foam, seaweed, and slick water that must have been familiar to sailors since ancient times. Systematic, three-dimensional observations were begun by Langmuir (1938) after whom the phenomenon was subsequently named. He studied it in the surface layers of Lake George, New York, with the aid of tracers like leaves, corks, and submerged umbrellas buoyed with light bulbs. More recent, and more expensive, studies include satellite SAR observations, release of dye clouds (Thorpe et al., 1994), surface velocity measurements with side-looking radar and sonar and subsurface sonar observations of bubble clouds (Zedel and Farmer, 1991; Thorpe, 1992). These clouds are produced by breaking waves. They are then transported horizontally and pulled down below the convergent regions of the Langmuir rolls to a depth where the friction of the downwelling water is balanced by their buoyancy. That depth, therefore, is related to the circulation intensity. Quantitative measurements of the three-dimensional velocity field were obtained by Weller and Price (1988), who reported vertical and horizontal circulation velocities in excess of $0.2 \, \mathrm{m \, s^{-1}}$ on several occasions during their experimental period. To the human eye, wind-rows often suggest some repetitive regularity, but quantitative studies indicate that the distances between them occupy a relatively broad spectral band. Sonar observations reported by Thorpe (1992) show that bubble bands often merge. This suggests a gradual growth of the scale of Langmuir circulation with time. The effect may be related to a simultaneous increase in the mixed-layer depth $h$.

A perturbation wavelength $\lambda$ is specified by the width of two adjacent counter-rotating rolls. Observations suggest that a major spectral peak is often associated with aspect ratios in the range $1 < \lambda/h < 3$ (i.e., with rolls that have about the same height and width). Atmospheric observations and numerical simulations suggest a similar configuration. The spacing between wind-rows can therefore be expected to give some indication of the layer-depth that is affected by these perturbations.

A relevant set of sea-surface observations has been obtained by Smith and Pinkel during March 1990 in the California Current (Jerome Smith, personal communication). Figures 6.3–6.5 reproduce part of their results. Figure 6.3 indicates a rapid change in windspeed and direction at about 7:20 A.M. at the observation site. The wind dropped again slowly after 8:00 A.M., but, at a mean speed of about $10 \, \mathrm{m \, s^{-1}}$, it remained stronger than it had been the night before. Figure 6.4 represents simultaneous sonar observations of the sea surface velocity component in the cross-wind direction. The figure shows the onset of Langmuir circulations after the wind change. It also suggests an increase in their size or wavelength after the initial burst. This development is demonstrated more clearly by the spectral plot in Fig. 6.5. The figure shows how the mean wavelength of the perturbations increased after the initial forcing change. This increase in wavelength or decrease in wavenumber actually occurred during a period when the wind and its surface stress were slowly decreasing. It therefore could not have been produced directly by the change in stress. It seems

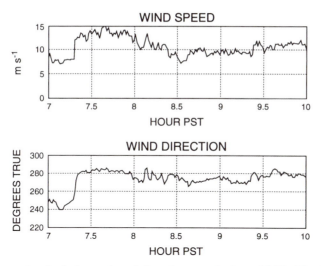

**Fig. 6.3.** Records of wind speed and direction on platform FLIP, March 4, 1990. Courtesy J. A. Smith and R. Pinkel, SIO.

more likely that the general increase in wind, stirring after 7:00 A.M., together with the associated enhancement of surface cooling, caused a gradually deepening of the mixed-layer. It was the greater layer-depth that allowed the growth of larger longitudinal rolls. The development stabilized presumably when a deeper layer had become established.

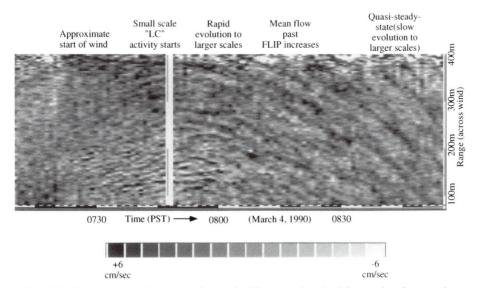

**Fig. 6.4.** Sonar record of near surface velocities associated with growing Langmuir circulations. Courtesy J. A. Smith and R. Pinkel, SIO.

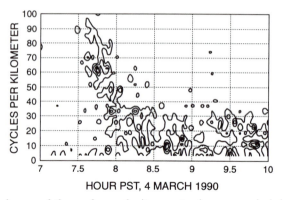

HOUR PST, 4 MARCH 1990

**Fig. 6.5.** The change of the surface velocity spectra in a crosswind direction during initial growth of Langmuir circulations. Contour interval is 30 $(cm\,s^{-1})^2$ per cycle $m^{-1}$. Courtesy J. A. Smith and R. Pinkel, SIO.

Langmuir circulations carry water with near-surface properties downward below the surface convergence zones. Plankton, sea weeds, surface detritus, and dissolved atmospheric gases tend to be concentrated in these sinking currents. On the other hand, water, different in temperature and salinity and often rich in nutrients, is brought up from the bottom of the mixed-layer toward the surface. In general, the presence of coherent structures causes the horizontal variance of conservative properties to be of the same order as the vertical variances across the layer-depth. This requires a careful specification of mean values for these properties.

### 6.2.2   Observations of atmospheric longitudinal rolls

The structure of longitudinal rolls in the marine PBL was first studied by Woodcock (1940) in a classic paper that was based on the soaring pattern of sea gulls. This was followed by Kuettner's (1959, 1967) work with soaring planes. Since then, these perturbations have been studied intensively over a wide range of locations from the tropics (LeMone, 1973) to the high arctic (Brown, 1990). Free-floating, constant-level balloons (tetroons) were used by Angell et al. (1968) to trace a helical progression downwind. The weather service routinely uses satellite observations of cloud-streets, which often indicate the presence of PBL rolls, to estimate wind direction over the sea. A good example of cloud-streets, as seen from a satellite, is shown in Fig. 6.6. The signature of atmospheric rolls in the surface wave field has been studied by Gerling (1986). Recent studies involving quantitative velocity measurements with dual Doppler radar and lidar are listed in the review by Etling and Brown (1993).

LeMone (1973) found aspect ratios of order 2.0–3.5 in most of her

**Fig. 6.6.** Satellite picture (NOAA 7, November 20, 1983) of cloud streets over the Greenland Sea. Courtesy D. Etling.

atmospheric data. Similar ratios were observed in many subsequent atmospheric studies. However, longitudinal perturbations in nature are not a monochromatic phenomenon. The wavenumber spectrum is usually continuous with several minor peaks. Walter and Overland (1984) observed cloud-streets in weakly stratified marine boundary layers that suggested aspect ratios of order 4–6 and 10–20. Kelly (1984) and Muira (1986) noted longitudinal vortices in cold air outbreaks with aspect ratios of order 10–15. Similar broad cloud bands were also observed by LeMone and Meitin (1984) in the marine tropical boundary layer. These circulations can evolve into even broader perturbations with aspect ratios of about 30 (Atlas et al., 1986).

Mourad and Brown (1990), who summarized many of these observations, define two classes of longitudinal perturbations. One has a single dominant horizontal length-scale, which is related to rolls of approximately circular cross-section. The other is made up of several elementary

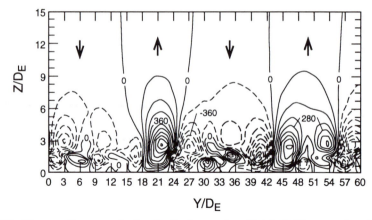

**Fig. 6.7.** The nonlinear vertical velocity for a class II triad. The wavelengths are $\lambda_1 = 8D_E$, $\lambda_2 = 11D_E$ and $\lambda_3 = 28D_E$. There are strong, relatively narrow updrafts with a horizontal separation of about $30D_e$ separated by broad, weaker downdrafts. This configuration would lead to cloud streets with an aspect ratio of about 4. After Mourad and Brown (1990).

perturbations and involves a number of different scales. Figure 6.7 shows a snapshot of the vertical velocity field associated with the multiscale eddy configuration. It demonstrates that multiscale boundary layers may allow different single-scale interpretations. Sensors on towers or low-flying aircraft may pick up strong spectral signals from up- and downdrafts with aspect ratios of about 1–4. These strong circulations receive their energy mainly from buoyancy and bulk vorticity in the lower boundary region. The broader cloud-streets seem to be associated with wave–wave interactions and with a mechanism that couples processes within the inversion layer nonlinearly with the energetically favored roll vortices.

Rising air currents, with anomalous potential temperature, humidity, and salt nuclei concentrations, have been observed by research aircraft above the surface convergence of atmospheric rolls. It has also been shown by LeMone (1976) that the transport of small-scale turbulent eddies by the roll circulation distorts the TKE field in an analogous manner. As in the oceanic case, the phenomenon causes the horizontal variance of conservative properties to be of the same order as the vertical variance across the layer-depth, with corresponding dynamic effects.

### 6.2.3 *Laboratory experiments*

Longitudinal perturbations were generated in rotating water tanks by Faller (1963) and by Tatro and Mollo–Christensen (1967). To study the onset of instability, the rotation rate was gradually increased. When the

Reynolds number reached a critical value of 55, the flow became disturbed by waves with long crests that were nearly parallel or had a small clockwise deflection from the geostrophic velocity vector, $U_g$. At somewhat higher Reynolds numbers a second, slower propagating wave appeared. It has closely spaced crests that deviated counterclockwise from the geostrophic flow. Although the widely spaced waves appear at a lower critical Reynolds number, they were designated type II, because Faller discovered them only after he had found the more ubiquitous, closely spaced, counterclockwise waves that were designated as type I. At higher Reynolds numbers both perturbations loose their banded character and become turbulent. For type I waves, the normalized wavelength $\lambda/D_E >$ 10, while for type II waves $\lambda/D_E > 20$, where $D_E$ is the Ekman depth.

The experiments were carried out with neutrally stratified fluid. Lilly (1966) argued that the type II instability is of a viscous character since it vanishes at higher Reynolds numbers. The type I waves are essentially inviscid. As of now, it has not been established whether either wave-type acts as a precursor of PBL roll vortices in the ocean or in the atmosphere. Their relative wavelength or inverse aspect ratio is certainly much larger than that of the commonly observed natural phenomenon. Their alignment and width is more indicative of the broader cloud-streets, but the evidence, as summarized by Mourad and Brown (1990), does not suggest that these can be generated by some primary, shearing instability.

### 6.2.4   Numerical simulations

The character of PBL perturbations depends on the mean velocity and density fields. The presence of coherent structures in turn affects the mean distribution of the bulk variables and the boundary fluxes. Although the finite-amplitude problem can only be solved numerically, one can use analytical methods to search for instabilities or growth criteria.

During the early growth of small disturbances, the boundary fluxes and mean fields may be considered constant. One can then represent small, a priori specified, perturbations by a set of waves with time-dependent amplitudes. Etling (1971) demonstrated that the complete set of the governing non-linear equations can be reduced in this way to a single eighth-order differential equation for the density anomaly. Free convection without shear and forced convection in an unstratified fluid may then be treated as special cases of this general formulation.

In particular, neglect of rotation reduces the governing equation from eighth to sixth order. If one further stipulates an upward buoyancy flux as the sole source of perturbation energy, then one gets free convection. The governing equation for this case was first integrated by Jeffreys (1926). His work was followed and extended by Pellew and Southwell (1940). A regular, hexagonal Benard cell pattern was obtained analytically by Christopherson (1940) from the superposition of three perturbation wave trains with equal amplitudes. In nature this occurs rarely if ever. All these papers dealt essentially with steady-state conditions. Relevant time-dependent developments were addressed by Ruby Krishnamurti (1975).

A different simplification of the complete problem involves neutrally stratified fluids. Shear flows are then the only possible source of energy for developing perturbations. Velocity shears in boundary layers are caused primarily by stress along the bounding surface. In rotating geophysical fluids they tend to involve directional turning. Replications of the rotating tank experiments with nonstratified fluids were carried out by Faller (1965) and by Lilly (1966). Finite-amplitude secondary flows, as distinct from infinitesimal disturbances, were simulated numerically by Faller and Kaylor (1966) as an initial value problem and by Brown (1970) as an equilibrium steady-state problem. Later investigations of sheared flows, some of which included stratification effects, were published by Etling (1971), Brown (1972), Asai and Nakasuji (1973) and Shirer (1987). Additional references can be found in Brown's (1980) review.

Figure 6.8 indicates a concentration of directional shear in the upper part of the PBL. This is caused by the upward transport of horizontal momentum in the rising branch of the rolls. The overall effect of these transports on a horizontally averaged velocity profile is not unlike that shown schematically in Fig. 6.2. The difference between the various curves in Fig. 6.8, however, indicates the pitfalls that may affect the sampling of velocity profiles in the planetary boundary layer and the specification of horizontal averages.

Most of the quoted investigations are two-dimensional, with a stipulated, uniform, subgrid-scale eddy viscosity $K$. The Ekman depth, $D_E = (2K/f)^{0.5}$, is then also fixed a priori and remains unaffected by the presence of finite-amplitude perturbations. The simulated developments were to some extent further biased by the imposition of preconceived perturbation patterns. These restrictions can be avoided or reduced by the complete simulation of all possible velocity distributions in the PBL.

Numerical integrations of the three-dimensional, nonlinear Navier–Stokes equations, as applied to the PBL, were carried out first by Deardorff (1970a,b,c, 1972). Even though no particular perturbation pattern was prescribed a priori, structures with a general alignment in the direction of the low level shear did in fact evolve. In contrast to the two-dimensional models of longitudinal rolls discussed earlier, these structures did not form equally spaced parallel bands. Moreover, they appeared only in the presence of some upward buoyancy flux. Deardorff's numerical studies were extended and refined by Moeng (1984), Mason and Thompson (1987), Mason (1989), and others. All of these simulations resolved the motion pattern of the larger perturbations and parameterized small-scale turbulence with eddy viscosity or eddy diffusivity coefficients. Calculations which were based directly upon the unadulterated Navier–Stokes equation, without appeal to the always somewhat arbitrary eddy exchange coefficients, were carried out by Coleman et al. (1990). Their paper contains a comprehensive reference list of earlier PBL studies.

Most of the later three-dimensional numerical simulations confirmed Deardorff's original conclusion that the existence of longitudinal rolls in nature requires both a bulk velocity shear and a mild, upward buoyancy flux. The model by Coleman et al. (1990) actually did begin to develop longitudinal vortices in neutrally or stably stratified Ekman layers, but these were subsequently wiped out again by turbulent interactions when initial conditions included small disturbances at all resolved wavelengths. This is probably always true for real boundary layers above and below the sea surface, which are usually very turbulent, with correspondingly high Reynolds numbers. The numerical studies suggest that longitudinal rolls cannot persist under these conditions, when shearing instability is the sole source of energy. On the other hand, their organized structure disappears also when convection becomes too vigorous. This conclusion is in agreement with LeMone's (1973) atmospheric observations. With $L$ and $h$ denoting the Obukhov length and the inversion height, she found that rolls occurred only when $3 \leq -h/L \leq 10$. Wipperman et al. (1978) arrived at similar results on the basis of dimensional and analytical arguments. A relatively weak, upward buoyancy flux is common immediately above and below the sea surface. We are not aware of any PBL observations that would indicate the presence of longitudinal rolls under truly neutral conditions.

### 6.2.5   Energetics of longitudinal rolls

The relative importance of an upward buoyancy flux is confirmed by studies of the kinetic energy balance. An energy equation for longitudinal rolls was apparently first derived by LeMone (1973). A simplified version of that equation has the form

$$\frac{1}{2}\frac{\partial}{\partial t}\overline{(w^2 + v^2)} = -\overline{vw}\frac{\partial V}{\partial z} + \overline{b'w} - f\overline{uv} + f_c\overline{uw}\cos\phi + D_1$$

$$\frac{1}{2}\frac{\partial}{\partial t}\overline{u^2} \qquad = -\overline{uw}\frac{\partial U}{\partial z} \qquad\qquad + f\overline{uv} - f_c\overline{uw}\cos\phi + D_2$$

$$(6.14)$$

where $U$, $V$ are the bulk velocity components parallel and normal to the roll axis and $u$, $v$, $w$ are the longitudinal, lateral and vertical perturbation velocities. The left-hand term in the first equation represents the change in roll kinetic energy. The first term on the right-hand side represents, as usual, a transfer of energy from a sheared mean. The following term indicates the contribution by the buoyancy flux. The terms involving the vertical $(f = 2\Omega \sin \varphi)$ and the meridional $(f_c = 2\Omega \cos \varphi)$ components of the Coriolis parameter cause a transfer of energy from the longitudinal to the transverse flow. The angle between the roll axis and the E–W direction is denoted by $\phi$. It is interesting to note that for small $\phi$ at low latitudes the effect of $f_c$ can be as large as, or larger than, that of $f$. $D_1$ and $D_2$ represent pressure and triple correlation terms in the standard TKE equation (1.46).

Hein and Brown (1988) measured perturbation spectra at various heights in cloud-streets of different width or wavelength over the Bering Sea. They used these data to evaluate the magnitude of the terms in (6.14). Some of their results have been reproduced later in Table 6.1. The streets had an approximate N–S direction on that occasion (i.e., $\phi \approx \pi/2$) and hence the terms including $f_c$ vanish. The table shows that the contribution of the standard Coriolis term was also relatively small at all heights and times. Most of the perturbation energy came from the buoyancy term. The contribution of the cross-wind or cross-roll shearing term was somewhat less than that of the buoyancy flux, but it was still large enough to have a significant influence upon the energy balance of the cloud-streets. When they analyzed broader rolls, they found the contribution of this term to be negative. In that case, this implies that energy released by buoyancy was fed through the shearing term from the perturbation into the mean flow.

The relative magnitude of the various terms in Table 6.1 may be fairly typical for conditions in any outbreak of cold air over warmer water. In the trade-wind region, the effect of the Coriolis term is likely to be even

**Table 6.1.** Contribution of Different Terms ($m^2 s^{-3}$) to the Kinetic Energy Balance of Cloud-Streets with a Wavelength of 2.3 km Observed by Hein and Brown (1988) at 57°N 171°W

| $z$ | $-\overline{vw}\dfrac{\partial V}{\partial z}$ | $\overline{b'w}$ | $-f\overline{uv}$ | $-\overline{wu}\dfrac{\partial U}{\partial z}$ |
|---|---|---|---|---|
| 56 m | $6.7 \times 10^{-6}$ | $2.7 \times 10^{-4}$ | $-4.8 \times 10^{-6}$ | $3.7 \times 10^{-5}$ |
| 215 m | $2.0 \times 10^{-4}$ | $5.2 \times 10^{-4}$ | $-3.8 \times 10^{-6}$ | $8.8 \times 10^{-5}$ |
| 531 m | $-2.5 \times 10^{-5}$ | $8.0 \times 10^{-5}$ | $1.7 \times 10^{-5}$ | $3.4 \times 10^{-6}$ |

weaker than over the Bering Sea. Equation (6.14) was originally applied to longitudinal perturbations in the atmosphere, but it is equally relevant for Langmuir rolls in the upper ocean.

### 6.2.6  Physical concepts and theories

The development of coherent PBL structures breaks any pre-existing horizontal symmetry. It produces different velocity profiles in the rising and sinking branches of the perturbation, as illustrated by Fig. 6.8, as well as differential transports of matter, energy, and small-scale turbulence. The phenomenon is essentially chaotic, with a relatively broad, continuous

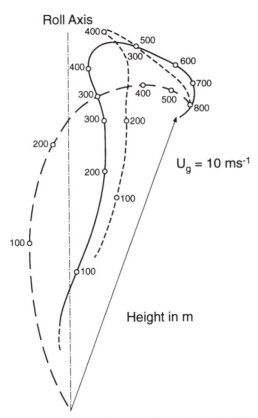

**Fig. 6.8.** Long dashes: unmodified Ekman spiral for $D_E = 215.5$ m. Solid line and short dashes: hodograms modified by longitudinal rolls, which make an angle of $20°$ with the geostrophic wind and which have a wavelength equal to $4\pi D_E = 2.7$ km. The two lines are characteristic for two different locations within the rolls. After Brown (1970).

wavenumber-frequency spectrum. It is not possible to predict whether a left-handed or right-handed roll will appear at any particular place and time.

Laboratory experiments and numerical simulations indicate that longitudinal perturbations can form at low Reynolds numbers in sheared, neutrally, or stably stratified fluids. They tend to be aligned nearly—but not quite–parallel to the geostrophic flow. Initial development is favored under these conditions near the profile inflection level of the normal or spanwise Ekman velocity component. In a laminar PBL, this inflection occurs at a level that is equal to $\pi$ times the Ekman depth. However, in fluids of finite depth, at Reynolds numbers above about 200, any laminar Ekman profile is rapidly distorted by the developing secondary flows. This tends to destroy the instabilities that generated these flows. It appears that the self-organization of boundary layer flows into coherent rolls can persist only if disequilibrium is maintained by an upward flux of buoyancy (see for example Coleman et al., 1990).

Convection starts when the Rayleigh number (1.81), exceeds the critical value of 1708. Convective perturbations seem to be favored when they have similar vertical and horizontal dimensions. High and thin perturbations can be wiped out easily by lateral mixing. On the other hand, squat and mainly horizontal flow patterns may not receive sufficient convective energy to overcome the adverse effects of vertical mixing and surface friction. Theoretical studies by Mourad and Brown (1990) suggest that structures with an aspect ratio of order 2 develop most rapidly. However, these structures are themselves unstable. Non-linear wave–wave-type interactions can subsequently transfer some of their energy to perturbations with lower wavenumbers. The spectrum evolves in this way with time, even if the external forcing remains constant. Relevant observations in the ocean have been described by Thorpe (1992).

Convection and mean shear flows may not be the only source of energy for longitudinal rolls. A theory that connects Langmuir circulations in the ocean with the surface wave field has been described by Craik and Leibovich (1976) and by Leibovich et al. (1989). It was extended by Huang (1979) to include the action of the Coriolis force. The concept has been favored by many oceanographers (see, e.g., Phillips, 1977), but it cannot explain longitudinal perturbations in the atmosphere, although these involve similar physical principles as their oceanic counterparts.

The same objection applies to hypotheses that stipulate a possible effect of sea surface contaminations upon the growth of Langmuir rolls. Welander (1963) suggested that the wind blows faster over the surface-active material that forms the smooth streaks of the wind-rows. The vorticity component in the direction of the surface wind is therefore affected by a stretching term of the form $(\partial w/\partial x - \partial u/\partial z)\partial u/\partial y$. The quasi-periodic variations of $\partial u/\partial y$ across the wind-rows can supply vorticity and energy through this term to counterrotating rolls—both in the atmosphere and in the ocean. The mechanism would imply rolls with the same width above and below the interface. This has not been observed.

A related, but somewhat different, mechanism has been proposed by Kraus (1967). As the skin drag of the wind is smaller over surface films than over uncontaminated water, slick streaks cannot have been drawn out directly by the wind as sometimes has been suggested. They could be produced, however, by the convergence of capillary or short gravity waves. The lobe-shaped directional distribution of these waves was discussed in subsection 4.4.2. Their absorption by patches of filmy material produces an unbalanced radiation stress, which drives the patch downwind and squeezes it out into a streak along the wind direction. The Stokes drift of the small waves in the slick-free lanes transports mass and momentum toward the streaks. Together, these processes can generate a helical circulation. The suggested mechanism skims the continuously renewed wave momentum from the surface and transports it downward below the streaks. Though the process seems plausible, its magnitude has not been assessed.

The surface processes discussed above may contribute to the forcing of longitudinal rolls in the ocean, but they cannot affect the large atmospheric roll vortices. A growing body of observations and numerical results suggest that wind and an upward buoyancy flux (i.e., kinematic and convective instability) are both necessary prerequisites for the generation and persistence of longitudinal perturbations in the atmospheric and oceanic PBL.

## 6.3 Parametric representations of the PBL fluxes and profiles

It was shown in Chapter 5 that the lowest part of the boundary layer (the surface layer) can be described with considerable accuracy by established universal parameterization schemes. Even there we still found unsolved problems, particularly below the interface, when surface waves are present. The much deeper PBL is affected by larger horizontal inhomogeneities. This limits our ability to establish horizontally averaged profiles of the bulk properties or of their vertical transports. It is probably fair to say that researchers have as yet not found a recipe for realistic, universally valid parameterizations of the distribution of these first- and second-order quantities.

The problem is caused partly by the coherent structures that are embedded in the PBL. As seen in the previous section, explicit numerical models of these deep perturbations usually parameterize subgrid-scale effects with the aid of separate eddy exchange coefficients. This approach is valid only if the large structures are separated by a spectral gap or valley from the small-scale turbulent eddies. Such a gap may well exist, but this implies that it is difficult, if not impossible, to represent the whole coherence spectrum with a single, universal parameterization scheme.

The present section deals with the merits and limitations of various schemes, which try to represent either the vertical fluxes, or the horizontally averaged PBL profiles as functions of the bulk or mean variables. A quite different approach makes use of the fact that well-mixed boundary layers do exist. The characteristics of these mixed layers and their relation to external forcing will be discussed in Section 6.4.

### 6.3.1  *Diffusive models*

Diffusive models deal essentially with down-gradient transports of conservative properties. Gradients are local features of scalar fields. In numerical simulations, they have to be approximated by differences between adjacent gridpoints. Models that are designed to reproduce diffusive processes therefore have an essentially 'local' character.

The oldest, and still most widely used, form of diffusive modelling involves eddy exchange coefficients, $K$, as introduced by (1.48). Some functional relations for $K$, valid for the surface layer, were discussed in Section 5.2. In Prandtl's and G. I. Taylor's pioneering studies, the eddy viscosity was parameterized in the form

$$K \propto l_m q. \tag{6.15}$$

The mixing length, $l_m$, represents the distance which a fluid parcel travels before it mixes completely with its new environment. The mean turbulence velocity, $q$, can in principle be derived from the TKE equation, (1.46). Mixing-length values for the neutrally stratified atmospheric PBL have been parameterized in the form

$$l_m = kz\left(1 + c_1 \frac{zf}{U_g}\right) \quad \text{or} \quad l_m = kz\left[1 + c_2\left(\frac{zf}{u_*}\right)^n\right] \tag{6.16}$$

The first of these expressions was proposed by Blackadar (1962) and the second by Lettau (1962). The constants $c_1, c_2$ and $n$ were chosen empirically; $k$ is the von Karman constant; $U_g$ the geostrophic wind; and $u_*$ the surface friction velocity. Starting from the TKE equation, Yamada (1983) developed a prognostic equation of the type

$$\frac{\partial l_m}{\partial t} = b\left[e_1 - e_2\left(\frac{l_m}{l_r}\right)^2\right]$$    (6.17)

where $e_1, e_2$ are two other empirical constants and the parameter $l_r$ depends both on the stratification and on the distance from the boundary. Another prognostic equation, based on energy dissipation rates, was developed by Duynkerke and Driedonks (1987).

Transport rates in turbulent fluids are affected by bulk gradients, as assumed in $K$ theories, but also by changes in the products or squares of the fluctuating quantities. To allow for these relations and to predict stochastic quantities, such as the turbulence kinetic energy or the temperature variance, one has to formulate *higher-order closure schemes*. These may include the TKE equation and other nonlinear predictive equations of the same type. Commonly used simplifications make the buoyancy a unique function of the temperature, and pressure fluctuations are neglected. The three velocity components and the temperature are then the only fluctuating variables. From these four quantities one can form 10 different mean products and squares, which can be identified as the six components of the Reynolds stress tensor, three components of the temperature flux vector, and the scalar temperature variance. Predictive equations for these mean second-order products and squares involve triple or third-order moments and correlations. The third-order moments in turn can be predicted only with the aid of fourth-order quantities, and so on. To form a closed system, it is always necessary to parameterize the highest-order moments as a function of lower-order terms.

A system of parameterized second-order equations can be further simplified and truncated in various ways. Levels of simplification were assessed systematically by Mellor and Yamada (1974). They scaled the basic set of equations with a parameter that measured deviations from isotropy. The progressive neglect of terms that were considered small on this scale gave them four different levels of simplification. Their highest level 4 involves all 10 second-order, predictive differential equations, with triple correlations expressed in simplified or parameterized form. On level 3, the three velocity variance equations collapse into a single predictive equation for isotropic TKE. The only other predictive equation is that for the temperature variance. The turbulent momentum and temperature transports are now derived from diagnostic equations. Levels 2 and 1 involve only diagnostic equations for the stochastic variables. They are therefore related to the various parameterization schemes for the eddy viscosity or the mixing length.

A somewhat different hierarchy of models can be based on the order of the predicted quantities. First-order models predict only changes in the mean or bulk variables. This includes parameterizations in terms of eddy transfer coefficients, $K$. Mellor and Yamada's level 1 or 2 models and most large-scale circulation models fall into this category. Second-order models attempt to predict at least some transients of variances and covariances. Second-order models for the upper ocean have been described among others by Worthem and Mellor (1979), and by Klein and Coantic (1981), and have been applied to the cloud-topped marine PBL by Koračin and Rogers (1990) and Rogers and Koračin (1992).

The second-order closure schemes cannot evaluate third-order processes, such as the turbulent flux of TKE. A third-order, level 4 model, which was designed specifically to assess the turbulent transport of TKE down to the base of the ocean mixed layer, has been described by André and Lacarrère (1985). Their approach involves only two bulk variables: a horizontal (one-dimensional) velocity, $U$, and the temperature, $\bar{T}$. That reduces the scheme to two first-order predictive equations for these variables and to six second-order equations for the products and squares of $u$, $w$, and $T'$. In addition to these equations, nine third-order predictive equations were needed for triple correlation terms such as $\overline{wu^2}$ or $\overline{T'^3}$. This gave them 17 prognostic equations that had to be integrated at each grid level at each time step. Results, derived from a hypothetical forcing situation, look realistic. Such realism is perhaps not surprising in a scheme that involves 14 empirical and, to some extent, tunable constants.

Diffusive models are particularly useful for the study of transports under neutral or stable conditions. Their validity becomes debatable under unstable conditions, when countergradient transports may come into play. The original $K$ theories were physically transparent and easy to understand, but their application to the PBL tended to yield only approximate results. The higher-order closure schemes can be used to obtain more realistic simulations in special cases, but it seems that this can be achieved only at the cost of less generality and less physical insight.

### 6.3.2   *The transilient scheme*

Diffusive models of all levels of complexity are limited essentially to numerical interactions between variables at neighboring grid points. They are based on the assumption of a continuous structure in the limit of a sufficiently fine grid scale. A different approach, involving advective transports over different distances by a set of differently sized eddies, has been developed by Fiedler (1984) and by Stull (1984, 1986). Stull called this type of parameterization *transilient turbulence.*

The one-dimensional transilient model divides the vertical fluid column into $N$ discrete cells that are stacked on top of each other. The distribution of some conservative property, $S$, within this array of cells, can then be represented by a vector with components $S_j$, where $j = 1, 2, \ldots, N)$. Eddies of various sizes carry fluid mass and specific concentrations of $S$ between the cells. The concentration $S_i$ in cell number $i$ at time $t + \Delta t$ is therefore a function of the $S_j$'s in all the cells at an earlier time $t$. The resulting change in the distribution of $S$ during the time interval $\Delta t$ can be specified by the matrix equation.

$$S_i(t + \Delta t) = c_{ij} S_j(t) \qquad (6.18)$$

The dimensionless *transilient coefficients* $c_{ij}$ represent the fraction of fluid in cell $i$ that has come from $j$ during $\Delta t$. Together the $c_{ij}$ compose a square matrix with $N^2$ elements. In the absence of mixing, each of the diagonal elements $c_{ii} = 1$ and all other coefficients are zero. Conservation of the total mass of fluid and the total amount of $S$ requires that

$$\sum_{j=1}^{N} c_{ij} = \sum_{i=1}^{N} c_{ij} = 1. \qquad (6.19)$$

The particular values of the various $c_{ij}$ have been parameterized in terms of bulk Richardson numbers $Ri_{ij}$ that are functions of the momentum and density at cells $i$ and $j$, and of the distance between them. As an alternative, the available energy that would be released by mixing between these two cells has also been used.

The transilient scheme can be characterized as a nonlocal, first-order or linear representation of the turbulent exchange process. It contains the diffusive $K$ theory as a special case, that is represented by a matrix in which only the diagonal elements $c_{ii}$ and the elements next to the diagonal $c_{i,i+1}, c_{i,i-1}$ are allowed to be different from zero. In contrast to classic mixing length theory, it involves an array of fixed lengths. These are determined by the distances between the different cells.

The transilient scheme involves action at a distance within the PBL. This means that it can deal explicitly with the effects of eddies of any desired size. Without a priori assumptions about the sources of turbulence, it can simulate shear-generated and convective mixing at all levels of the domain. Its weakest feature is the somewhat arbitrary stipulation of the transilient coefficients. Being first-order, it also neglects the triple correlations, which denote turbulent transports of second-order quantities such as TKE or temperature variance. Though developed originally for the atmospheric PBL, the transilient scheme was later applied to the upper ocean by Stull and Kraus (1987), Kraus (1987), and Gaspar et al. (1988).

### 6.3.3  *Parametric representation of PBL profiles*

A different line of studies has sought universal parameterizations for the vertical distribution of the first-order bulk properties of the PBL. The associated eddy exchange coefficients can then be derived indirectly from the form of the calculated profiles.

The universal form of the viscous, laminar boundary layer in a rotating system is the Ekman profile, which was discussed in Section 6.1. Though the concept is fundamental to our understanding, perfect Ekman spirals are practically never observed in the atmsopheric or oceanic PBL. The fluids there are turbulent and, in the atmosphere at least, the scale of the turbulent eddies is known to increase with distance from the interface. If this increase is linear, it results in the logarithmic profile that was discussed in Section 5.1. The velocity vector within this logarithmic region was shown in (5.17) to be a function of the surface roughness length $z_0$.

The influence of the surface roughness upon the turbulence must diminish or disappear at a height where $z/z_0 \gg 1$. It is often assumed that $u_*/f$ is the appropriate scale-length for this outer, deeper part of the PBL under neutral conditions, as given in (6.13). Following Gill (1968) or Blackadar and Tennekes (1968) one can express the velocity distribution in the following general form

$$\frac{\mathbf{U}}{u_*} = \mathbf{f}_1\left(\frac{z}{z_0}\right) \quad \text{inner constant stress layer,} \tag{6.20}$$

$$\frac{\mathbf{U}}{u_*} = \mathbf{f}_2\left(\frac{zf}{u_*}\right) + \frac{\mathbf{U}_g}{u_*} \quad \text{outer Ekman layer.} \tag{6.21}$$

The vector functions $\mathbf{f}_1$ and $\mathbf{f}_2$ were required to match at an intermediate level $z = z_1$, where it was assumed that the velocity, together with its vertical derivatives, could be expressed equally well by the functions appropriate either to the upper or the lower region

$$\mathbf{f}_1\left(\frac{z}{z_0}\right) = \mathbf{f}_2\left(\frac{z}{z_0} \cdot \frac{z_0 f}{u_*}\right) + \frac{\mathbf{U}_g}{u_*} \quad \text{for} \quad z = z_1. \tag{6.22}$$

This procedure is identical to the matching of the viscous and turbulent layers in Section 5.1.2. We introduce the new independent variables $z/z_0 = \zeta$ and $z_0 f/u_* = \xi$. Differentiation of the last equation by each of these two new variables yields

$$\mathbf{f}_1' = \xi \, \mathbf{f}_2',$$

$$0 = \zeta \, \mathbf{f}_2' + \frac{\partial}{\partial \xi} \frac{\mathbf{U}_g}{u_*}. \tag{6.23}$$

The prime denotes here the derivative of a function with respect to its argument. Elimination of $\mathbf{f}_2'$ from (6.23) yields

$$\zeta \, \mathbf{f}_1' = -\frac{\partial}{\partial \xi} \frac{\mathbf{U}_g}{u_*}. \tag{6.24}$$

The left-hand side of (6.24) is a function of $\zeta$ only, while the right-hand side depends only on $\xi$. Each side is therefore a constant. Blackadar and Tennekes (*loc. cit.*) equated this constant to $\mathbf{i}/k$, where $\mathbf{i}$ is a unit vector in the direction of the surface stress and $k$ the von Karman constant. With $\mathbf{j}$ denoting another horizontal unit vector at right angles to $\mathbf{i}$, we get, by separate integration of each side of (6.24),

$$\frac{\mathbf{U}_g}{u_*} = (\mathbf{i} \cos \alpha_0 + \mathbf{j} \sin \alpha_0)\frac{U_g}{u_*} = \frac{1}{k}\left[\mathbf{i}\left(\ln \frac{u_*}{f z_0} - A\right) - \mathbf{j}B\right]. \tag{6.25}$$

The two components of the last vector equation can be solved graphically or numerically for the two unknowns $u_*/U_g$ and $\alpha_0$, the angle between the surface stress and the isobars. The result can also be expressed as a function of the so-called surface Rossby number $Ro_s = U_g/f z_0$. An equation similar to (6.25) had been derived earlier from different premises, involving the eddy viscosity, by investigators at the Shirshov Institute of Oceanology in Moscow (see Zilitinkevich et al., 1967). For the constants in (6.25), Blackadar and Tennekes (1968) quote values $A \sim 1.7$; $B \sim 4.7$. Deardorff's (1970c) numerical computations of neutral boundary layers yield $A = 1.3$; $B = 3.0$. The value of $u_*/U_g$ is not very sensitive to variations in these constants. Under maritime conditions ($z_0 \approx 0.01$ cm), values from Blackadar and Tennekes and from Deardorff yield an almost identical result that can be approximated very closely by the simple relation

$$\frac{u_*}{U_g} = 0.025 + 0.00035(9.0 - \log_{10} Ro_s). \tag{6.26}$$

This approximation is valid for $Ro_s > 10^8$, a condition that is practically always realized over by sea. Sin $\alpha_0$ can be seen from (6.25) to be equal to $u_*/U_g$ multiplied by the factor $-B/k$. The geostrophic drag coefficient $C_g = (u_*/U_g)^2$ under neutral conditions turns out to be roughly half the coefficient $C_{10}$, which affects the wind at the 10 m level. For a geostrophic wind of $10\,\mathrm{m\,s^{-1}}$ at a latitude $\varphi = 44°$, we have $C_g \sim 0.6 \times 10^{-3}$. The corresponding value of $\alpha_0 \sim -11°$ if $B = 3.0$ or $\alpha_0 \sim -17°$ if $B = 4.7$.

In the sea, $U_w \sim (\rho/\rho_w)U$ while $u_{*w} = (\rho/\rho_w)^{1/2}u_*$. Therefore, the ratio $(u_*/U)$ is always much larger in water than in air. The whole procedure of matching a quasi-constant turbulent stress-layer with an outer region of the planetary boundary layer is hardly applicable in water, where the depth of the wave zone usually exceeds the theoretical depth of the constant stress layer. In fact, if we want to restrict our attention to features and scales that are large compared to the mean wave height, the velocity is considered to be discontinuous across the interface. The stress does remain continuous. This means that under stationary conditions the integrated ageostrophic transports in the two media must be equal and opposite as indicated by (6.10).

If the stratification of the atmosphere is not neutral, the quantities $A$ and $B$ in the solution, (6.25), become variable. They can then be conceived as functions $A(S')$ and $B(S')$ of a stability parameter, $S'$, which can be equated to the Rayleigh number (1.81). Alternatively, Clarke (1970) stipulated $S'$ to be a function of the Obukhov length (5.31), with the form

$$S' = \frac{u_*}{fL}. \tag{6.27}$$

A matching procedure, analogous to the one used in deriving (6.25), yields in this case an expression for the (virtual) potential temperature difference $\Delta\Theta_v$ across the planetary boundary layer,

$$u_*(\overline{wT_v'})_0^{-1}\Delta\Theta_v = \frac{1}{k}\left[\ln\frac{u_*}{fz_0} - C(S')\right]. \tag{6.28}$$

Valiant attempts to assess the value of three functions $A(S')$, $B(S')$, and $C(S')$ from observational determinations of $U_g, u_*$, and $\overline{wT'}$ over land have been described in Zilitinkevich et al. (1967). Later determinations by Clarke (1970) indicate a sharp fall-off in magnitude of the quantities $A(S')$ and $C(S')$ when the stratification is stable. These results were obtained from relatively weak winds under conditions of marked diurnal variability. The variations of hydrostatic stability were in this case associated with an

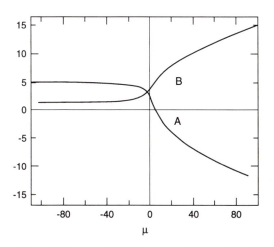

**Fig. 6.9.** Values of $A$ and $B$ versus stability parameter $\mu \equiv ku_*/fL$ from two-layer similarity modelling. After Brown (1982). Reprinted by permission of Kluwer Academic Publishers.

almost fifty-fold range of the geostrophic drag coefficient, much of it occurring close to neutral equilibrium.

Based on these earlier results Brown (1982) formulated a comprehensive parameterization scheme that describes the stability dependence of the similarity functions $A$ and $B$. He also considered the effects of baroclinity, secondary flows, humidity, latitude, surface roughness, and characteristic height scale. Figure 6.9 represents the functions $A$ and $B$ Brown recommends.

In the case of free convection in the PBL, Tennekes (1970) argues that $U_g/u_*$ does not depend on the Obukhov length but that it must be similar to the relation under neutral conditions, see (6.25). This seems to be confirmed by Clarke's observational data that indicate no significant changes in stress with increasing instability after a free convection regime has become established. On the other hand, the vertical heat flux continues to increase as conditions become more unstable. The virtual potential temperature difference between the surface and the top of the PBL under these conditions was found by Tennekes to be specified by

$$u_*(\overline{wT_v'})_0^{-1}\Delta\Theta_v \approx \frac{1}{k}\ln\frac{-L}{30z_0}. \tag{6.29}$$

The corresponding characteristic turbulent velocity, $q$, is indicated by

$$\frac{q}{u_*} \approx 0.5\left(\frac{u_*}{fL}\right)^{1/2}. \tag{6.30}$$

It is not clear how relevant all these deductions are for marine conditions. In particular, the ratio $|L/z_0|$ can be very large, and its logarithm usually provides no more than an approximate estimate. The specification of the temperature difference $\Delta\Theta_v$ is also ambivalent in neutrally stratified mixed layers that are bounded by thermoclines and inversions. The meaning of (6.28) and (6.29) becomes uncertain under these circumstances.

Mixed layers are usually characterized by an upward flux of heat, that is, by a downward flux of density anomaly $(\overline{\rho'w} < 0)$. The depth of the PBL does not depend on the Coriolis parameter under these circumstances. This is indicated by the absence of variations of the oceanic mixed-layer depth with latitude as well as by Deardorff's (1970c) numerical simulations. These show that even a very slight degree of thermal instability causes the vertical mixing in the atmospheric boundary layer to be limited only by the height of the inversion base. It follows that the relevant scale-length for the statistics of turbulence in the unstable case is provided by the mixed-layer depth $h$. An appropriate indicator of static instability is then

$$S' = -\frac{h}{L} \tag{6.31}$$

instead of (6.27). Above a level $z \approx -L$, the characteristic turbulent velocity scale was found by Deardorff to be independent of $f$, and to be of order

$$q \sim u_* \left( -\frac{h}{L} \right)^{1/3}. \tag{6.32}$$

Increasing instability (i.e., increasing values of $-h/L$) were associated in numerical experiments with a drastic reduction of the vertical wind shear. For a ratio $-h/L = 45$, a value often exceeded in the unstable marine environment, the cross-isobaric flow became virtually uniform with height. Immediately above $h$, a relatively large shear frequently occurs, which adjusts the wind to geostrophic equilibrium.

Static instability tends to generate the three-dimensional coherent structures that were discussed in the preceding section. These contribute to the mixing of potential temperature, momentum, and fluid composition. The rate of mixing will not necessarily be the same for all these properties. In fact, heat and momentum are generally transported at different rates. However, the end result is the same. It is characterized by an approach to a vertically uniform statistical distribution throughout the mixed layer. The generality of Deardorff's (1970c) deductions are limited only by the a priori prescription of the layer-depth. In reality, $h$ is time-dependent and a function not only of the instantaneous and local conditions, but also of the previous history of the fluid. That topic is treated in the next section.

## 6.4  Mixed-layer models

Observations in the ocean show that temperature and salinity are usually well mixed through a shallow layer below the sea surface. This *mixed layer*, therefore, has a near-neutral stratification ($N \approx 0$). In the atmosphere above the sea surface, one finds also at many places, much of the time, a similar well-mixed layer, with vertically uniform heat and moisture content and hence uniform equivalent-potential temperature (see Section 2.31). Oceanic mixed layers tend to be bounded below by discontinuities or highly stratified pycnoclines. The corresponding features in the atmosphere are inversions or series of inversions with potentially warmer and drier air aloft. Temperature differences across these stable strata are often larger than the temperature difference across the air–sea interface, as shown, for example, by Fig. 6.10.

A vertically uniform distribution of conservative properties within the mixed layers can be produced by large eddies and coherent structures of the type discussed in Section 6.2. It was indicated there that small-scale

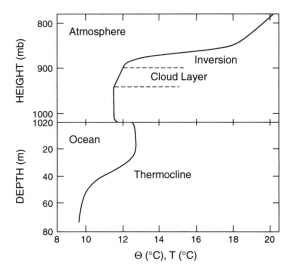

**Fig. 6.10.** Typical profiles of potential temperature in the atmosphere and temperature in the ocean measured during the Joint Air Sea Interaction (JASIN) experiment on Aug. 25, 1978, at 59°N and 12°W.

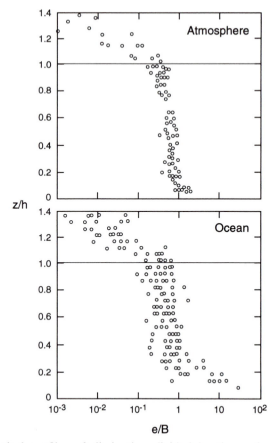

**Fig. 6.11.** Vertical profiles of dissipation divided by the surface buoyancy flux ($\epsilon/B$). Sea surface is at the bottom of each plot. For the region $0.4 < z/h < 1.0$ the mean value of $\epsilon/B \approx 0.45$, both above and below the air–sea interface. After Shay and Gregg (1984).

turbulence is also advected by these perturbations. The resulting uniform distribution of energy dissipation, both above and below the sea surface, is illustrated in Fig. 6.11.

Integral mixed-layer models are based on the assumption that vertical gradients of conservative properties are negligibly small within the layer. All of these models have to deal one way or other with the physical processes of fluid 'entrainment' from the adjacent stable layers. Entrainment always thickens mixed layers, because turbulent regions tend to expand into more quiescent fluid. If there is not enough energy available for entrainment, a new shallower mixed layer may form.

The basic principles of mixed-layer modelling are the same above and below the air–sea interface. Oceanic models are somewhat simpler than

their atmospheric counterparts because sea water compressibility is negligible in the upper ocean. We will therefore start our discussion in
subsection 6.4.1 with the oceanic case. Many of the equations developed
there can be adapted directly to deal with processes in the cloud-free
atmospheric mixed layer, which will be considered in subsection 6.4.2. The
added complication of clouds in the upper part of the atmospheric mixed
layer is the topic of subsection 6.4.3.

### 6.4.1 *The oceanic mixed layer*

The depth to which a fluid layer can be thoroughly stirred depends on the
amount of kinetic energy that is available for this purpose. Within the
stirred layer, the stratification is neutral. A mixed-layer theory, based on
these simple precepts, was conceived originally by K. Ball (1960). It was
further developed by Kraus and Turner (1967). Later refinements of the
theory, in the framework of oceanic and atmospheric applications, all
involve the same principles.

In our coordinate system $z = 0$ at the ocean surface and decreases
downward. The level of mixed-layer boundaries is indicated by $z = h$, with
$h < 0$. Local changes in layer depth can be represented by

$$\frac{\partial h}{\partial t} = -h \, \nabla \cdot \mathbf{U} + w_e \equiv w_h + w_e, \tag{6.33}$$

where $\mathbf{U}$ denotes here the mean horizontal velocity in the layer and $w_h$ is
the change in layer thickness associated with processes such as Ekman
pumping (see subsection 7.2.4) or internal gravity waves (see subsection
7.2.2). The *entrainment velocity*, $w_e$, indicates the rate of layer thickening
that is caused by the turbulent entrainment of fluid across $h$ and by its
incorporation into the layer.

Thorough stirring tends to produce a vertically uniform distribution of
the $T$ and $s$. Being determined entirely by $T$ and $s$ in the upper ocean, the
vertical density gradient must then also vanish. We shall use the Boussinesq approximation (see Section 1.2.2) throughout this section. It implies
that the gravitational acceleration and the variations in density occur only
in the combination

$$b = -g \frac{\rho - \rho_r}{\rho_x} \approx g[\alpha(T - T_r) - \beta(s - s_r)], \tag{6.34}$$

which describes the buoyancy per unit volume. The quantities with a subscript $r$ are reference values. If they are close enough to the observed temperature and salinity, then the coefficients of thermal expansion $\alpha(T, s)$ and haline contraction $\beta(T, s)$ can be considered constant. The buoyancy $b$ can be interpreted as the acceleration that would be experienced by a fluid element of density $\rho$ that is embedded in an ambient fluid of density $\rho_r$.

With horizontal variations assumed to be negligibly small, the conservation equation for $b$ has the form

$$\frac{\partial b}{\partial t} + \frac{\partial \overline{wb'}}{\partial z} = -\frac{g\alpha}{\rho c}\frac{\partial F_{is}}{\partial z} \equiv \frac{\partial S_b}{\partial z}, \qquad (6.35)$$

where $F_{is}(z)$ is the penetrating part of short-wave irradiance, as specified by (3.19); $c$ is the specific heat; $\partial S_b/\partial z$ is the source of buoyancy caused by the internal absorption of radiation. With $\partial b/\partial z = 0$ within mixed layers, integration of (6.35) from 0 to $h$ yields

$$h\frac{\partial b}{\partial t} = (\overline{wb'})_0 - (\overline{wb'})_h - S_b(0) = B - w_e\,\Delta b - S_b(0). \qquad (6.36)$$

The very small amount of radiation $F_{is}(h)$ that traverses the mixed layer to penetrate into the water below has been neglected in this equation. The term $(\overline{wb'})_0 = B$ indicates the buoyancy flux at the water side of the air–sea interface. It can be represented by

$$B = \frac{g}{\rho}\left[\frac{\alpha}{c}Q_w + \beta s(E - P)\right]. \qquad (6.37)$$

$Q_w$ is the total surface heat flux as specified in (5.62), and $E - P$ is the evaporation–precipitation difference (i.e., the net fresh water flux per unit area and unit time). The surface buoyancy flux has different numerical values but usually the same sign in the ocean and in the overlying atmosphere. The term

$$(\overline{wb'})_h = w_e(b - b_h) \equiv w_b\,\Delta b, \qquad (6.38)$$

represents the power that is needed to entrain relatively dense water from below and to incorporate it into the less dense mixed layer. The buoyancy of the water immediately below the layer boundary is denoted here by $b_h$ and $\Delta b$ is the buoyancy discontinuity at $z = h$. As noted in the introduction, entrainment causes the layer boundary to migrate away from the air–sea interface. In the ocean it cools the surface mixed layer and lowers its buoyancy. In our coordinate system, with $z = 0$ at the sea surface, $h$ and $w_e$ are negative in the ocean, in contrast to the atmosphere where these variables are positive. If the adiabatic velocity $w_h \approx 0$, the change of $b_h$ is determined by $w_e$ and by the gradient of the buoyancy $b(z)$ at a depth lower than $h$. Thus

$$\frac{\partial b_h}{\partial t} = w_e \frac{\partial b(z)}{\partial z}.$$

To obtain a second equation for $w_e$, we integrate (1.46) vertically upward across the mixed layer. As the rate of TKE change is always relatively small compared to the production and dissipation terms, we can set $d(\overline{u_j u_j})/dt = 0$. Velocity shears tend to be concentrated in the surface layer and at the outer mixed-layer boundary. In the surface layer $\overline{wu} \equiv -u_*^2$ and $\partial U/\partial z$ is a function of $u_*$ as indicated by (5.16). At the outer layer boundary $(\overline{wu})_h = w_e \Delta U$, where $\Delta U$ denotes the velocity change across $h$. Based on these considerations we can parameterize the vertical integral of the shear production term in (1.46) by

$$\int_0^h \overline{uw} \frac{\partial U}{\partial z} dz \approx m u_*^3 - \frac{1}{2} w_e (\Delta U)^2, \qquad (6.39)$$

where $m$ is a constant of proportionality.

The vertical integral of the second right-hand term in (1.46) represents the change in potential energy that is caused by a vertical transport of buoyancy. One can obtain explicit values for $\overline{wb'}$ at any intermediate depth $-h' < -h$ within the layer, by vertical integration of (6.35) from 0 to $h'$. After re-arrangement this yields

$$(\overline{wb'})_{h'} = B - h' \frac{\partial b}{\partial t} + S_b(h') - S_b(0). \qquad (6.40)$$

The change in potential energy is obtained by integration of (6.40) over $h'$ from $h$ to 0. After subsequent elimination of the term $\partial b/\partial t$ between the result of this integration and (6.36), one can set $h' = z$. The result may be expressed in the form

$$\int_h^0 \overline{wb'} \, dz = -\frac{1}{2} h[B - S_b(0)] + \int_H^0 S_b \, dz - \frac{1}{2} h w_e \, \Delta b = w_*^3 + \frac{1}{2} w_e c_1^2, \qquad (6.41)$$

where $w_*$, which can be either positive or negative, has been introduced as a convective velocity scale and

$$-h \, \Delta b = hg \frac{\rho - \rho_h}{\rho_r} \approx c_1^2 \qquad (6.42)$$

is the squared velocity of internal gravity waves along the density discontinuity at $h$, as derived below in (7.30).

The transport and pressure terms in (1.46) do not produce TKE, and they vanish after integration over the mixed layer when there is no interaction with the thermocline. In reality turbulent pressure fluctuations may set up gravity waves in the stable layer. Energy is needed to generate turbulence in the entrained quiescent fluid, which is given by

$$-\frac{1}{2}\int_0^h \frac{\partial}{\partial z}(wq^2)\, dz = \frac{1}{2}w_e q^2, \qquad (6.43)$$

where $q^2 \equiv \overline{u_j u_j}$, the mean squared turbulence velocity.

The dissipation $\varepsilon$ tends to be distributed uniformly within the layer, as is shown in Fig. 6.11. We can therefore write

$$\int_h^0 \epsilon\, dz = -\epsilon h \equiv D_*. \qquad (6.44)$$

Introduction of (6.39)–(6.44) into the vertical integral of (1.46) yields, after re-arrangement of terms,

$$-\frac{1}{2}w_e(c_1^2 + q^2) + D_* = mu_*^3 + w_*^3 - \frac{1}{2}w_e(\Delta U)^2. \qquad (6.45)$$

The terms on the left-hand side of (6.45) represent TKE consuming processes. Those on the right-hand side generate TKE, provided their sum is larger than zero. This will not be the case when the buoyancy flux is downward into the ocean ($B < 0$), and when the resulting negative value of $w_*^3$ exceeds the other two terms on the right-hand side of (6.45).

The definition of $w_*^3$ in (6.41) contains two terms that include $S_b$. As a first approximation, one can assume that the short-wave radiation flux $F_{is}$, and hence $S_b$, is attenuated exponentially. The relevant terms can then be assessed by

$$\int_0^h S_b\, dz + \frac{h}{2}S_b(0) \approx -\frac{g\alpha}{\rho c}\left(\lambda + \frac{h}{2}\right)F_{is}(0), \qquad (6.46)$$

where $\lambda$ denotes the scale-depth or approximate e-folding depth of the penetrating short-wave irradiance, $F_{is}$. Radiation heating at depth can increase $w_*^3$ and contribute to convective stirring if $\lambda > -h/2$. It will consume TKE, tending to stabilize the layer in the opposite case.

To evaluate the dissipation integral $D_*$ in (6.45), we may argue that the three TKE generating processes specified by the terms on the right-hand side of (6.45) tend to produce eddies with different characteristic scales. As a first approximation, we assume that the contribution of these processes to the total dissipation is proportional to the three generating rates. Based on this assumption, we may rewrite (6.45) in the form

$$-\frac{1}{2} w_e (c_1^2 + q^2) = m_1 u_*^3 + m_2 w_*^3 - \frac{1}{2} m_3 w_e (\Delta U)^2. \qquad (6.47)$$

The proportionality factors $m_1$, $m_2$, and $m_3$ measure the fractions of TKE which, after generation by the three indicated processes, can be converted into mean potential energy. In other words, these are the fractions that are available for entrainment after dissipation has been taken into account. Numerical values of these factors will be discussed below. Before doing so, it is useful to consider the physical implications of the preceding equations.

Together, (6.33), (6.36), and (6.47) may be used to derive the three unknowns $h$, $b$, and $w_e$ as functions of the external forcing and of the state of the fluid below the pycnocline. When (6.47) is solved for $w_e$ we obtain

$$w_e = -2 \frac{m_1 u_*^3 + m_2 w_*^3}{c_1^2 + q^2 - m_3 (\Delta U)^2}. \qquad (6.48)$$

Entrainment occurs when the denominator and the numerator in (6.48) are both positive. This is always the case when $w_*^3 > 0$ (i.e., when the buoyancy flux is upward). In this case it acts as a source of TKE and its effects are therefore subject to dissipation. On the other hand, a downward buoyancy flux ($w_*^3 < 0$) consumes TKE and is not involved in dissipation. This implies that $m_2$ should be unity when $w_*^3 < 0$. Note that entrainment can still occur in this case, as long as the numerator in (6.48) remains positive. To allow for this asymmetry between upward and downward buoyancy transports, one may express the relevant term in (6.47) or (6.48) in the form

$$\frac{1}{2} [(1 + m_2) w_*^3 - (1 - m_2) |w_*^3|]. \qquad (6.49)$$

When the buoyancy flux is upward (6.47) and (6.48) remain as written. When it is downward, (6.49) causes the factor $m_2$ to be replaced by unity.

This means that TKE is decreased by the total amount of energy needed for the downward buoyancy flux.

In the case that $(m_1 u_*^3 + w_*^3) < 0$, no energy is available for entrainment and $w_e$ must vanish. The negative value of $w_*^3$ adjusts itself automatically until the TKE, that is consumed by the downward transport of buoyancy, is balanced exactly by the work of the wind stress minus dissipation [i.e., until $(m_1 u_*^3 + w_*^3) = 0$]. Equation (6.41) indicates that $w_*^3$ is a linear function of the layer-depth $h$. The adjustment may therefore involve the establishment of a new mixed layer with a shallower depth. To demonstrate this process, we assume that $S_b$ and $\Delta U$ are both negligible. It follows from (6.41) that

$$w_*^3 = \frac{1}{2} hB. \tag{6.50}$$

Elimination of $w_*^3$ between (6.50) and the condition $(m_1 u_*^3 + w_*^3) = 0$ yields

$$h = \frac{2m_1 u_*^3}{B} \propto L \quad \text{(balanced layer-depth)} \tag{6.51}$$

where $L$ is the Obukhov length (5.31). Equation (6.51) may be readily generalized to include $S_b$ and $\Delta U$.

The phenomenon of mixed-layer shallowing or *detrainment* has received much less attention in the literature than the entrainment process, although the topic is just as important. Incorporation of the detrainment formulation (6.51) into large-scale circulation models can involve numerical and conceptual problems. A way to circumvent these has been described by Bleck et al. (1989).

Deardorff (1983) argued that (6.48) can involve a singularity. The entrainment velocity $w_e$ becomes infinitely large when the denominator in (6.48) goes toward zero. This singularity cannot occur when an entrainment hypothesis developed by Pollard et al. (1973) is used. They argued in essence that the ratio $(c_1/\Delta U)^2$ has the character of a Richardson number. As conditions become more unstable, this Richardson number becomes progressively smaller. An entrainment event always occurs when it approaches a critical value $Ri_v(\text{crit})$. The resulting mixing decreases the velocity difference $\Delta U$ and tends to stabilize conditions, at least temporarily. It follows that $(c_1/\Delta U)^2$ cannot decrease significantly below $Ri_v(\text{crit})$. This causes the denominator in (6.48), and hence $w_e$, to remain finite at all times, provided $Ri_v(\text{crit}) > m_3$.

We now return to a discussion of the three proportionality factors in

(6.47) and (6.48). Kraus and Turner (1967) did not consider the processes represented by the terms containing $q^2$ and $(\Delta U)^2$ in (6.48). They also assumed that $m_2 = 1$ or, more precisely, that dissipation of convectively generated TKE can be disregarded. This disregard had been justified earlier by Ball (1960) with the argument that molecular viscosity would not affect the relatively large, coherent perturbations that are generated by an upward flux of buoyancy. Since then we have learned that most energy is transferred from the large energy-containing eddies to smaller scales, where it can ultimately be dissipated. In spite of the inherent simplifications of the original Kraus–Turner approach, their results were in reasonable agreement with laboratory experiments by Turner and Kraus (1967) and with observations in the ocean. In particular, the theory indicated that ocean surface mixed layers should have minimum depth at the time of the summer solstice, when solar irradiation has a maximum. The highest ocean surface temperature can only be reached later during the heating season. This phase difference and the resulting hysteresis loop are typical for the real ocean as illustrated by Figs. 6.12 and 6.13. They are

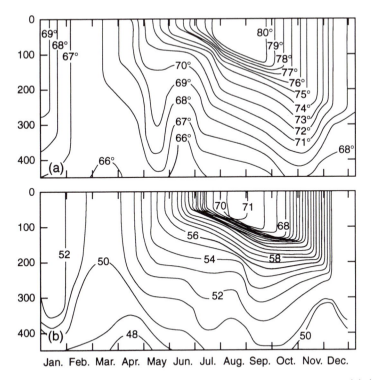

**Fig. 6.12.** The seasonal temperature cycle in °F of the upper ocean: (a) in the Bermuda area: (b) in the central N. Pacific. From Technical Report No. 6, N.D.R.C., Washington (1946).

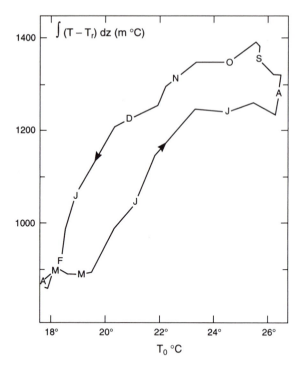

**Fig. 6.13.** Heat content as a function of surface temperature at Ocean Weather Station Echo (35°N, 48°W). The integral extends from a fixed depth of 250 m to the surface. The reference temperature $T_r$ represents the mean of the temperatures at 250 m and 275 m. After Gill and Turner (1976).

an inevitable result of the model equations (6.33), (6.36), and (6.48), which indicate that $w_e$ and $\partial b/\partial t$ (or $\partial T/\partial t$) cannot be zero simultaneously. This makes it impossible for mixed layers in the ocean to reach a stationary state with constant depth and temperature.

An equation of the form (6.47) was derived by Niiler and Kraus (1977). Numerical values of $m_1, m_2$, and $m_3$ are uncertain and controversial, in spite of numerous, relevant investigations. Niiler and Kraus quoted laboratory and oceanographic determinations of $m_1$ that vary from 0.6 to 4.0. Later studies suggested lower numbers. A choice of $m_1 = 1.0$ produced the best results in a large-scale circulation model with an explicitly resolved mixed layer (Bleck et al., 1989). Oceanic studies by Price et al. (1986) indicated $m_1 \leq 0.3$. Possibly the best documented estimate of $0.4 \leq m_1 \leq 0.5$ was obtained by Davis et al. (1981) from continuous measurements during 20 days in the North Pacific. Gaspar (1988) found later that the slightly higher range of $0.55 \leq m_1 \leq 0.7$ provided a good fit for four years of routine observations at station P in the Gulf of Alaska.

Observed or stipulated values of $m_2$ range all the way from $m_2 = 1$, as assumed implicitly by Ball, to $m_2 = 0.036$ for a deepening mixed layer under the ice of a frozen lake (Farmer, 1975). Gaspar (1988) published a useful table, summarizing many different assessments of $m_2$, mainly in the atmosphere, that cluster around a value $m_2 = 0.2$.

The last factor, $m_3$, has been studied less than the other two, partly because the very existence of a velocity discontinuity $\Delta U$ is in question. In their original exposition, Niiler and Kraus (1977) tacitly assumed that $m_3 = 1$. Price et al. (1986) recommended $m_3 = 0.65$.

It is probably unrealistic to treat these proportionality factors as universal constants. The numerical value of $m_1$, and possibly also of $m_2$, is likely to be affected by static stability, characteristics of the surface wave field, shear currents, and perhaps by other nonlocal influences. They also may vary with depth. The working of the wind stress is always positive ($u_*^3 > 0$). Gill and Turner (1976) proved that this causes a monotonic increase of potential energy in the model over several heating cycles. To avoid the resulting long-term drift, Resnyanskiy (1975), Garwood (1977), and Wells (1979) all stipulated a decrease of $m_1$ and $m_2$ with the layer-depth $h$. The same stratagem was used later by Woods and Barkmann (1986) and by Gaspar (1988). A decrease with depth is plausible for the wind-generated TKE, and hence for $m_1$, but convective perturbations tend to penetrate the entire layer. There is no physical reason why $m_2$ should decrease with depth and why convectively generated energy should not be available for entrainment at the layer bottom.

It seems obvious that there is at present no single, clear-cut answer to the dissipation problem. Fortunately, the numerical values of $m_1$, $m_2$, and $m_3$ have a greater influence upon the rate of mixed-layer development than they do upon the final state. In the simulation of cyclical or seasonal developments, the maximum and minimum values of $h$ remain relatively insensitive to the particular individual choice of these factors. If the wind stress is the the only TKE-generating force, it follows from (6.48) and (6.42) that $m_1 \propto h^2 \, \partial h / \partial t$. On the other hand, if the buoyancy flux is the only source of TKE, then we find similarly that $m_2 \propto h \, \partial h / \partial t$. Using these relations, estimates of $m_1$ and $m_2$ may be obtained from observations of the rate of layer-deepening and of the surface wind stress and the surface buoyancy flux. The relevant rates of layer-deepening are given in (6.60) and (6.61).

An important aspect of the model is the discontinuity at the layer base. Deardorff (1983) considered this discontinuity unrealistic and replaced it with a finite transition layer of thickness $\Delta h$ across which buoyancy and velocity vary continuously. He introduced a second-order closure scheme for the transition layer to complete the model. This type of modelling has not improved the description of the mixed layer. In fact, observations of the transition layer suggest that a discontinuity is a good approximation. A striking example is given later in Fig. 6.18 for the cloud-topped convective boundary layer. More realistic results may be obtained by *large eddy simulation* (LES) models, as reviewed by Garratt (1992).

### 6.4.2   *The cloud-free atmospheric mixed layer*

The mixed-layer concept was adapted specifically to atmospheric conditions by Lilly (1968), Betts (1973), Tennekes (1973), Stull (1976), Tennekes and Driedonks (1981), and others. When there are no phase changes, atmospheric mixing conserves the potential temperature, $\Theta$, and the specific humidity, $q$. Following (2.24) we combine the effect of both these properties on the density or potential density by introduction of a virtual potential temperature

$$\Theta_v = \Theta(1 + 0.6q). \tag{6.52}$$

In high latitudes where $q$ is always relatively small, one can generally assume that the actual potential temperature $\Theta$ is conserved during mixing. However, in the tropics near the sea surface, the humidity effect becomes important, causing the potential density to be inversely proportional to $\Theta_v$.

The equations developed for the oceanic mixed layer in the preceding subsection are directly applicable to the cloud-free atmosphere if the buoyancy $b$, as defined in (6.34), is replaced by $g\Theta_v/\Theta_v(0)$. This implies choice of the virtual potential temperature $\Theta_v(0)$ at the sea surface as a reference value. The use of a fixed reference value may introduce an unnecessary inaccuracy. Usually the average value of $\Theta_v$ is used, as determined at the level where the buoyancy flux is considered.

The buoyancy flux immediately above the sea surface is now specified by

$$(\overline{wb'})_0 \equiv B = \frac{g}{\Theta_v(0)} (\overline{w\theta_v})_0 \tag{6.53}$$

The buoyancy flux across the upper layer boundary at $z = h$ is indicated by

$$(\overline{wb'})_h = w_e \, \Delta b = w_e \frac{g}{\Theta_v(0)} \Delta\Theta_v. \tag{6.54}$$

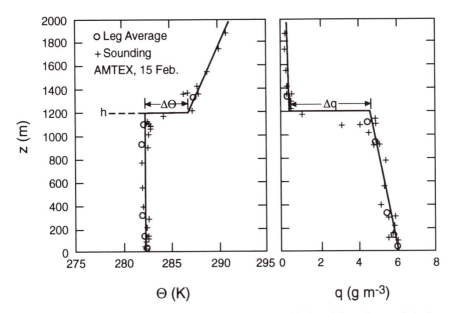

**Fig. 6.14.** Profiles of potential temperature and specific humidity observed during the Air Mass Transformation Experiment (AMTEX). After Lenschow et al. (1980).

When sign differences of $h$ and $w_e$ are allowed for, all the equations of Section 6.4.1, including (6.36), (6.41), and (6.48) remain valid also for the cloud-free atmosphere.

If the buoyancy flux is the dominant source of TKE, with the shear production terms in (1.46) or (6.47) negligibly small in comparison, the mixed layer is also called the *convective boundary layer*. Conditions of this type are common in the marine boundary layer. As an example, Fig. 6.14 presented a convective boundary layer during a cold air outbreak. It is clear that the potential temperature is uniform with height, but that the humidity still shows a gradient, which will be discussed shortly. If absorption of radiation within a relatively dry air layer can be neglected, the convective velocity $w_*$ assumes the form (6.50). With $q^2$ also considered relatively small, the atmospheric version of (6.48) reduces to

$$w_e = m_2 \frac{\overline{(w\theta_v)}_0}{\Delta\Theta_v}. \tag{6.55}$$

The value of $m_2$ is difficult to determine because the entrainment flux cannot be measured directly. Observations of the growth of convective boundary layers suggest that $m_2 \approx 0.2$. If all buoyant energy was used for entrainment $m_2 = 1$. The other extreme, $m_2 = 0$, implies that all buoyant TKE is dissipated before it reaches the top of the boundary layer, or that $\Delta\Theta_v = 0$. In this case we speak of encroachment.

In order to determine $w_e$ we need to know the magnitude of $\Delta\Theta_v$. The rate of change for this quantity depends on the lapse rate $\gamma$ above the inversion and may be written as

$$\frac{\partial(\Delta\Theta_v)}{\partial t} = \gamma w_e - \frac{\partial\Theta_v}{\partial t}. \tag{6.56}$$

With $S_b = 0$, the atmospheric version of (6.36) has the form

$$h\frac{\partial\Theta_v}{\partial t} = \overline{(w\theta_v)}_0 + w_e\,\Delta\Theta_v. \tag{6.57}$$

Following Betts (1973) we eliminate $\partial\Theta_v/\partial t$ from (6.56) and (6.57). The resulting equation and (6.56) may be solved for $w_e$ and $\Delta\Theta_v$. This yields

$$w_e = \frac{1 + 2m_2}{\gamma h}\overline{(w\theta_v)}_0 \tag{6.58}$$

and

$$\Delta\Theta_v = \frac{m_2\gamma h}{1 + 2m_2}. \tag{6.59}$$

From these two equations we see that $w_e$ is inversely proportional to $h$ and that $\Delta\Theta_v$ is proportinal to $h$. Furthermore, if we integrate (6.58), assuming that $\overline{(w\theta_v)}_0$ is constant and that $w_h = 0$, we find that

$$h(t) - h(0) = \frac{1 + 2m_2}{\gamma}\overline{(w\theta_v)}_0 t^{1/2}. \tag{6.60}$$

In the special case that shear production is the only source of TKE, common for the oceanic mixed layer, (6.47) may be written in the form

$$\frac{1}{2} g h \frac{\Delta \Theta_v}{\Theta_v} \frac{\partial h}{\partial t} = m_1 u_*^3 .$$

In this case the solution for $\Delta \Theta_v$ is similar to (6.59) (i.e., $\Delta \Theta_v \propto h$) which, combined with the previous equality, yields

$$\frac{\partial h}{\partial t} \propto h^{-2} \quad \text{and} \quad h(t) - h(0) \propto t^{1/3} . \tag{6.61}$$

This is in contrast with the convective case given in (6.60), where $h(t) - h(0) \propto t^{1/2}$.

In reality, the structure of the convective boundary layer is not as simple as has been sketched earlier. There usually is large-scale convergence or divergence and $w_h \neq 0$. The fluxes of sensible and latent heat at the surface are not uniform in place and time. The coefficient $m_2$ is probably not a constant. Nevertheless, this assumption gives satisfactory results. The absorption and emission of radiation can cause generation or consumption of TKE even in cloud-free layers. Wind shear near both mixed-layer boundaries must play a significant role on many occasions. In particular, it has been demonstrated by Brost et al. (1982) that TKE production by shear at the top of the boundary layer may enhance entrainment by a substantial amount.

The convective boundary layer has been extensively studied with LES models. This type of modelling was pioneered by Deardorff (1970a) and applied to the convective boundary layer (Deardorff, 1972). LES studies confirmed that the simple one-dimensional parameterization of the convective boundary layer is an adequate model for practical use. Further studies by Wyngaard and Brost (1984) and Moeng and Wyngaard (1984) dealt with the fluxes originating at the bottom and the top of the boundary layer. They introduced the concepts *bottom-up* and *top-down* convection. The difference between these two types of convection relates to the fact that the bottom is a rigid surface and the top is a free surface. Top-down convection may start with rather large eddies whereas bottom-up convection is restricted at the bottom by the rigid surface. In this context even the sea surface is a rigid surface for atmospheric convection. In the ocean mixed layer we have mainly top-down convection. By separating these two fluxes, it is possible to introduce eddy transfer coefficients $K$ for the various scalar fluxes. This is especially useful for the description of passive scalar quantities, such as water vapour. In Fig. 6.14 we saw an example of a water vapour gradient in the mixed layer. The vapour flux may be obtained by multiplying this gradient by the eddy transfer coefficient, determined with the LES model.

### 6.4.3   *The cloud-topped convective marine boundary layer*

As soon as the top of the well-mixed boundary layer reaches beyond the *lifting condensation level* (LCL), the structure of the boundary layer changes dramatically. A stratus layer forms in the upper part of the boundary layer with the cloud top coinciding with the top of the boundary layer. Strong radiative divergence, associated with strong cooling, occurs at the cloud top. This destabilizes the boundary layer and enhances entrainment. Models of the cloud-topped mixed layer have been developed by Lilly (1968), Deardorff (1976), Stage and Businger (1981), Kraus and Leslie (1982), and others.

The properties that are conserved when condensation occurs are the entropy (2.30) [or the closely related equivalent potential temperature $\Theta_e$ (2.31)] and the total water content, $q_w$, which is given by

$$q_w = q + q_1,$$

where $q$ is the specific humidity and $q_1$ is the specific humidity for liquid water. The value of $\Theta_e$ can be readily adjusted to allow for the virtual temperature effect. The cloud base (i.e., the lifting condensation level), $z_c = z_c(q_w, \Theta_0)$.

The governing equations (6.33) and (6.36) remain valid also in the cloud-covered case. Condensation at cloud base causes a discontinuity in the buoyancy flux as is illustrated in Fig. 6.15. This figure also presents typical profiles of the other quantities, such as $q_w$ and $\Theta_e$. The derivation of the equations describing the cloud-topped mixed layer is straightforward but somewhat cumbersome and can be found in the papers quoted above. However, modelling of entrainment requires special attention, because entrainment is a major factor in determining the mixed-layer height and has significant influence on the heat and moisture of the mixed layer.

For the description of the dry mixed layer it is not critical whether the entrained buoyancy flux $\overline{(wb')}_h$ is distributed only over the upper part of the boundary layer to the level where $\overline{wb'} = 0$, or over the entire boundary layer.

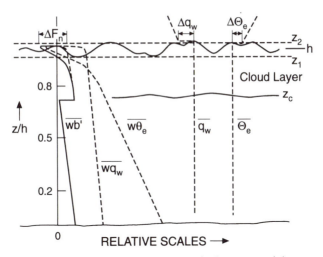

**Fig. 6.15.** Profiles of total water content $q_w$, equivalent potential temperature $\Theta_e$, buoyancy flux $\overline{wb'}$, moisture flux $\overline{wq_w}$, and flux of equivalent potential temperature. At cloud top there are jumps in $q_w$, $\Theta_e$, and net irradiance $\Delta F_n$. The cloud base is indicated by $z_c$.

The energetics of these two cases are compared in Fig. 6.16. In both cases it is assumed that $m_2 = 0.2$. In the first case (a), the consumed buoyant energy, $N$, is only 4 per cent of the production, $P$. This means that only 4 per cent of the production is used for the given entrainment. Stull (1976), Randall (1980b) and others came to this conclusion. However, at

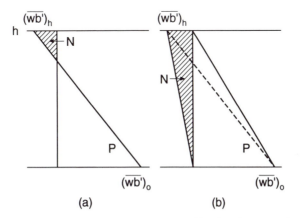

**Fig. 6.16.** Mean buoyancy flux profile with partitioning into production $P$ and consumption $N$: (a) the conventional partitioning; (b) as proposed by Stage and Businger (1981).

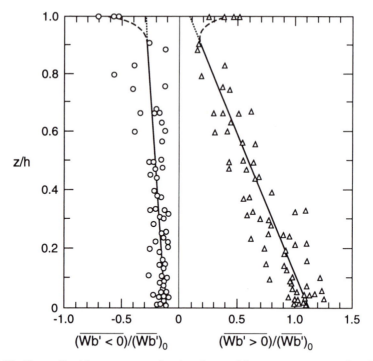

**Fig. 6.17.** Normalized buoyancy production flux and buoyancy consumption flux as a function of $z/h$. After Wilczak and Businger (1983).

the level where $\overline{wb'} = 0$, there are eddies with both positive and negative contributions to the buoyancy flux; turbulence is more exuberant than the averages reflect. A more realistic entrainment hypothesis suggested by Stage and Businger (1981) is illustrated in Fig. 6.16(b). In this case the rate of consumption over production of TKE is

$$-\frac{N}{P} = m_2 = 0.2.$$

To test whether or not this formulation is better than the one given by Fig. 6.16(a), Wilczak and Businger (1983) analyzed data obtained at the Boulder Atmospheric Observatory in 1978. The result, given in Fig. 6.17, shows that case (b) gives a reasonable approximation, superior to case (a), and that the real world is more complex than the best of the simple models presented here. The same figure also gives a glimpse of bottom-up and top-down convection. This entrainment model is needed to describe the interaction of the entrained air with the cloud layer. The cloud physics in this interaction still requires additional work. It is an area beyond the scope of this book.

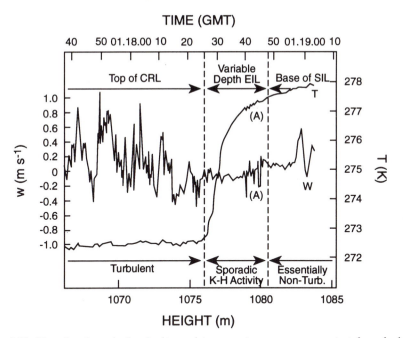

**Fig. 6.18.** Details of vertical velocity and temperature measurements taken during a traverse of the entrainment interfacial layer (EIL) when this layer was relatively quiescent. The feature (*A*) may be a consequence of Kelvin–Helmholtz instability in a region where the Richardson number may have been reduced to around 0.25 by wind shear. CRL denotes 'cloud radiative layer' and SIL 'subsidence inversion layer'. After Caughey et al. (1982).

The cloud top is uneven and varies between heights $z_1$ and $z_2$, as shown in Fig. 6.15. This suggests that there is a transition region between the boundary layer and the free atmosphere. However, when we take a close look at the interfacial region between cloud and no-cloud, we see a very sharp transition in temperature. Figure 6.18 shows this transition measured with a probe, which was lifted by a tethered balloon. The sharpness of the transition is marked by the temperature, which increases by 5°C over less than 5 m. Over the same height interval the vertical velocity signal changes from turbulent to quiescent except for a patch of activity (A) where possibly Kelvin–Helmholtz instability occurs. A striking feature of these observations is that the *entrainment interfacial layer* is only a few meters thick. Another example is given in Fig. 6.19, where this layer is about 8-m thick but where possibly active entrainment takes place. Especially feature B in the temperature trace is suggestive of this possibility.

Because the top of the cloud layer is able to emit long-wave radiation to space, the entire boundary layer loses energy. Also, because the loss

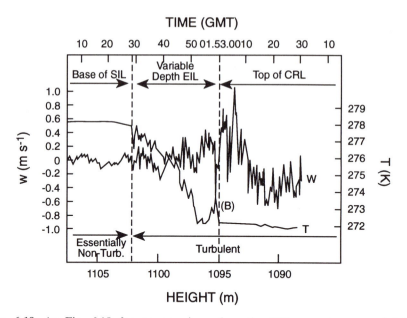

**Fig. 6.19.** As Fig. 6.18 for an occasion when the EIL was deeper and fully turbulent. The feature $(B)$ may be a volume of air in process of entrainment. After Caughey et al. (1982).

occurs at the top of the boundary layer, it tends to destabilize the entire layer and to enhance convection. The emittance of long-wave radiation appears to be mainly a function of the liquid water content of the cloud only; it depends relatively little on the droplet spectrum, as shown in Fig. 3.10. The short-wave radiation is not so easily treated in clouds. To model this quantity, both liquid water content and droplet spectra are needed. Whereas the long-wave radiation is emitted near the top of the cloud, the short-wave radiation is absorbed over a considerably deeper layer of the cloud. Figure 6.20 presents measurements of the absorption of short-wave radiation as well as the long-wave emission. Long-wave radiation loss still tends to destabilize the cloud while short-wave absorption has a stabilizing effect. Further information on the transfer of short-wave radiation in clouds can be found in Joseph et al. (1976).

The number of cloud particles that form in a cloud depends to some extent on the pre-existing aerosol. In the marine boundary layer the air is usually clean, and the number of condensation nuclei and consequently of droplets in the cloud is relatively small, allowing them to grow. Once droplets have attained a diameter greater than 1 or 2 $\mu$m, they will accumulate virtually all the water that condenses out of the air while rising, and no new droplets will form. This takes place because the saturation

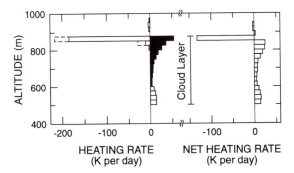

**Fig. 6.20.** Theoretical longwave (open), shortwave (solid) and net heating rate profiles from the Roach and Slingo (1979) and Slingo and Schrecker (1982) schemes for profiles measured at midday in the JASIN area. The longwave heating rates from the Schmetz and Raschke (1981) scheme indicate a deviation at cloud top where they are shown dotted. After Slingo et al. (1982b).

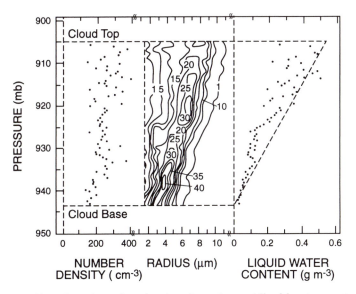

**Fig. 6.21.** Profiles of total number density, drop size and liquid water content. The drop-size spectra are presented as contours of the percentage normalized spectral density, such that at any height the sum of the values in each 1-$\mu$m interval is 100 per cent. The adiabatic liquid water content profile is shown as the dashed line. After Slingo et al. (1982a).

vapour pressure over the larger droplets is very close to the saturation vapour pressure over a flat surface. Figure 6.21 shows that, indeed, the number density of the droplets tends to be constant with height. As expected, this figure also shows that the average droplet radius increases with height, and therefore the average liquid water content increases as well. The liquid water content shows considerable variability with height. The average liquid water content is somewhat less than the adiabatic value, probably due to entrainment of dry air at the top of the cloud. However, it is also of interest to note that some values are larger than the adiabatic liquid water content. This suggests that the radiative cooling, which is due to the emission of long-wave radiation by the droplets, enhances condensation onto these droplets. This may result locally in conditions of a higher than adiabatic liquid-water content.

Figure 6.22 shows the net upward radiative flux and the heating rate as a function of height. We see that at cloud base a small amount of heating occurs due to radiation from below. Near the top of the cloud, the cooling reaches a pronounced maximum. This layer of strong radiative divergence is about 50–70-m thick, which is an order of magnitude greater than the thickness of the layer shown in Fig. 6.19. Furthermore, a small amount of cooling takes place above the cloud top. This will cool the air above the boundary layer and may lead to so-called *direct entrainment*. The figure shows that this does not appear to be an important process. By and large, the radiative divergence occurs within the cloud layer below the EIL. This issue became somewhat controversial over the years since Lilly (1968) introduced a model of the cloud-topped boundary layer where radiative cooling resulted in direct entrainment. Deardorff (1976) introduced a model where both direct and indirect entrainment took place, and Kahn and Businger (1979) argued that all radiative cooling took place in the cloud layer and that entrainment was a result of convection due to radiative cooling. Further discussions on this topic are given by Nieuwstadt and Businger (1984).

The persistence of the marine stratus or stratocumulus and its widespread occurrence suggest that, once formed, the cloud deck is quite stable. We have seen that at midlatitude the long-wave cooling over a 24-hour period is stronger than the short-wave heating (Fig. 6.20). A cloud cover tends to cool the boundary layer. This results in a thickening of the cloud layer, when other conditions are kept the same.

In the preceding discussion, the radiative cooling near the top of the cloud was assumed to set up mixed-layer convection. This is the case when the radiative cooling is balanced by a sensible heat flux from below. Brost et al. (1982) pointed out that the radiative cooling may also be balanced by the release of latent heat or by the entrainment of sensible heat from

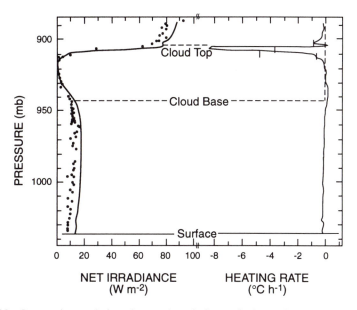

**Fig. 6.22.** Comparison of the observed and theoretical net long-wave irradiance and heating rates for profile 1, Nov. 19–20, 1976. The continuous lines are from the radiation scheme, using the balloon temperatures and humidities. The dots represent the corrected measurements from the upper and lower radiometers, respectively. Only every fifth value has been plotted. After Slingo et al. (1982a).

above. In these two cases the radiative cooling may not contribute to convection of the mixed layer and to entrainment. The downward flux of sensible heat may be the result of shear-driven entrainment. Observations by the NCAR Electra on June 13 and June 17, 1976, in the cloud-topped mixed layer off the California coast show strong evidence of shear-driven entrainment. The boundary layer was characterized by high windspeeds, near-neutral stability, and a jump in wind direction across the inversion at the top. The shear-driven entrainment mixes dry air from above the cloud top into the cloud layer. It can be a mechanism for cloud dissipation.

We have seen that a quasi–steady-state may occur in the dry convective boundary layer, when $\frac{\partial h}{\partial t} = w_e + w_h = 0$. In this case $h$ is constant, but $\Theta_v$ must increase with time. On the other hand, the cloud-topped convective boundary layer may develop a local steady-state, when the radiative cooling at the cloud top is balanced by the heating at the surface and, due to entrainment, when $w_e = -w_h$, and when the evaporation at the surface is equal to the entrainment of dry air at the top.

The cloud layer also may develop an existence independent from the lower boundary layer. This situation may occur when the boundary layer

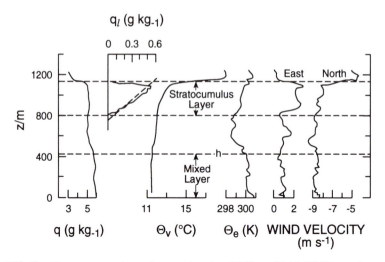

**Fig. 6.23.** Profiles measured on descent by the C130 at 11.15 GMT on August 7, 1978, during JASIN. The dashed line on the $q_1$ trace represents the adiabatic liquid water content for a cloud base of 750 m. After Nicholls et al. (1983).

is advected over cooler water. The convection in the lower part of the boundary layer is then suppressed and may no longer reach the cloud base. It is easy to see in Fig. 6.15 that, when the heat flux at the surface decreases, the buoyancy production of TKE may not reach the LCL. As a result the layer of stratocumulus is disconnected from the surface and maintains its own existence. An example is shown in Fig. 6.23.

A large number of models have been developed to study the cloud-topped convective boundary layer. These include simple one-dimensional models, such as mentioned earlier, as well as much more sophisticated LES models, such as Deardorff (1980, 1981). Garratt (1992) provides an overview of the state of LES modelling.

In conclusion, a few remarks concerning the break-up of a stratocumulus deck. The most common mechanism is the presence of sufficiently dry air above the boundary layer. When this air is dry enough, $\Delta\Theta_e$ may become negative at the top of the cloud, and *entrainment instability* may set in. A parcel of entrained air will cause droplets from the surrounding cloud to evaporate and cool until it reaches saturation level. At that stage the parcel will be cooler than its environment and tend to be negatively buoyant. In order for this process to cause entrainment instability, $\Delta\Theta_e$ must be sufficiently negative to overcome the liquid water loading in the cloud. Deardorff (1980) and Randall (1980a) considered this effect and calculated the threshold value of $\Delta\Theta_e$ at which entrainment instability sets in. In this context it is of interest to note that the average profiles for the

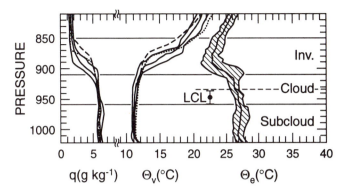

**Fig. 6.24.** Mean profiles of specific humidity, virtual potential temperature and equivalent potential temperature for radiosonde flights between 3.00 and 20.00 GMT, August 31, 1978, from the Endurer, Hecla, and Meteor (ships participating in JASIN). After Taylor and Guymer (1983).

daytime of August 31, 1978, show a $\Delta\Theta_e < 0$ at the top of the boundary layer (Fig. 6.24). This type of situation is presumably characteristic of a broken cloud deck (i.e., patches of Sc and individual Cu clouds).

The cloud deck also tends to break up when the boundary layer air is advected over gradually warmer water in the subtropics. The boundary layer tends to become deeper and entrain rather dry air from above. This process lifts the condensation level sufficiently to break up the deck of stratocumulus into isolated cumulus, which are characteristic for the trade winds. The modelling of broken clouds raises additional difficulties, particularly with the radiation field.

## 6.5 Discussion and evaluation

Each of the various approaches to PBL modelling has its advantages and drawbacks. None of them is suitable for all occasions. The least controversial and most realistic representations are obtained from a detailed, three-dimensional numerical simulation of all the fluctuating quantities. PBL studies that followed this line of attack have been listed at the end of Section 6.3 and can be found in Garratt (1992) or Coleman et al. (1990). The approach requires relatively large computing resources. It also generates a great deal of information that is simply not needed in the study of large-scale phenomena. Much effort continues to go, therefore, into the search for simpler, universal parameterization schemes. The complete numerical models have a potentially invaluable role in this search. They permit experimentation with a practically unlimited range of boundary conditions, external forcing, latitudes and so forth. As long as carefully selected experiments are carried out to test them, the models may reduce the need for a corresponding set of real atmospheric and oceanic observations.

Parametric models entail necessarily a trade-off between specific realism and universal validity. The number of empirical proportionality constants in PBL models increases with the order and complexity of the modelling scheme. However, it is difficult to derive statistically significant values of more than one nondimensional ratio from observational records of noisy and complex natural systems. Schemes, which contain several empirical and partly adjustable parameters, can produce realistic simulations of specific, observed features; however, those empirical parameters tend to reduce their universal validity and significance.

The diffusive approach is in the direct line of traditional turbulent transport studies. It is supported by a large body of laboratory experiments, field tests, and theoretical investigations. The concept of eddy diffusivity or eddy viscosity has been found very useful for the representation of fluxes through relatively small, localized systems, such as the 'constant stress' surface layer. It tends to become more arbitrary when applied to the much deeper PBL, where complex variations of $K$ have to be stipulated. LES models may be used to establish simpler parameterizations. This procedure has been proposed by Wyngaard and Brost (1984), to introduce simple $K$-models in a more complex context.

Diffusive models are most appropriate for the simulation of transports in neutral or stable conditions. Their validity becomes questionable in unstable conditions, when countergradient transports can come into play. Vertical countergradient transports occur necessarily during the development or intensification of inversions and thermoclines. The diffusive approach also becomes awkward when turbulent transports of different properties are coupled dynamically. Coupled systems often involve double diffusion phenomena, for example, a downward transport of heat in the upper ocean that is driven by an unstable salinity gradient.

The transilient scheme differs from the diffusive approach, by being able to deal explicitly with the effects of eddies of any desired size. It can deal with surface layers that are not mixed completely and with transients in the stable strata below. It can also be adopted to reproduce differential rates of transport and dissipation for different properties. This might make it useful for the investigation of biological transport processes in the sea and for studies of the equatorial ocean, where the near surface vertical profiles of momentum and density are often quite different. The weakest feature of the transilient method is the arbitrary stipulation of the transilient coefficients. Being first-order, the scheme neglects also the triple correlations, which represent the transport of TKE and other

second-order quantities by the turbulent eddies. These third-order transports can be parameterized only by higher-order closure schemes.

The various matching models for the parametric representation of
vertical profiles of bulk variables in the PBL suffer from the same
drawback as the diffusive models. In fact, these models involve implicitly a
continuously and monotonically changing eddy viscosity. They are therefore not able to account for the sharp changes in the velocity gradient that
occur characteristically in inversions or thermoclines.

The formation of inversions and thermoclines is an essential feature of
the mixed-layer models. None of the other techniques can account for this
phenomenon as simply or directly. The models stimulate the different rates
of sea surface temperature change during heating and cooling periods
reasonably well. Self-consistent mixed-layer models treat all conservative
properties as being homogeneous throughout the layer-depth. This implies
not only a vertically uniform distribution of buoyancy and concentrations,
but also a horizontal displacement of the whole layer like a solid slab
without mean internal shears. The models cease to be applicable when the
layer is not in a well-mixed state. Their representation of the bounding
inversions and thermoclines as discontinuities is in remarkable agreement
with observations, such as presented in Fig. 6.16. Entrainment is clearly an
intermittent phenomenon. The detailed structure of these intermittent
events is very difficult to observe. Kelvin–Helmholtz instabilities may play
an important role but much research is still needed in this area.
Experiments and ideas that deal with the physics of the entrainment
process have been reviewed by Fernando (1991).

# 7

# ATMOSPHERICALLY FORCED
# PERTURBATIONS IN THE OCEANS

Kinetic energy flows almost exclusively downward, from the atmosphere into the ocean. The upward flux of energy is thermal, and that will be the topic of our concluding chapter. In the present chapter, we shall deal with the effects of kinetic energy inputs into the ocean. Although this requires some discussion of the different types of oceanic perturbations, our treatment of these topics is necessarily brief and incomplete. We are not concerned with details of the motion pattern in the deeper ocean or with processes involving friction and non-adiabatic mixing in the interior. These processes are essential for an understanding of ocean circulations. They are treated in general oceanographic textbooks and in many monographs that deal specifically with these subjects.

Surface stress and air pressure variations produce surface wind waves along with a variety of other wave forms. Most of these waves are relatively slow, with periods that can be measured in hours, days, or even years. The amplitude of internal gravity waves in the oceans is often much larger than that of surface waves and their wavelengths tend to be in the kilometer range. The square of the amplitude-wavenumber product is usually a very small quantity. This makes first order approximations appropriate for many purposes. It justifies use of the hydrostatic approximation and of the linear equations (4.1) as a basis for the following discussion. To do so, it is necessary to represent the unspecified forcing terms on the right-hand side of those equations in a linearized form. An algorithm for the inclusion of the various atmospheric inputs as a linearized boundary condition in the equations of motion for the ocean is discussed in Section 7.1.

Section 7.2 describes a two-layer ocean model. Systems of this type are convenient for the conceptual consideration of atmosphere–ocean interactions, because the wind affects the ocean primarily through action upon the surface mixed-layer. Internal waves, the topic of Section 7.3, are ubiquitous both in the ocean and in the atmosphere. Essentially, sea surface gravity waves can be viewed as internal waves at the interface between two fluids of very unequal density. The atmospheric forcing of internal waves leads to a discussion in Section 7.4 of oceanic perturbations produced by traveling storms. Section 7.5 provides a brief survey of wind-driven coastal upwelling and of coastal Kelvin and shelf waves. Under suitable conditions, these phenomena can develop into disastrous storm surges along relatively flat coasts. Planetary or Rossby waves have a meridional extent that is large enough for them to be affected by variations in the Coriolis parameter. They are considered in Section 7.6. In Section 7.7 we deal with the general character of equatorial perturbations and their sensitivity to changes in the wind stress pattern.

## 7.1 Perturbations of a shallow, homogeneous ocean

### 7.1.1 The different types of atmospheric forcing

Let $\partial \zeta_F / \partial t$ denote the change in sea surface elevation that would be forced by the atmosphere in the absence of internal oceanic adjustments. This forcing can be caused by a number of separate, independent processes. To describe them, we set

$$\frac{\partial \zeta_F}{\partial t} = \frac{\partial (\zeta_A + \zeta_B + \zeta_M + \zeta_E)}{\partial t}. \tag{7.1}$$

The deformation $\zeta_A$ is directly proportional to transients of the departure $p_A$ of the surface air pressure from its horizontal mean

$$\frac{\partial \zeta_A}{\partial t} = -\frac{\partial p_A / \partial t}{g \rho}. \tag{7.2}$$

Called the *inverse barometer depression*, $\zeta_A$ indicates the change that would be registered by a water barometer as a function of air pressure changes—about 1 cm per millibar. Air pressure gradients cannot drive currents if the actual surface level $\zeta$ conforms to $\zeta_A$.

The surface buoyancy flux $B$ was specified in (6.29). It can force sea level changes by causing expansion or contraction of the water column below. The resulting deformation rate of the surface of a shallow ocean with vertically invariant temperature $T$ and salinity $s$ is indicated by

$$\frac{\partial \zeta_B}{\partial t} = \left( \alpha \frac{\partial T}{\partial t} + \beta \frac{\partial s}{\partial t} \right) D = \frac{B}{g}. \tag{7.3}$$

The quantity

$$\frac{\partial \zeta_M}{\partial t} = \frac{P - E}{\rho} \tag{7.4}$$

stands for a sea surface elevation change, which could be produced by the precipitation–evaporation difference $(P - E)$, expressed as an addition of water mass per square meter per second. On a global scale, $(P - E)$ tends to be negative in the subtropics. It is usually positive in polar and equatorial regions. The resulting modulations of the density-driven *thermo-haline circulation* can be very significant, but they are of no concern in the present discussion, which deals with the mechanical forcing of the oceans by the atmosphere.

The last term in (7.1) represents the change of surface deformation that is forced by convergence of a wind-driven Ekman transport $\mathbf{M}_E$. To derive a relevant expression, one forms both the curl and the divergence of (6.10). Neglecting meridional variations of the Coriolis parameter in the present context, one gets, after elimination of the curl $\nabla \times \mathbf{M}_E$ from the resulting two equations,

$$\frac{\partial \zeta_E}{\partial t} = \nabla \cdot \mathbf{M}_E \approx \rho^{-1} \left[ f(\nabla \times \boldsymbol{\tau}) + \frac{\partial(\nabla \cdot \boldsymbol{\tau})}{\partial t} \right]. \tag{7.5}$$

As the surface stress vector $\boldsymbol{\tau}$ is always horizontal, its curl $(\nabla \times \tau) \equiv (\nabla \times \tau) \cdot \mathbf{n}$ where $\mathbf{n}$ is an upward-pointing unit vector. To prevent equations from becoming too long, we have identified $(\nabla \times \boldsymbol{\tau})$ with its vertical component throughout this chapter.

### 7.1.2   *The forced shallow water equation*

Gill (1982) showed that for motion in a 'shallow' homogeneous $(N = 0)$ ocean, the linear equations (4.1) can be transformed so that all forcing terms appear only in the continuity equation. To demonstrate this, we let $\zeta'$ represent departures from an 'adjusted' sea surface level, which does not contribute to horizontal pressure gradients in the water below. This adjusted level is not necessarily horizontal. Surface slopes can adjust themselves to air pressure gradients as well as to local rainfalls, local heating, and local upwelling in a way that leaves the pressure horizontally uniform in the water below. With the notation used earlier, changes in $\zeta'$ are indicated by

$$\frac{\partial \zeta'}{\partial t} = \frac{\partial \zeta}{\partial t} - \frac{\partial \zeta_F}{\partial t}. \tag{7.6}$$

Horizontal velocities and the horizontal pressure gradient $\nabla p = g\rho\nabla\zeta'$ cannot vary significantly with depth, if the horizontal perturbation scale is much larger than the water depth. This would make the equation for forced motion,

$$\left(\frac{\partial}{\partial t} + f\mathbf{n}\times\right)\mathbf{U} = -g\nabla\zeta', \tag{7.7}$$

valid at any depth. We shall consider changes of the ocean depth $D$ in Section 7.6, but assume a flat bottom ($\nabla D = 0$) for the time being. If $D \gg \zeta$ one can represent the continuity equation for the whole fluid column in the form

$$\nabla\cdot[(D+\zeta)\mathbf{U}] \approx D\nabla\cdot\mathbf{U} = -\frac{\partial\zeta}{\partial t}. \tag{7.8}$$

To obtain a single equation in $\zeta'$ we use the well-known vector calculus formulas

$$\nabla\times(\mathbf{n}\times\mathbf{U}) = \nabla\cdot\mathbf{U} \quad \text{and}$$

$$\nabla\cdot(\mathbf{n}\times\mathbf{U}) = -(\nabla\times\mathbf{U})\cdot\mathbf{n} = -\left(\frac{\partial V}{\partial x} - \frac{\partial U}{\partial y}\right) \equiv -\eta, \tag{7.9}$$

where $U$ and $V$ are the zonal and meridional components of the horizontal velocity vector $\mathbf{U}$, and $\eta$ denotes again the vorticity. Meridional variations of the Coriolis parameter $f$ can be represented by

$$\beta \equiv \frac{\partial f}{\partial y} = R^{-1}\frac{\partial f}{\partial\varphi} \tag{7.10}$$

($R$ = earth's radius, $\varphi$ = latitude). With aid of (7.8) and (7.9), one gets, after multiplication by $D$, the following expressions for the curl and the divergence of (7.7)

$$D\frac{\partial\eta}{\partial t} - f\frac{\partial\zeta}{\partial t} + D\beta V = 0$$

$$\tag{7.11}$$

$$-Df\eta - \frac{\partial^2\zeta}{\partial t^2} + D\beta U = -gD\nabla^2\zeta'$$

with $\nabla^2 = (\partial^2/\partial x^2 + \partial^2/\partial y^2)$.

Elimination of $\eta$ by cross-differentiation of the two equations (7.11) yields, with consideration of (7.7) and (7.6),

$$
\left[\frac{-\partial^3}{\partial t^3} + (gD\nabla^2 - f^2)\frac{\partial}{\partial t} + gD\beta\frac{\partial}{\partial x}\right]\zeta' = \left(\frac{\partial^2}{\partial t^2} + f^2\right)\frac{\partial\zeta_F}{\partial t}
$$

$$
= \rho^{-1}\left\{\left(\frac{\partial^2}{\partial t^2} + f^2\right)\left[\left(\rho B_0 - \frac{\partial p_A}{\partial t}\right)g^{-1} + (P - E)\right] + f(\nabla\times\tau) + \frac{\partial(\nabla\cdot\tau)}{\partial t}\right\}
$$

(7.12)

In the last equality $\partial\zeta_F/\partial t$ was replaced by the explicit forcing functions that were specified in the preceding section.

Gill called (7.12) the *forced shallow-water equation*. It can be used to describe the atmospheric forcing of a shallow, homogeneous ocean through time-dependent changes of the surface boundary conditions. Similar equations can be used to deal with 'modes' of motion in a deep stratified ocean. This will be discussed in somewhat more detail in Section 7.2.

The second term on the left-hand side of (7.12) involves the convergence of an *isallobaric flow* with velocity

$$
\mathbf{U} = -\frac{1}{f^2\rho}\nabla\frac{\partial p}{\partial t} = -\frac{g}{f^2}\nabla\frac{\partial\zeta'}{\partial t}.
$$

(7.13)

This expression was first obtained by Brunt and Douglas (1928). It can be derived by solving (7.7) for its components $U$ and $V$ with the assumption that $\partial/\partial t \ll f$. The isallobaric velocity is directed from regions of increasing pressure into regions of decreasing pressure.

### 7.1.3  *Perturbations of different extent and duration*

To assess the relative importance of the different terms in (7.12), we consider that the right-hand side of that equation becomes zero for unforced perturbations. We express $\zeta'$ as in (4.2) by

$$
\zeta' = Ae^{i(lx+my-\omega t)} = Ae^{i(\mathbf{k}\cdot\mathbf{x}-\omega t)}
$$

or by a sum of terms of this form. When this expression is introduced into the left-hand side of (7.12), one gets, after division by $-i\zeta'$,

$$
\omega^3 - gD(k^2 + a_0^2)\omega - gD\beta l = 0.
$$

(7.14)

The symbol

$$a_0^2 = \frac{f^2}{gD} \qquad (7.15)$$

is the inverse of the squared (barotropic) *Rossby radius of deformation*. The deformation radius is a fundamental horizontal scale-length in fluids that are affected by both gravity and rotation. At a latitude of 30° and an ocean depth $D = 4000$ m, the Rossby radius $a_0^{-1} \approx 3750$ km.

For very slow perturbations, characterized by a relatively small $\omega$, the second and third term must approximately balance each other. The resulting truncated frequency equation is

$$\omega = \frac{-\beta l}{k^2 + a_0^2} \quad \text{for} \quad \omega^3 \ll gDl\beta. \qquad (7.16)$$

This is the dispersion relation of *planetary or Rossby waves*, which will be discussed in Section 7.6. On the other hand, if $\omega$ is relatively large, one gets the dispersion relation for horizontally-propagating *inertio-gravity* waves

$$\omega^2 = gD(k^2 + a_0^2) \quad \text{for} \quad \omega^3 \gg gDl\beta. \qquad (7.17)$$

If further $k^2 \gg a_0^2$ (i.e., if the wavelength of the perturbation is much smaller than the Rossby radius), (7.17) approaches the dispersion relation (4.21) of plane, long surface gravity waves. On the other hand, if $k^2 \ll a_0^2$, then the frequency $\omega \approx a_0(gD)^{1/2} = f$. This is the case of inertial oscillations, an example of which was given in Section 6.1. The actual, selective generation of these different wave types depends on the time and space scales of the forcing function $\zeta_F$.

## 7.2   The two-layer ocean model

### 7.2.1   *The governing equations*

Over much of the world, particularly in tropical and temperate latitudes, the interior ocean is separated by a very stable thermocline from a shallow upper ocean layer. As a first approximation, therefore, one can simulate actual conditions by a hypothetical ocean of constant depth $D$, which is divided by a frictionless discontinuity at depth $z = -h = -h(\mathbf{x}, t)$ into two strata. Each of these strata is assumed to have uniform density and vertically uniform horizontal velocity. It is usually also stipulated that the turbulent downward transport of horizontal momentum decreases linearly from $\tau$ at the surface $z = 0$ to zero at $z = -h$, and that this is the only external driving force. An ocean model of this type appears to have been used first by Charney (1955). A more general discussion of perturbations in such a system has been published by Crépon (1974). The two-layer ocean model may also give a useful introduction to the concept of barotropic and baroclinic modes and to the important process of Ekman pumping.

Meridional changes of the Coriolis parameter will be considered in Sections 7.6 and 7.7. Until then it will be assumed that these changes can be neglected and that (7.17) holds. We shall denote properties pertaining to the layers above and below the internal interface by the suffices $a$ and $b$. The horizontal component of the hydrostatic pressure gradient at any level in the lower layer is then indicated by

$$\nabla p_b = g[\rho_a(\nabla\zeta - \nabla h) + \rho_b\nabla h] = \rho_b\left[g\frac{\rho_a}{\rho_b}\nabla\zeta + g'\nabla h\right]$$

$$\approx \rho_b(g\nabla\zeta + g'\nabla h), \quad (7.18)$$

where $g' = g(\rho_b - \rho_a)/\rho_b$ as in (1.26). The last approximate equality in (7.18) is based on the relative smallness of ocean density differences $(g'/g \ll 0.01)$ and $(\rho_a/\rho_b) = 1 - g'/g \approx 1$. Allowing for (7.18), the linearized equations of motion for the two layers are

$$\left(\frac{\partial}{\partial t} + f\mathbf{n} \times \right)\mathbf{U}_a = -g\nabla\zeta + \frac{\tau}{\rho_a}h \qquad (a)$$

$$\left(\frac{\partial}{\partial t} + f\mathbf{n} \times \right)\mathbf{U}_b = -g\nabla\zeta - g'\nabla h \qquad (b)$$

$$(7.19)$$

Departures of the sea surface and of the internal interface from their horizontal equilibriums tend to be small compared to the mean layer-depth $\bar{h}$. The continuity equations for the two layers can therefore be approximated by

$$\bar{h}\nabla \cdot \mathbf{U}_a \approx -\left(\frac{\partial\zeta}{\partial t} + \frac{\partial h}{\partial t}\right) \equiv -W(0) + W(h) \qquad (a)$$

$$(D - \bar{h})\nabla \cdot \mathbf{U}_b \approx \frac{\partial h}{\partial t} \equiv -W(h) \qquad (b)$$

$$(7.20)$$

where $W(0)$, $W(h)$ are the vertical displacement rates of the sea surface and of the internal interface.

Elimination of the horizontal velocities from these equations by the same procedure used for the derivation of (7.12) yields the following equations for $W(0)$ and $W(h)$

$$\left(\frac{\partial^2}{\partial t^2} + f^2 + \bar{h}g\nabla^2\right)W(0) - \left(\frac{\partial^2}{\partial t^2} + f^2\right)W(h) = f\left(\nabla \times \frac{\boldsymbol{\tau}}{\rho}\right) + \frac{\partial}{\partial t}\left(\nabla \cdot \frac{\boldsymbol{\tau}}{\rho}\right) \qquad \text{(a)}$$

$$-(D - \bar{h})g\nabla^2 W(0) + \left[\frac{\partial^2}{\partial t^2} + f^2 - (D - \bar{h})g'\nabla^2\right]W(h) = 0 \qquad \text{(b)}$$

$$\text{(7.21)}$$

One can represent the velocity field also in terms of *barotropic* and *baroclinic* modes rather than by $\mathbf{U}_a$ and $\mathbf{U}_b$. *Barotropy* is defined as a state in which isopycnic surfaces of constant density coincide with isobaric surfaces. To obtain approximate equations for the barotropic velocity $\mathbf{U}_0$ one neglects internal density differences ($\rho_a \approx \rho_b \approx \rho$). One then multiplies the equations for the upper layer by $\bar{h}/D$, those for the lower layer by $1 - \bar{h}/D$, and adds the products. For the sake of future reference we will express the resultant equations in component form

$$\frac{\partial U_0}{\partial t} - fV_0 + g\frac{\partial \zeta_0}{\partial x} = \frac{\tau_x}{\rho D}$$

$$\frac{\partial V_0}{\partial t} + fU_0 + g\frac{\partial \zeta_0}{\partial y} = \frac{\tau_y}{\rho D} \qquad \text{(7.22)}$$

$$\frac{\partial U_0}{\partial x} + \frac{\partial V_0}{\partial y} = \frac{-W_0(0)}{D}$$

where $W_0(0) = \partial\zeta_0/\partial t \approx \partial\zeta/\partial t$ represents the contribution of the barotropic mode to surface elevation changes. After elimination of $U_0$ and $V_0$ from these equations one gets

$$\left(\frac{\partial^2}{\partial t^2} + f^2 + Dg\nabla^2\right)W_0(0) = f\left(\nabla \times \frac{\boldsymbol{\tau}}{\rho}\right) + \frac{\partial}{\partial t}\left(\frac{\nabla \cdot \boldsymbol{\tau}}{\rho}\right). \qquad \text{(7.23)}$$

The velocities of the baroclinic mode can be obtained from the difference between the actual velocities in the two layers and the barotropic velocities, or alternatively from the velocity difference $\Delta\mathbf{U} = \mathbf{U}_a - \mathbf{U}_b$. We denote the baroclinic contribution to the vertical velocity by

$W_1$. Relevant component equations have the form

$$\frac{\partial(\Delta U)}{\partial t} - f(\Delta V) + g'\frac{\partial h_1}{\partial x} = \frac{\tau_x}{\rho D}$$

$$\frac{\partial(\Delta V)}{\partial t} + f(\Delta U) + g'\frac{\partial h_1}{\partial y} = \frac{\tau_y}{\rho D} \qquad (7.24)$$

$$\bar{h}\left[\frac{\partial(\Delta U)}{\partial x} + \frac{\partial(\Delta V)}{\partial y}\right] = -W_1(0) + \frac{W_1(h)}{1 - h/D} \approx -W_1(0) + W_1(h)$$

The approximate equality involves the assumption $\bar{h} \ll D$. The baroclinic part of the interface displacement is usually much larger than the barotropic part $[W_0(h) \ll W_1(h)]$. The opposite holds at the sea surface $[W_0(0) \gg W_1(0)]$. If $W_1(0)$ is negelected, we may derive an equation for $W_1(h) = \partial h_1/\partial t \approx \partial h/\partial t$ by elimination of $\Delta U$ and $\Delta V$ from the set (7.24). This yields

$$\left(\frac{\partial^2}{\partial t^2} + f^2 + \bar{h}g'\nabla^2\right)W_1(h) = f\left(\nabla \times \frac{\boldsymbol{\tau}}{\rho}\right) + \frac{\partial}{\partial t}\left(\frac{\nabla \cdot \boldsymbol{\tau}}{\rho}\right). \qquad (7.25)$$

### 7.2.2  Gravity waves at an internal density discontinuity

Mathematical studies of internal gravity waves by Lord Rayleigh and other nineteenth-century scientists preceded their discovery in the ocean and in the atmosphere. Nansen later noticed that boat crews in the Arctic and fishermen in Norwegian fjords had to row exceptionally hard when the salty sea was covered by a thin layer of fresh meltwater. The existence of such a *dead-water* phenomenon had been known since antiquity. In his *Natural History*, Pliny the Elder (A.D. 23–79) suggested that the effect was caused by some mollusk or fish that attached itself to the keel of Roman ships. After Nansen's return from his arctic drift cruise (see Section 6.1), it was suggested by V. Bjerknes that the rowers had so much difficulty because they had to put additional energy into the generation of invisible internal waves at the saltwater–freshwater interface. Systematic analytical and experimental studies of the phenomenon were published subsequently by the young Ekman (1904).

We introduce a new variable,

$$c_j^2 = \frac{\omega_j^2 - f^2}{k^2} \qquad (j = 0, 1). \qquad (7.26)$$

If $\omega_j^2 \gg f^2$, then the value of $c_j$ approaches that of a phase velocity $\omega_j/k$. We seek a wave-like solution of (7.21) and represent the periodic vertical velocities again by

$$w_j = \sum_j W_j(z)e^{i(k \cdot \mathbf{x} - \omega_j t)}.$$

Substitution of these expressions and of (7.26) into (7.21) yields

$$[-(c_j^2 - \bar{h}g)W_j(0) + c_j^2 W_j(-h)]k^2 e^{i(k \cdot \mathbf{x} - \omega_j t)} = \frac{f(\nabla \times \boldsymbol{\tau})}{\rho} + \frac{\partial}{\partial t}\left(\frac{\nabla \cdot \boldsymbol{\tau}}{\rho}\right) \qquad \text{(a)}$$

$$\{(D - \bar{h})gWj(0) - [c_j^2 - (D - \bar{h})g']Wj(h)\}k^2 e^{i(k \cdot \mathbf{x} - \omega_j t)} = 0 \qquad \text{(b)}$$

$$(7.27)$$

For free waves, the right-hand side of (7.27a) is zero. In that case, solutions for $W_j(0)$ and $W_j(h)$ exist only if the determinant of their coefficients is zero. Neglecting terms that contain the factor $g'/g$ one gets

$$\begin{vmatrix} c_j^2 - g\bar{h} & -c_j^2 \\ -g(D - \bar{h}) & [c_j^2 - g'(D - \bar{h})] \end{vmatrix} \approx c_j^4 - gDc_j^2 + gg'\bar{h}(D - \bar{h}) = 0. \quad (7.28)$$

The last equation has two roots, which correspond to the two different wave modes. The larger root

$$c_0^2 = gD\left(1 - \frac{g'\bar{h}(D - \bar{h})}{gD^2} + \cdots\right) \approx gD. \qquad (7.29)$$

If $\bar{h} \ll D$, the smaller root can be represented with good approximation by

$$c_1^2 = g'\bar{h}\left(1 - \frac{\bar{h}}{D}\right)\left(1 + \frac{g'\bar{h}(D - \bar{h})}{gD^2} + \cdots\right) \approx g'\bar{h} \equiv gD_1. \qquad (7.30)$$

The approximate equalities could have been obtained directly from (7.23) and (7.25). It follows that $c_0$ and $c_1$ can be identified with phase velocities that correspond to the barotropic and the baroclinic modes, provided $\bar{h} \ll D$.

The barotropic velocity $c_0$ is equal to the speed of long surface gravity waves (see Section 4.23) on an ocean of depth $D$; while $c_1$ can be interpreted as being equal to the speed of long waves that propagate on a sheet of water with 'equivalent depth' $D_1 = g'\bar{h}/g$. With $D \approx 4000$ m, $\bar{h} \approx 200$ m, and $g'/g \leq 2 \times 10^{-3}$ one gets $c_0 \approx 200$ m s$^{-1}$, $c_1 \leq 2.0$ m s$^{-1}$ and $D_1 \leq 0.4$ m.

Following (7.26) the frequencies associated with the two modes are

$$\omega_j^2 = f^2 + gD_j k^2 = gD_j(a_j^2 + k^2). \qquad (7.31)$$

For $j = 0$, this equation is identical with (7.17). In analogy with (7.15), a *baroclinic deformation radius* $a_1^{-1}$ is defined by

$$a_1^2 \equiv \frac{f^2}{gD_1} = \left(\frac{f}{c_1}\right)^2. \qquad (7.15')$$

With the numbers given above $a_1^{-1} \approx 30$ km (i.e., more than 100 times shorter than the barotropic radius). Free perturbations characterized by $k^2 \gg a_j^2$ (i.e., perturbations with a horizontal extent smaller than the Rossby radius), tend to behave like plane progressive gravity waves, which propagate with a speed $\omega_j/k \approx c_j$. On the other hand, perturbations that are long compared to the Rossby radius ($k_2 \ll a_j^2$) will be practically stationary and will have near-inertial frequencies ($\omega^2 \approx f^2$).

Numerical values of the amplitudes $W_0$ and $W_1$ depend on initial conditions. The right-hand side of (7.27a) is not equal to zero, if the upper layer is driven externally. In this case the motion can be represented by a sum of the general solution for the free system, plus a particular solution of the forced system. The amplitude of the forced perturbation depends on the degree of matching or resonance between the free values of $\omega_j$ and $k^2$ on the one hand, and the time and space scales of the external forcing on the other hand. A general qualitative discussion of atmospheric forcing of internal waves can be found in Section 7.3.3.

### 7.2.3   The rigid-lid approximation

Allowing for the hydrostatic approximation, the horizontal pressure gradients and velocities do not vary with depth within each layer. It follows that $w$ is a linear function of $z$ within each layer. Introduction of (7.29) and (7.30) into the first equation (7.27) indicates that in the absence of external forcing

$$\frac{W_0(0)}{W_0(h)} \approx \frac{D}{D - \bar{h}} \qquad \text{(barotropic)}$$

$$\frac{W_1(0)}{W_1(h)} \approx \frac{-g'}{g - g'} \approx \frac{-g'}{g} \qquad \text{(baroclinic)}$$

(7.32)

The expressions (7.32) are not limited to wavelike perturbations—they also apply to all types of barotropic and baroclinic motions. Barotropic vertical displacements vary linearly with depth. For $\bar{h} \ll D$, therefore, one gets $W_0(0) \approx W_0(h)$. In the baroclinic mode, surface displacements are always negligibly small compared to internal interface displacements. This permits representation of the continuity equation (7.20a) for the upper layer by the *rigid-lid* approximation

$$\bar{h}\nabla \cdot \mathbf{U}_a \approx \bar{h}\nabla \cdot \Delta\mathbf{U} \approx W_1(h). \qquad (7.33)$$

When this approximation is applicable, it follows from (7.20a,b) that

$$\rho_a \bar{h}\mathbf{U}_a + \rho_b(D - \bar{h})\mathbf{U}_b \approx 0, \qquad (7.34)$$

which indicates that baroclinic perturbations involve nearly zero volume transports. The horizontal velocities have opposite signs in the two layers. If $D \gg h$, the lower layer will be practically at rest. The horizontal velocity that is induced by internal waves in the upper layer tends to be correspondingly larger. Therefore, internal waves can affect high-frequency surface waves, like alternating advective currents (see Section 4.4). The resulting bands of rougher and smoother water can be seen in satellite pictures.

For simplicity's sake we shall from now on use the symbol $W_0$ for $W_0(0)$ with the understanding that the actual barotropic velocity varies linearly from its surface value to zero at the ocean bottom. Likewise, we shall denote the baroclinic vertical velocity at the internal interface simply by $W_1$ instead of $W_1(h)$. With these conventions and with allowance for (7.29) and (7.30), one can express (7.23) and (7.25) by a single expression of the form

$$\left(\frac{\partial^2}{\partial t^2} + f^2 + c_j^2\nabla^2\right)W_j = f\left(\nabla \times \frac{\boldsymbol{\tau}}{\rho}\right) + \frac{\partial}{\partial t}\left(\frac{\nabla \cdot \boldsymbol{\tau}}{\rho}\right). \qquad (7.35)$$

### 7.2.4   Ekman pumping

If local changes remain very small during the time of a half-pendulum day, transients involving $\partial/\partial t$ can be neglected in (7.35), which then assumes the form

$$(f^2 + c_j^2\nabla^2)W_j - f^2(1 + a_j^{-2}\nabla^2)W_j = f\left(\nabla \times \frac{\boldsymbol{\tau}}{\rho}\right). \tag{7.35'}$$

We assume that the horizontal distribution of the stress curl can be described by a sum of exponential or trigonometric functions with length-scales $k_\tau^{-1}$, so that $\nabla^2(\nabla \times \boldsymbol{\tau}) = \pm k_\tau^2(\nabla \times \boldsymbol{\tau})$. Solutions of (7.35') are then represented by

$$W_j = \frac{\nabla \times \boldsymbol{\tau}/\rho}{f(1 \pm k_\tau^2/a_j^2)}. \tag{7.36}$$

The values of $W_0$ amd $W_1$ depend on the horizontal scale of the atmospheric perturbation and on the ocean depth. In general, for synoptic scale atmospheric perturbations over the open ocean, one gets

$$W_1 \approx \frac{\nabla \times \boldsymbol{\tau}}{\rho f} \quad \text{for} \quad \left(\frac{k_\tau}{a_1}\right)^2 \ll 1 \tag{7.37}$$

and

$$W_0 \approx \frac{\pm(\nabla \times \boldsymbol{\tau}/\rho)(a_0/k_\tau)^2}{f} \ll W_1 \quad \text{for} \quad \left(\frac{k_\tau}{a_0}\right)^2 \gg 1. \tag{7.38}$$

The case represented by $(k_\tau/a_1)^2 > 1$ is not of practical interest, because synoptic-scale perturbations are always larger than the oceanic baroclinic radius. On the other hand, the scale of atmospheric perturbations over shallow coastal regions can exceed the oceanic barotropic radius. The motion tends to become purely barotropic if $(k_\tau/a_0)2 \ll 1$.

The so-called Ekman pumping equation (7.37) relates the vertical velocity at $z = -h$ to the curl of the wind stress. When this curl is cyclonic, the pumping velocity is positive and hence upward. Ekman pumping is due to the divergence of the Ekman transport. It follows from (6.10) that this divergence has the same magnitude, but opposite signs, in the atmosphere and the ocean. The flow is always upwards in both fluids when the stress curl is positive or cyclonic. It is downward when the curl is anticyclonic.

Ekman pumping can be affected not only by horizontal wind variations, but also by surface current gradients. As indicated by (5.53), the surface stress is a function of the velocity difference between the wind and the current. The contribution of currents to the stress magnitude is usually neglected, because their speed is negligible compared to the windspeed. However, the horizontal scale of ocean currents tends to be much smaller than the scale of the wind field. Their horizontal gradients can contribute significantly to the stress curl.

To demonstrate this effect, which has been investigated in some detail by Rooth and Xie (1992), we stipulate an $x$-axis in the direction of a steady surface current $U_w$ so that $\partial U_w/\partial x = 0$. For the special case of a horizontally-uniform wind with $x$-component $U_a$ we obtain from (5.53)

$$\frac{\nabla \times \boldsymbol{\tau}}{\rho} = 2C_D \frac{\partial (U_a - U_w)^2}{\partial y} \approx 2C_D \frac{\partial U_w}{\partial y} |U_A|. \tag{7.39}$$

This expression is always positive, regardless of wind direction, near the right bank of the current where $\partial U_w/\partial y > 0$. It is negative near the left bank. In the northern hemisphere, this implies a cyclonic stress curl with positive upward Ekman pumping on the right bank and downward pumping on the left bank. The opposite holds in the southern hemipshere. In either case, the stress curls at the current edge tend to decelerate the current. This is illustrated schematically by Fig. 7.1.

When a wind blows over an ocean eddy, the surface eddy current will be decelerated by this process. However, the cyclonic curl on the inner (right-hand) side of an anticyclonic eddy current will also generate some upwelling there. In a warm anticyclonic eddy or Gulf Stream ring, this tends to bring cold water from below toward the surface. It therefore contributes to the spin-down of the eddy. The opposite happens near the surface of a cold-core cyclonic eddy. Similar processes affect the stress curl at the edge of narrow boundary currents.

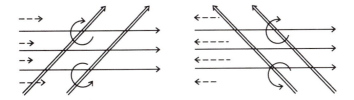

**Fig. 7.1.** Schematic diagram of surface current effects on the wind stress: The current is represented by full arrows and the wind by double arrows. The wind-stress component parallel to the current is indicated by the short stippled arrows. In the northern hemisphere, the stress curl, as depicted by the curved arrows, is cyclonic along the right bank of the current and anticyclonic along the left bank, regardless of the wind direction.

Ekman pumping is the dominant process by which the atmosphere affects the interior ocean on synoptic scales. Because of its importance, we repeat the limitations involved in the derivation of (7.36). These include neglect of the $\beta$ effect, the assumption of steady motion or very slow transients $(\partial^2/\partial t^2 \ll f^2)$, of $\bar{h} \ll D$ and of $(a_0/k_\tau)^2 \ll g'/g$. The last inequality is most likely to be satisfied in tropical and subtropical deep ocean regions where $f^2$ and $a_0^2$ are relatively small. In our two-layer system, the maximum vertical velocity occurs at the internal interface. In the real ocean, with continuous stratification, the depth over which surface friction is distributed is not as easily specified. Ekman pumping can draw up water in this case in a slantwise direction along sloping isopycnal surfaces. In spite of its limitations, the Ekman pumping equation (7.36) has been widely used to obtain an indication of interior ocean forcing by the wind.

## 7.3  Internal inertio-gravity waves

The present section provides a brief outline of a topic that is the subject of a very extensive literature. A thorough treatment can be found in monographs by Eckart (1960), Krauss (1966), or Lighthill (1978). More recently, a concise review has been published by Munk (1981).

### 7.3.1  *Internal waves in a continuously stratified ocean*

Throughout the previous section it was assumed that the wind stress acts like a vertically uniform body force upon the upper layer of a two-layer system, characterized by uniform density $(N = 0)$ within each layer. In the real ocean, density and current velocities tend to vary continuously with depth. Perturbations can then propagate vertically. It follows from (4.5) that waves in the frequency range $N^2 > \omega^2 > f^2$ have real vertical wavenumbers. Waves within that frequency range propagate in a direction that makes an angle $\alpha$ with the horizontal. This permits us to express (4.5) and the dispersion relation (4.6) in the form

$$\frac{N^2 - \omega^2}{\omega^2 - f^2} = \left(\frac{n}{k}\right)^2 = \tan^2 \alpha \qquad (7.40)$$

and

$$\omega^2 = \frac{N^2 k^2 + f^2 n^2}{k^2 + n^2} = N^2 \cos^2 \alpha + f^2 \sin^2 \alpha. \qquad (7.41)$$

With $(k^2 + n^2) \equiv k_*^2$ and with an $x$-axis in the direction of the horizontal wavenumber component, the horizontal and vertical propagation of the wave phase is specified by

$$c(x) = \frac{\omega}{k} = \frac{\omega}{k_* \cos \alpha} \quad \text{and} \quad c(z) = \frac{\omega}{n} = \frac{\omega}{k_* \sin \alpha}. \tag{7.42}$$

The absolute phase velocity $\mathbf{c} = \omega/\mathbf{k}$ has components

$$c_x = \frac{\omega}{k_*} \cos \alpha \quad \text{and} \quad c_z = \frac{\omega}{k_*} \sin \alpha. \tag{7.43}$$

The corresponding group velocity components are

$$c_{gx} = \frac{\partial \omega}{\partial k} = \frac{N^2 - f^2}{\omega k_*} \cos \alpha \sin^2 \alpha$$

$$c_{gz} = \frac{\partial \omega}{\partial n} = -\frac{N^2 - f^2}{\omega k_*} \cos^2 \alpha \sin \alpha \tag{7.44}$$

It follows from (7.44) and (7.43) that $\mathbf{c} \cdot \mathbf{c}_g = 0$ (i.e., the group and phase velocity vectors are normal to each other).

It can be readily confirmed that the following expressions for the perturbation velocity components and the perturbation pressure satisfy (7.41) and the basic equations (4.1–4.3)

$$u = -\sin\alpha \, A\omega \cos(nz) \sin(kx - \omega t) \tag{7.45}$$

$$v = \sin\alpha \, Af \cos(nz) \cos(kx - \omega t) \tag{7.46}$$

$$w = \cos\alpha \, A\omega \sin(nz) \cos(kx - \omega t) \tag{7.47}$$

$$\frac{p'}{\rho} = \cos \alpha \, \frac{N^2 - \omega^2}{n} A \cos(nz) \sin(kx - \omega t). \tag{7.48}$$

These equations describe a banded velocity structure, which is uniform in the $y$-direction. The ridges and troughs of the perturbation pressure are aligned parallel to the group velocity and normal to the phase velocity. Individual fluid particles move anticyclonically along inclined ellipses with axes proportional to $\omega$ and $f$.

The elliptical orbits become very narrow when $\omega \gg f$. In the limit, they approach the rectilinear orbits of ordinary gravity waves in nonrotating fluids. On the other hand, when $\omega \approx f$ the orbits become nearly circular. Perturbations acquire then the character of *inertial oscillations* as discussed in Section 6.1.2. The existence of circular horizontal orbits when $\omega = f$ can also be derived directly from the first two equations (4.1), which are satisfied in this case by

$$u = A^*f \sin (ft) + U_g \qquad v = A^*f \cos (ft) + V_g \qquad (7.49)$$

with $A^*$ dependent on initial conditions and external forcing and with the geostrophic velocity components $U_g$, $V_g$ balanced by the pressure gradient.

Internal waves cannot propagate into layers where $N^2 \leq \omega^2$. It follows from (7.40) that $\tan^2 \alpha$ decreases when $N^2 - \omega^2$ becomes smaller (i.e., when waves enter less stably stratified layers). The phase speed vector is then bent toward the horizontal and the group velocity becomes more vertical. Total reflection occurs when $N^2 - \omega^2 = 0$ and hence $\tan \alpha = \sin \alpha = 0$. The effect of a monotonically varying $N(z)$ is illustrated in Fig. 7.2 (a). The little patch of parallel lines in that figure represents crests in a wave packet. As indicated earlier, the group velocity and hence the path of the packet is parallel to these crests. This path is shown as a solid line. When the group reaches its *turning level* where $N^2 = \omega^2$, it is turned back into the more stably stratified region. The phase speed becomes horizontal at the turning level.

The value of $N^2$ is much larger in the thermocline than it is in the layers above and below. The turning process, therefore, can inhibit the vertical propagation of some waves into the less stably stratified upper and lower ocean. The thermocline acts in this way as a horizontal waveguide for waves with relevant frequencies. Horizontally propagating, baroclinic waves on a density discontinuity can now be seen to represent a limiting case for waves in a thermocline waveguide, characterized locally by very high values of $N$.

Internal waves are also constrained by a *turning latitude* where $\omega^2 = f^2$. When waves propagate poleward, $f^2$ becomes larger while $\omega^2$ remains constant. Equation (7.40) indicates that $\tan \alpha$ increases with $f$ when $N$ is horizontally uniform. At the turning latitude $\cos \alpha = 0$ and $\sin \alpha = 1$. The wavenumber vector becomes vertical ($k^2 = 0$) and (7.44) shows that energy cannot be radiated beyond this latitude. The turning latitude can trap

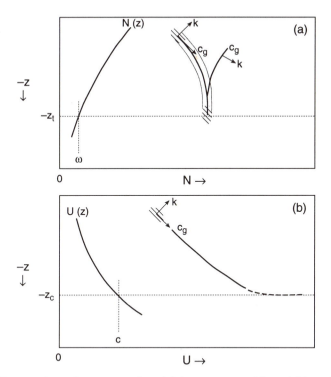

**Fig. 7.2.** Propagation of a wave packet: (a) in an ocean with variable stability and without shear. The packet is turned back at a level $z_t$ where $\omega = N(z_t)$; (b) in an ocean with constant stability and current shear. At the critical depth, $z_c$, the wave propagation equals the current velocity and wave energy is absorbed. After Munk (1981).

meridionally propagating inertio-gravity waves in a zonal *equatorial wave-guide*, which corresponds to the horizontal thermocline waveguide for vertically propagating waves.

A vertically changing advective current $U(z)$ can constrain wave propagation in a different way. It is obvious that waves cannot transport energy against a current that exceeds the group velocity $\mathbf{c}_g$. Something different happens when waves travel downstream [see Fig. 7.2 (b)]. Their frequency is Doppler-shifted to

$$\omega' = [c(x) - U]k = \omega - Uk, \qquad (7.50)$$

where $c(x)$ is the phase velocity in the direction of the current. As the wave propagates vertically into a region of faster current speed, $\omega'$ decreases and finally becomes zero at a 'critical level' where $c(x) - U = 0$. Wave energy and momentum are converted into current energy and

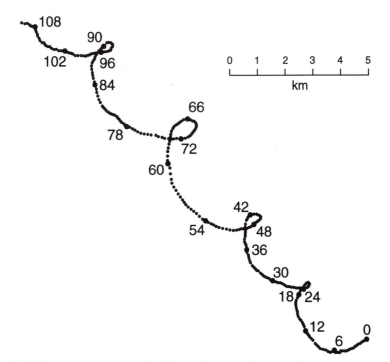

**Fig. 7.3.** Current meter record, July 1964, at 29°11′ N, 68°21′ W, depth 617 m. Plotted in 20-min increments; time shown every 6 h; inertial period 23.6 h. Courtesy F. Webster.

momentum at that level. The process is the inverse of the Miles mechanism (see Section 4.4.3), which involved a transformation from mean current to wave momentum above the air–sea interface, at a critical matching level where $c - U = 0$. It is illustrated by Fig. 7.3.

### 7.3.2 *Long waves*; *normal modes*

It follows from the last equation (4.2) that (4.3) can also be expressed as

$$\frac{\partial^2 w}{\partial z^2} + n^2 w = 0. \tag{7.51}$$

With the rigid-lid approximation, this equation has to satisfy the sea surface and sea bottom boundary conditions

$$w_{z=0} \approx 0 \quad \text{and} \quad w_{z=-D} = 0.$$

These conditions can be met only if the vertical wavenumber $n$ is restricted to a set of discreet values $n_j$. If $N$ is constant, then these must have the form

$$n_j = \pm j\frac{\pi}{D} \qquad j = 1, 2, 3, \ldots \tag{7.52}$$

Use of these values instead of $n$ in (7.45) shows that the horizontal velocity components change sign $j$ times along the vertical. This can be identified with $j$ orthogonal, baroclinic modes. Without loss of generality, we can also set

$$n_j^2 \equiv \frac{N^2}{c_j^2} \equiv \frac{N^2}{gD_j} \equiv \left(\frac{N}{f}\right)^2 a_j^2, \tag{7.53}$$

where $c_j$ is a baroclinic velocity fully specified by (7.52) and (7.53), $D_j$ is a corresponding equivalent depth and $1/a_j$ is the associated baroclinic Rossby radius.

In this subsection, we consider perturbations with horizontal length-scales larger than the ocean depth $(kD < 1)$. It follows from (7.52) that

$$\frac{k^2}{n_j^2} = \cos^2 \alpha_j \ll 1. \tag{7.54}$$

This means that the vertical phase propagation is very much slower than the horizontal phase speed. It also permits us to approximate the dispersion relation (7.41) by

$$\omega_j^2 \approx \left(\frac{Nk}{n_j}\right)^2 + f^2 = (kc_j)^2 + f^2 = c_j^2(k^2 + a_j^2). \tag{7.55}$$

If the horizontal extent of the disturbance is much larger than the baroclinic Rossby radius, then the frequency $\omega$ exceeds the Coriolis frequency $f$ by only a very small amount. One can then write

$$\frac{\omega^2}{f^2} = 1 + \epsilon^2 \quad \text{with} \quad \epsilon^2 \ll 1 \quad \text{if} \quad k^2 \ll a_j^2. \tag{7.56}$$

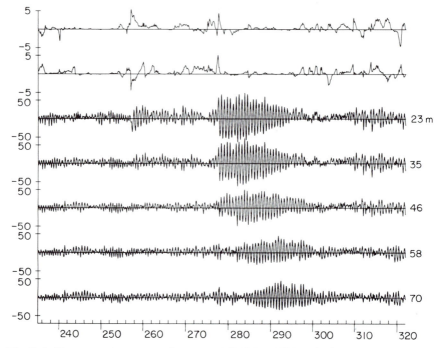

**Fig. 7.4.** Upper ocean current changes at 47.5°N, 135°W. Abscissa is the year–day of 1987. Plots from top to bottom are eastward wind stress (dynes centimeters squared), northward wind stress, and eastward components of current velocities (centimeters per second) measured at depths indicated along the right margin. Mixed-layer depth ranged from 30 to 50 m. Courtesy of R. E. Davis and A. Toruella.

Perturbations with near-inertial frequencies ($\omega \approx f$) are probably the most common and best-documented oceanic response to mesoscale atmospheric forcing. Their presence in Ekman layers has been discussed in Section 6.1 [see (6.12)]. Relevant observations in a somewhat deeper stratum are reproduced in Fig. 7.4. Oscillations with an exact inertial frequency $f = \omega$ are possibly only if $k = 0$, but (7.56) indicates that $\omega \approx f$, whenever $k^2/a_1^2 \ll 1$. This condition tends to be realized within a broad spectral band. For example, the graph for $(\omega_1)_g$ in Fig. 7.10 shows that $\omega \approx f$ for all waves longer than about 400 km. The baroclinic deformation radius $1/a_j$ decreases with the mode number $j$. Higher modes ($j > 1$), therefore, have near-inertial frequencies within a still broader wavenumber range. The prevalence of inertial or near-inertial frequencies in oceanic records is in keeping with these arguments.

The preceding equations were based on the assumption of $N$ being constant. In general $N = N(z)$, the vertical wavenumber $n_j$ as given by

(7.53) is then also a function of depth. To deal with this case one can express (7.51) in the form

$$\frac{\partial^2 w}{\partial z^2} + \left(\frac{N(z)}{c_j}\right)^2 w = 0. \tag{7.57}$$

This equation is satisfied by an infinite sequence of eigenvalues of $c_j$, which can be computed from observed or stipulated vertical profiles of $N^2$. Explicit descriptions of the relevant algorithm can be found in the monographs quoted at the beginning of this section.

### 7.3.3  Atmospheric forcing of internal gravity waves in the ocean

Momentum can be transferred downward from the sea surface by pressure forces, turbulence and internal wave radiation. The radiation mechanism tends to be inhibited by the neutral or near-neutral stratification of the surface layers, which suppresses vertically propagating internal waves, as indicated in Section 7.3.1. Large-scale surface pressure gradients are transmitted with the velocity of sound through the whole water column. We can expect them to drive oceanic motion that is predominantly barotropic, if one neglects the relatively small change of density with depth. This is obviously not true for pressure changes of relatively small horizontal extent. In particular, pressure perturbations caused by surface waves on deep water are known to decrease downward at a rate proportional to $\exp(kz)$. They may perturb the bottom boundary of a surface mixed layer, but that perturbation can have a significant amplitude only if the layer-depth $h$ is smaller than the wavelength.

Sufficiently long surface waves can deform the lower mixed layer boundary. To resonate with internal waves, the surface wave frequency would have to be equal to or lower than the value of $N$ in the thermocline. Relatively few wind-generated, linear surface waves are likely to satisfy this requirement. Groups of surface waves propagate more slowly and have a longer composite wavelength. They therefore have a better chance to resonate with the sluggish internal waves. Watson et al. (1976) used the computed exponential growth rate of an internal wave spectrum, to derive the energy transfer to that spectrum, from a spectrum of wind-generated surface waves. Assuming $N(z)$ to have a maximum immediately below $h = 100$ m, they computed growth rates as low as two hours for the lowest internal wave mode. The resulting power transfer from surface to internal waves was estimated to be of the order $3 \times 10^{-3}$ W m$^{-2}$. Garrett and Munk (1972) estimated that internal wave energy is dissipated at an average rate of about $7 \times 10^{-3}$ W m$^{-2}$. Together, these numbers might suggest that nearly half the internal wave energy is derived from surface waves. The mean global wind energy input into surface waves is almost certainly less than $20 \times 10^{-3}$ W m$^{-2}$. It would follow that at least 15 per cent of this wind input is used ultimately to generate internal waves. This may be a rather high estimate.

Internal waves could also be generated by travelling air pressure fluctuations through a resonance process that is identical with the one discussed for surface waves in Section 4.4. If an atmospheric disturbance with horizontal scale $1/k_a$ is advected with a velocity $U_a$, it could give rise to internal waves with a frequency $\omega = k_a U_a$. Advected air pressure fluctuations with significant amplitudes usually have horizontal dimensions that are large compared to even the lowest-mode baroclinic deformation radius in the ocean ($k_a \leq a_1$). It follows from (7.56) that perturbations which are forced by this mechanism in the ocean, are likely to have near-inertial frequencies.

A turbulent downward flux of horizontal momentum $\tau(z) = -\rho\overline{uw}$ is probably the most important link in the mechanical forcing of oceanic motion by the atmosphere. Unfortunately, we have no valid universal relationships for the distribution of this flux in the upper ocean. Observational data concerning turbulence in the surface layer are sparse. Theories about changes with depth of the relevant length-scale below a waving interface, or about the interactions between waves, Langmuir circulations and small-scale turbulence are still controversial. We seem to know more about the situation below about 100 m. Turbulence kinetic energy there has been observed at many places to be dissipated at an average rate of $0.1 - 1.0 \times 10^{-9}$ W kg$^{-1}$ (Gargett et al., 1981; Crawford, 1982; Lueck et al., 1983; Oakey, 1985). The corresponding eddy viscosity was observed by Moum and Osborne (1986) to be inversely proportional to $N$. It is obvious that this relation cannot be extrapolated to the surface mixed layer where $N \approx 0$.

To assess the interior forcing of the ocean, one could represent $\partial\tau/\partial z$ by a series of orthogonal functions that correspond to the different baroclinic modes. Wunsch and Gill (1976) have made an effort to do that. They expressed the convergence of the momentum flux in the form

$$\frac{\partial(\overline{uw})}{\partial z} = \Sigma_j \frac{\tau}{\rho} D_j^F. \tag{7.58}$$

The *equivalent forcing depth* $D_j^F$ determines the forcing of the corresponding mode $j$. For the lowest mode in the equatorial Pacific they calculated $D_1^F$ to be about 280 m. This is very different from the equivalent depth $D_1$, which was about 0.75 m. The mixed-layer depth $h$ in that area was almost certainly less than 100 m. This would suggest that only about one third of the stress was used to accelerate the mixed layer and that the remainder was used to drive the layers below. This conclusion is not in keeping with a common assumption that most or all stress is absorbed within the mixed layer.

In the absence of detailed observational data, the last-mentioned assumption has been used by many investigators. The particular assumption of a linear decrease of the stress ($\partial\tau/\partial z = \tau_0/h$), which was made in Section 7.2, identifies the forcing depth $D_1^F$ with the mixed-layer depth $h$ and with the depth of the Ekman layer, which are coincident in this case. The maximum pumping velocity $w_h$ occurs at the bottom of the layer and is independent of the layer-depth. The motion below is driven only by pressure forces. This approach, which implies that the surface wind stress acts instantaneously like a horizontal body force upon the mixed layer, seems to have been first used by Charney (1955). It forms the basis of most or all modelling studies of the oceanic response to moving storms.

The treatment of surface wind stress as a body force involves the assumption that the mixed layer response can be simulated by the behaviour of a rigid slab. This may not be unduly unrealistic. Davis and Torruella (personal communication) have obtained records of current changes that followed pronounced changes of the surface wind. Some of their data are reproduced in Fig. 7.4. It can be seen that current oscillations, which followed pronounced wind changes, had the same phase and amplitude at different depths within the surface mixed layer. In the stable strata below, their onset was delayed progressively with depth, suggesting a skewed downward propagation as inertio-gravity waves. The vertical wavenumber there must have been real and was perhaps relatively large ($N^2 \gg \omega^2$). Vertical coherence would not be observed under these circumstances. Increased local shears may have contributed to the perceptible attenuation of the oscillations below the mixed layer. The resulting transformation of wave energy into turbulence kinetic energy can cause enhanced local mixing and layer deepening. Statistical evidence for such a decrease of oscillation energy density with depth has been compiled by Pollard (1970).

If currents do not change much with depth within the mixed layer, the velocity shear must be concentrated at the outer layer boundary. A rationale for such a velocity distribution was suggested in the discussion of Fig. 6.2. Some support for it can be found in atmospheric observations by Riehl et al. (1951) and in Deardorff's (1970b) calculations. High shear flows in the upper thermocline immediately below the mixed layer is usually associated with relatively low values of the Richardson number. As discussed in Section 6.4, this was related by Pollard et al. (1973) to a mechanism of mixed-layer deepening. It may also cause the generation of internal gravity waves that can radiate energy and momentum downwards from the mixed-layer boundary.

The apparently very fast transmission of surface stress changes through the mixed layer presents a problem, which does not seem to have been solved quantitatively at this stage. It probably involves Langmuir circulations and other coherent boundary layer structures as discussed in Section 6.2. These structures may also contribute to the generation of internal waves in the interior fluid. The impact and overshooting of mixed-layer thermals can produce deformations of the density field in the thermocline and also in atmospheric inversions. These are then dissipated by internal wave radiation. Alternating stresses, caused by Langmuir rolls, could also conceivably act as wave generators. We do not know how much these processes contribute to the internal wave spectrum.

## 7.4   The response of the open ocean to moving cyclonic storms

### 7.4.1   *Observations*

Hurricanes and other cyclonic storms present a somewhat special case of atmospheric forcing. They transmit both energy and momentum, as well as vorticity, to the water below. The nature of the resulting, localized oceanic perturbation is a function of the intensity, extent, and translation speed of the storm. After the storm has passed, the oceanic perturbation will gradually be smoothed out by the emission of internal waves. Dissipation and the arrival of new perturbations tend to remove all traces of it. The present section deals with the characteristic sequence of these processes in the open ocean. The effect of shelving bottoms and lateral coastal boundaries will be considered in Section 7.5.

Figure 7.5 displays temperature and density data in the top 150 m, which were collected by Leipper (1967) after the passage of Hurricane Hilda across the Gulf of Mexico. It clearly shows a ridge of cold water, caused by upward Ekman pumping, along the hurricane's path. The temperature reduction along the trajectory was largest in the region where the storm translation speed was particularly slow. Cooling by evaporation and mixing must have contributed to the temperature reduction. These non-adiabatic processes probably played a secondary role in the upwelling region below the central part of the storm, but they must have been active over the whole area of high windspeeds. Together with the anticyclonic stress curl and the resulting downward Ekman pumping in the outer regions of the hurricane, this may account for an observed slight cooling and mixed-layer deepening there. The cold water ridge shown in Fig. 7.5 was observed to persist for several weeks. Leipper demonstrated that it

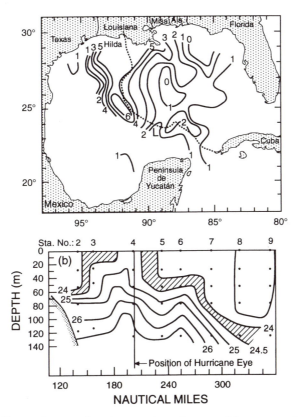

**Fig. 7.5.** (a) Observed sea surface temperature decrease (degrees centigrade) after passage of hurricane Hilda along dotted line trajectory. (b) Observed density distribution on section across path of hurricane Hilda. Courtesy D. F. Leipper.

was balanced geostrophically by currents of about $0.5\,\mathrm{m\,s^{-1}}$, which bounded the area of the coldest water. Later observations of the oceanic response to moving hurricanes, published by Price (1981), Sanford et al. (1987), Shay and Elsberry (1987), Church et al. (1989), and Brink (1989) indicated essentially similar configurations.

One can make a distinction between the oceanic perturbations that accompany or follow an intense storm during the first few days after its passage, and the long-term oceanic reaction. If $k$ denotes the inverse diameter or inverse length-scale of the storm, the short-term response involves packets of inertio-gravity waves, with frequencies $\omega \gg (c^2 k \beta)^{1/3}$. After dispersal of these waves, the $\beta$ term becomes relatively and progressively more important. The long-term response can be viewed, therefore, in terms of planetary waves, which will be discussed in somewhat more detail in Section 7.6 and which disperse only during a much longer period of time.

### 7.4.2   The simulated short-term oceanic response to moving storms

Oceanic perturbations, which can be associated with moving hurricanes, were first simulated theoretically by Geisler (1970). His analytical approach was later extended by Price (1981) and by Greatbatch (1983) with numerical, nonlinear models of hurricanes. All of these studies dealt with storms that did not change with time, except for a translation over the ocean surface with constant velocity $U_a$. This made it convenient to have the $x$-axis parallel to the storm translation and to introduce a moving coordinate system through the Galilean transformation $x' = x - U_a t$. The storm would appear to be stationary in this system, with the local time derivatives replaced by an advective change

$$\frac{\partial}{\partial t} = U_a \frac{\partial}{\partial x'}. \tag{7.59}$$

Using (7.59) one can transform (7.35) into

$$\left[ \left( \frac{U_a^2}{c_j^2} - 1 \right) \frac{\partial^2}{\partial x'^2} - \frac{\partial^2}{\partial y^2} - a_j^2 \right] w_j = \left[ \frac{\nabla \times \boldsymbol{\tau}}{\rho f} + U_a \frac{\partial}{\partial x'} \left( \frac{\nabla \cdot \boldsymbol{\tau}}{\rho} \right) \right] c_j^2. \tag{7.60}$$

An equation of this type was used by Geisler (1970) in his hurricane model. He also included forcing by air pressure changes on the right-hand side. Though this can be relevant for the barotropic mode, baroclinic motion in deep water is forced almost entirely by Ekman pumping with air pressure playing a negligible role.

   The solutions of (7.60) depend on the sign of the term $(U_a^2/c_j^2 - 1)$. If the storm propagates slower than the waves, then that sign is negative and the equation is elliptic. In that case the solutions represent perturbations that spread out from the storm center in all directions. The barotropic solution is always elliptic for hurricanes on the open ocean, because $c_0$, being of order $200 \text{ m s}^{-1}$, is always much larger than $U_a$. The baroclinic solution of (7.60) tends to be hyperbolic because $(U_a^2/c_1^2 - 1) > 0$ in most meterological situations. In this case it is not possible for the baroclinic disturbance to affect the water beyond a short distance ahead of the storm because energy cannot be radiated from the storm center by baroclinic waves faster than $c_1$. Behind the storm, the response is a mixture of an upwelling pattern on the scale of the storm diameter and a wake of inertio-gravity waves.

   In a fixed coordinate system, the frequency of waves, which propagate with the same phase speed as the moving storm, would be $U_a k_a$. On the

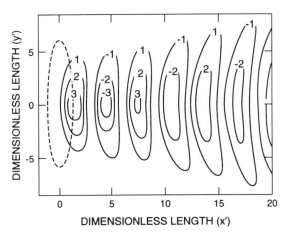

**Fig. 7.6.** Computed contours of vertical velocity (arbitrary units) at the bottom of the surface mixed layer under a hurricane, which moves toward the left at a speed $U_a = 5c_1$ ($\approx 7 \text{ m s}^{-1}$). The broken line denotes limits of forcing region. Unit of $x'$ is $5/a_1$ ($= 150 \text{ km}$). Unit of $y'$ is $1/a_1$. Courtesy J. Geisler.

other hand, this frequency must conform also with the dispersion relation (7.31). Elimination of $k_a$ between these two expressions yields

$$\omega_j^2\left(1 - \frac{c_j^2}{U_a^2}\right) = f^2. \tag{7.61}$$

Barotropic waves characterized by $c_0^2/U_a^2 \gg 1$, can be expected to have a frequency that is proportional to the storm translation speed ($\omega_0 \approx a_0 U_a$). This frequency is likely to be very low and the waves are correspondingly long. In baroclinic wake waves behind fast-moving storms $c_1^2/U_a^2 \ll 1$. These waves tend to have near-inertial frequencies ($\omega \approx f$). Figure 7.6 shows an idealized pattern of such baroclinic wake waves. The figure has been reproduced from Geisler's linear study. The nonlinear simulations by Price (*loc. cit.*) and by Greatbatch (*loc. cit.*) produced essentially similar, although less smooth and perhaps somewhat more realistic, configurations. To generate Fig. 7.6, Geisler stipulated a circular stress pattern that translates with a speed $U_a = 5c_1$ and that varies with distance $r$ from the storm center as specified by

$$\tau_0 \propto \frac{ra_1}{2} \exp\left(\frac{-ra_1}{2}\right)$$

If one selects $1/a_1 = 30 \text{ km}$, then the maximum stress curl occurs at 120 km from the center, and the curl becomes negative or anticyclonic beyond 180 km. Geisler considered the stress curl to be zero beyond this limit, which is marked by the dashed line in Fig. 7.6. The spreading angle of the wake becomes wider if the storm moves slower and covers a smaller area. It should be noted that the $y$-scale in the figure is five times longer than the $x'$-scale. The actual spreading angle of the baroclinic wake is smaller than apparent in the figure. A very small spreading angle was found also in the numerical results obtained by Price (1981), which were based on more realistic stress patterns.

A frictionless, steady-state solution of (7.60) is not possible for the eliptic case when $(U_a/c_j)^2 < 1$. The limiting case of a stationary storm $(U_a = 0)$ was investigated numerically by O'Brien and Reid (1967) as an initial value problem. They included friction at the internal interface and assumed the lower ocean to be infinitely deep, causing it to be at rest at all times. This assumption filters out the barotropic solution automatically, because it makes $a_0^2 \propto c_0^{-2} \approx 0$. Starting from rest, the O'Brien and Reid model produced an initial thickening of the upper layer. This was due to the inward directed, isallobaric component of the surface wind and the resulting convergence of the wind stress vector. After a few hours of simulated development this was replaced by Ekman transport divergence. This caused a slight lowering of the sea surface in the central part of the forced region ($w_0 < 0$) and pronounced upwelling below ($w_h > 0$). Similar results were obtained by Geisler in his elliptic solution for slow-moving storms.

The models in this subsection treat the Rossby radius $a_1$ and the baroclinic phase velocity $c_1$ as constants, although storms can cause dramatic changes in layer-depth and temperature. These can involve changes in $c_1^2 = g'h$ by a factor of 10 or more during the passage of a real hurricane. The neglected secondary effects are almost certainly important. The models, therefore, may be more of conceptual than of predictive interest.

### 7.4.3  *The long-term oceanic response to moving storms*

Persistent deformations of the initial density field are caused mainly by the stress component $\tau_x$, which is parallel to the storm track. The resulting Ekman flow away from the storm center lowers the sea surface along the path of the center and pumps cold water towards the surface from below. After a relatively short time, the inward-directed pressure force in the surface layer is balanced geostrophically by tangential currents. This balance involves two parts. The inward slope of the sea surface toward the trough along the hurricane path drives a barotropic current. According to Bjerknes' circulation theorem, the opposite slope of the internal isopycnic surfaces (see Fig. 7.5) is balanced by the pumping action of an upward

increasing Coriolis force. In a two-layer system, this implies a slope $-\nabla h$ of the internal interface, which is proportional to the baroclinic velocity difference and inversely proportional to the density difference between the two layers

$$-\nabla h = \frac{f}{g'} \mathbf{n} \times (\mathbf{U}_2 - \mathbf{U}_1). \qquad (7.62)$$

This pumping action prevents the warm surface water from flowing back over the cold water ridge that has risen from below.

The width of this ridge is approximately equal to the storm diameter. Its height and the strength of the associated geostrophic currents depend on the amount of vorticity that was imparted locally. The latter depends in turn upon the storm intensity and translation speed. If the storm moves slowly, Ekman pumping can draw water upward during a relatively long period of time. On the other hand, fast moving storms provide more energy to internal waves. The effect depends on the relative magnitude of the terms $(U_a/c_1)^2)\partial^2/\partial x'^2$ and $a_1^2$ in the baroclinic version of (7.60). It is illustrated in Fig. 7.7. For $(U_a/c_1)^2 = 25$, the response is almost entirely in the wave domain; for $(U_a/c_1)^2 = 2$, it involves a prominent mean rise of the internal interface.

The geostrophic currents, which balance the cold water ridge, are weakened gradually by the emission of planetary waves, as will be discussed in Section 7.6.2. The barotropic part of these currents is attenuated rather rapidly through emission of barotropic waves. This can allow significant filling of the surface trough within $4\pi a_0/\beta \approx 2\text{--}3$ days. As that happens, the baroclinic mode asserts itself in the lower ocean layer,

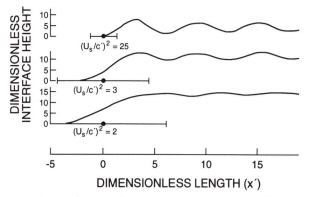

**Fig. 7.7.** Computed elevation of mixed-layer depth above initial value ($h_0 = 100$ m) along track of storm centre. Elevation in metres is approximately $0.4\,\tau_m$ times indicated dimensionless height ($\tau_m$ = max. stress in dynes per square cemtimeter). Horizontal bar indicates distance covered by the storm. Courtesy J. Geisler.

which then aquires an integral momentum that is opposite to that above the internal interface. The process does not affect the velocity difference between the two layers or the slope of the internal interface. The time-scale for the reduction of that slope and the dissolution of the baroclinic currents cannot be less than the period of the highest-frequency baroclinic Rossby waves, which is $4\pi a_1/\beta$ as indicated by (7.74) below. This can be many months (see, e.g., Fig. 7.10). The internal cold water ridge will then persist during a corresponding time interval.

The preceding discussion dealt mainly with an idealized, two-layer ocean. This allowed us to discuss the physical character of the oceanic response to moving storms without going into all the details and complications that characterize the real world. Some of these are discussed in the quoted papers. So far we have considered only conditions in the deep ocean. The changes produced by moving storms in shallow coastal waters are the topic of the next section.

## 7.5   The effect of lateral boundries on wind-forced perturbations

The existence of coastal boundaries determines the character of large-scale ocean circulations. The forcing of these circulations by the wind stress is buffered by the redistribution of energy and momentum within the bounded ocean basins. This is the central theme of physical oceanography, but it is not the topic of this book. The present section is limited to a discussion of locally forced currents and oceanic perturbations in coastal areas.

### 7.5.1   *Wind-forced upwelling and downwelling along a straight coast*

Cold, highly productive water masses tend to well up from below toward the surface along the subtropical eastern boundaries of all the ocean basins. The climate of these regions is characterized by surface winds that form the eastern branch of the great, quasi-stationary marine anticyclones. Blowing fairly persistently toward the equator, these winds drive westward Ekman transports away from the coast. The resulting horizontal divergence is compensated for by vertical convergence (i.e., by a thinning of the surface mixed layer or by entrainment of upwelling water). The process tends to lower the surface water temperature of the affected regions by several degrees below the global average for the same latitude. Ekman transports toward the coast have opposite kinematic effects. They cause a thickening of the surface layer, and hence downwelling, but no significant change in temperature of the layer. The climatological temperature effect of downwelling is therefore less pronounced.

To discuss the relevant physics we consider a two-layer ocean of constant depth $D$, which is bounded by a vertical cliff. Any wind stress parallel to the coast must cause an Ekman transport normal to the coast. This transport is divergent because it is zero at the coast itself. The divergence forces a slope of the sea surface and of the internal isopycnal surfaces normal to the coast. This slope in turn tends to be balanced by the development of quasi-geostrophic longshore currents.

The barotropic part of the velocity field can be derived from (7.22). We chose an $x$-axis parallel to the coast with $y = 0$ and $V = 0$ at the coast. The water is assumed to be at rest initially, but at time $t = 0$ a constant, spatially uniform wind stress $\tau_x$, parallel to the coast, is being turned on. It can be readily confirmed by substitution into (7.22), that the response has the form

$$U_0 = \frac{\tau_x}{\rho D} t e^{(-a_0 y)} + \sum_l A_{0l} \left(\frac{g}{D}\right)^{0.5} \cos(lx - \omega_0 t) e^{(-a_0 y)} \equiv U' + u_0$$

$$V_0 = \frac{\tau_x}{\rho D f} [1 - e^{(-a_0 y)}] \equiv V' \qquad\qquad (7.63)$$

$$W_0 = \frac{\tau_x}{\rho c_0} e^{(-a_0 y)} + \sum_l l A_{0l} \left(\frac{g}{D}\right)^{0.5} \sin(lx - \omega_0 t) e^{(-a_0 y)} \equiv W' + w_0$$

The solutions of (7.24) have the same structure and are given by

$$\Delta U = \frac{\tau_x}{\rho h} t e^{(-a_1 y)} + \sum_l A_{1l} \left(\frac{g'}{h}\right)^{0.5} \cos(lx - \omega_1 t) e^{(-a_1 y)} \equiv \Delta U' + \Delta u_1$$

$$\Delta V = \frac{\tau_x}{\rho h f} [1 - e^{(-a_1 y)}] \qquad\qquad (7.64)$$

$$W_1 = \frac{\tau_x}{\rho c_1} e^{(-a_1 y)} + \sum_l l A_{1l} \cos(lx - \omega_1 t) e^{(-a_1 y)} \equiv W_1' + w_1$$

These equations consist of two parts. The periodic component, which represents the general solution of the unforced part of (7.22) and (7.24), will be discussed in Section 7.5.2. The forced monotonic part of the solution is of interest here. It involves transports normal to the coast and vertical velocities that remain constant at all times $t > 0$. On the other hand the longshore currents increase as a linear function of time. It follows from $W_0' \approx \partial \zeta / \partial t$ and $W_1' = \partial h / \partial t$ that the coastal sea level anomaly and the slope of both the sea surface and the interface must also increase with time. A steady state cannot be achieved in this case.

The barotropic longshore transport and the mean current in the upper layer flow in the direction of the wind stress, but the flow in the lower layer must be opposite, because $U'/\Delta U' = \bar{h}/D \exp[(a_1 - a_0)yt] < 1$. The existence of this lower countercurrent has been confirmed by observations. It indicates that the sea surface and the internal interface must have opposite slopes. Wind-driven currents away from the shore cause both a drop of the coastal sea level and an upwelling of cold water from below.

The mean layer-depth $\bar{h}$ and hence $c_1$ and $a_1$ were considered constant in the derivation of (7.64). In reality, coastal upwelling is a rather rapid process. To consider an example, assume that $\tau_x/\rho = u_*^2 \approx 10^{-4} \, \text{m}^2 \, \text{s}^{-2}$, $g' \approx 0.005 \, \text{m s}^{-2}$ and $h \approx 50 \, \text{m}$. This makes $c_1 \approx 0.5 \, \text{m s}^{-1}$. The resulting monotonic rise of the interface at the coast ($y = 0$) would then be indicated by (7.64) to be $W_1' \approx 20 \, \text{m day}^{-1}$. It would become progressively faster as $h$ and hence $c_1$ decrease. After a relatively short time, $W_1't$ can become comparable to the initial value of $h$. It may even reach the sea surface, reducing $h$ to zero. Linear equations (7.24) and the resulting expressions (7.64) are then obviously no longer applicable.

The rapidity of this development is due to the large value of the Ekman transport divergence near the coast and the resulting concentration of upwelling within a narrow coastal zone. The approximate width of this zone is indicated by the internal Rossby radius. With the numbers quoted earlier, $1/a_1 = c_1/f \approx 10 \, \text{km}$ at a latitude of 30°. Ekman pumping produced by cyclonic winds over the open ocean tends to be less persistent, but it is almost always distributed over a wider area. It is therefore much less pronounced. Vigorous, persistent coastal upwelling of nutrient-rich water has provided some of the most productive fishing areas in the world.

The geographical concentration of persistent upwelling along the subtropical, continental west coasts is mainly due to the steadiness and strength of surface winds towards the equator in those regions. The poleward directed winds, on the western side of the quasi-stationary marine anticyclones near the continental east coasts, tend to be more variable. Coastal upwelling there is also more likely to be counteracted by downward pumping on the coastal side of the western boundary currents [see discussion following (7.39) and Fig. 7.1]. An exception is the coast of Somalia and southern Arabia, where very strong seasonal upwelling is forced during summer by the persistence and strength of the southwest monsoon.

## 7.5.2   Coastal Kelvin waves

The preceding discussion dealt with changes in sea level, or in the level of internal isopycnal surfaces that did not vary along the coast. In reality, some longshore variations are inevitably caused by the finite size of the forcing wind systems and by always present, coastline irregularities. The resulting differences in pressure, density, and velocity can give rise to perturbations that propagate parallel to the shore. Known as *edge waves,* these perturbations play a considerable role in the prediction of coastal storm surges and in large-scale atmosphere–ocean interactions and in climate studies.

Coastal Kelvin waves are edge waves, which travel along a boundary that resembles a quasi-vertical cliff. The case of equatorial Kelvin waves, where the lateral resistance of a rigid cliff is replaced by symmetrically opposed pressure gradients in the two hemispheres, will be considered in subsection 7.7.2. The sea surface topography and the velocities associated with barotropic, coastal Kelvin waves are specified by the oscillatory parts of (7.63) and (7.64). Pressure gradients normal to the coast are balanced in Kelvin waves by quasi-geostrophic flows parallel to the boundary, while long-shore pressure variations are balanced by longshore accelerations. There are no flows normal to the coast ($v \equiv 0$). The amplitudes of the longshore and vertical velocities have maxima at the coast where $y = 0$. The exponential seaward decrease of these amplitudes involves the Rossby radius as a length-scale. The barotropic surface velocity $u_0$ and the surface deformation

$$\zeta_0 = \int w_0 \, dt \approx -\omega^{-1} \Sigma A_{0l} \sin{(lx - \omega_0 t)} e^{(-a_0 y)} \tag{7.65}$$

are illustrated schematically in Fig. 7.8.

It can easily be confirmed by introduction of (7.63), (7.64) into (7.22), and (7.24) that phase and group velocities within each mode ($j = 0, 1$) are

$$\frac{\omega_j}{l} = (gD_j)^{1/2} = \frac{\partial \omega_j}{\partial l}. \tag{7.66}$$

Kelvin waves are therefore nondispersive. In the chosen coordinate system, with the $y$-axis pointing towards the sea, they can only propagate in the positive $x$-direction in the northern hemisphere. South of the equator, where $f < 0$ they must propagate in the direction of negative $x$. In other words, they always travel toward the equator along continental east coasts and poleward along west coasts.

### 7.5.3   Shelf waves

Shelf waves occur over sloping continental shelfs. Unlike Kelvin waves, they involve currents normal to the coast as an essential ingredient. Though they propagate in the same direction as Kelvin waves they have

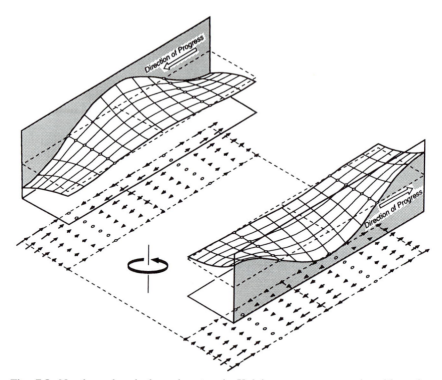

**Fig. 7.8.** Northern hemisphere barotropic Kelvin wave on opposite sides of an ocean basin. Current velocities (shown by arrows) are confined to vertical planes parallel to the coast. They are identical with perturbation velocities produced by irrotational ($f \approx 0$) long, internal gravity waves. Geostrophic balance is maintained by an exponential decrease of surface elevation from the coast. After Gill (1982).

different dynamics. To explain their behaviour, we invoke the principle of potential vorticity conservation (1.31). Within a fluid column, which is not affected by friction, the potential vorticity

$$\frac{\eta + f}{D} = \text{constant.} \tag{7.67}$$

The relative vorticity $\eta$ changes in shelf waves, due to variations of $D$ in the fluid column. The mechanism has been explained very clearly by Longuet–Higgins (1965). Consider a continental shelf that slopes down toward the west. The solid curve in Fig. 7.9 represents a vertical sheet of fluid that initially had been situated along a straight contour line, parallel to the coast, and that was displaced by some disturbance. Assume that $\eta$ was initially zero. The vertical thickness $D$ of water columns is decreased by their displacement toward the coast. They must therefore acquire

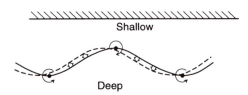

**Fig. 7.9.** The mechanism of shelf (and Rossby) waves. The solid line represents a string of particles whose undisturbed position lay along a depth contour (parallel circle). Conservation of potential vorticity implies that particles moving into shallower water (higher latitudes) acquire anticyclonic relative vorticity. Those moving into deeper water (lower latitudes) acquire cyclonic relative vorticity. Displacements (broad arrows) induced by the vorticity field, move string of particles to the position indicated by the dashed line. The sense of vorticity is shown for the northern hemisphere.

negative, anticyclonic vorticity for the ratio $(\eta + f)/D$ to remain constant. As shown in the figure, this tends to produce a seaward current on the poleward side of the perturbation and a current toward the coast on its equatorial side. The net effect of this local current pattern is a displacement of the perturbation along the shore, toward the right as seen from the land in the northern hemisphere (i.e., in the same direction as Kelvin waves).

We use the same coordinate system as in the preceding section and assume that the sea bottom slopes away from the coast in the $y$-direction. Continuity requires that the cross section of a water column that moves with velocity $v$ across the shelf slope, must change at a rate that is commensurate with the change of the column depth $\zeta + D$. If $\zeta \ll D$, the continuity equation is

$$(\zeta + D)\left(\frac{\partial u}{\partial x} + \frac{\partial v}{\partial y}\right) + v\frac{\partial(\zeta + D)}{\partial y} \approx D\left(\frac{\partial u}{\partial x} + \frac{\partial v}{\partial y}\right) + v\frac{\partial D}{\partial y} = 0. \qquad (7.68)$$

If the $\beta$ term is relatively small, then the conservation of potential vorticity can be expressed in linearized form by

$$\frac{d}{dt}\frac{\eta + f}{D} \approx D^{-1}\frac{\partial}{\partial t}\left(\frac{\partial v}{\partial x} - \frac{\partial u}{\partial y}\right) - D^{-2}fv\frac{\partial D}{\partial y}. \qquad (7.69)$$

These are the governing equations for unforced linear shelf waves. To express them explicitly, one has to know the variations of $D \equiv D(y)$ normal to the shore. Solutions based on a variety of topographic configurations, also including variations in the longshore direction, have been discussed by LeBlond and Mysak (1977) and by Csanady (1981). The effect of longshore variations in the shelf topography, which was not considered in (7.68) and (7.69), has been addressed specifically in papers by Wilkin and Chapman (1987), Middleton and Wright (1988), and Johnson (1989).

Shelf waves and Kelvin waves are both 'trapped' in the coastal zone. The surface displacement amplitude of shelf waves decreases with increasing water depth and surface height changes, therefore, become negligibly small at some distance from the shore. That distance depends on the characteristic length-scale of the shelf topography, which is often of the same order—about 30 km—as the local baroclinic radius of deformation. The water above a sloping shelf is also usually stratified. Hybrid forms that combine the characteristics of shelf and baroclinic Kelvin waves are therefore rather common.

### 7.5.4  Storm surges

Accumulations of high water that propagate along a shore can cause large losses of property and life. The physics of these surges involves all the processes discussed in subsection 7.5.3. The second equation (7.63) shows the mean flow towards the coast to be inversely proportional to the water depth $D$. The most devastating storm surges, therefore, occur in regions where the sea bottom slopes out very gently from the coast. On January 31, 1953, a North Sea storm produced a surge that flooded 25,000 km² of fertile land and drowned some 2000 persons in Britain and in the Netherlands. In 1989, a powerful Bengal cyclone is said to have claimed 200,000 lives in Bangladesh. The Gulf coast of the United States is among other areas at risk.

Rossiter (1954) collected data that showed how the peak of the 1953 storm surge propagated counterclockwise around the North Sea shores. This is the direction of edge wave propagation. A conceptual association of storm surges with coastal boundary waves was developed subsequently by Munk et al. (1956) and by Reid (1958). Depending on the coastal configuration, the travelling surges can have the character of shelf waves, Kelvin waves, or, most commonly, hybrid, intermediate forms. Fundamental frequencies and longshore wavenumbers of resonantly excited shelf waves tend to be of order

$$\omega_\alpha = g \frac{\sin \alpha}{U_{ac}} \qquad l_\alpha = g \frac{\sin \alpha}{U_{ac}^2} \qquad (7.70)$$

where $\tan \alpha$ is the bottom slope and $U_{ac}$ is the component of the storm translation speed parallel to the coast.

Storm surges are produced, in the first instance, by the localized piling up of water. This tends to be caused by convergent mass transports that are normal to the shore, but vary in the longshore direction. Equations (7.63) and (7.64) are therefore not applicable during the initial forcing process. If the longshore wind stress component is variable with time and with distance along the shore $[\tau_x \equiv \tau_x(x, t)]$, then the sea level change along a straight coast can be derived from the equation

$$\left(\frac{\partial \zeta}{\partial t} + c_0 \frac{\partial \zeta}{\partial x}\right)_{y=0} = \tau_x \rho c_0 \tag{7.71}$$

For constant forcing, (7.71) represents a balance between the stress and the pressure gradient. Theoretical models of idealized, forced Kelvin-type coastal surges have been described by Kajiura (1962), Thompson (1970), Keller and Watson (1981), and by Fandry et al. (1984). The amplitude of wind-forced shelf waves can be derived from an equation, which has the same structure as (7.71) and which can be found in Gill (1982).

These models of storm surges were based essentially on linear equations of the form (7.22) and (7.24). The actual rise of sea level that can be produced by a major storm involves nonlinear relationships. It is affected by tides and surface waves and therefore cannot be derived from local atmospheric forcing alone. Bottom friction and details of the local topography generally play a major role. For that reason, operational forecasting schemes are based on numerical models that permit a variety of localized inputs. A particular example is a model designed originally by Jelesnianski (1967), which now forms the basis of operational surge warnings by the U.S. National Hurricane Center in Miami. Analogous models have been developed for the North Sea and other, potentially endangered areas.

## 7.6 Rossby or planetary waves

Gravity waves tend to disperse available potential energy (APE). That happens, for example, on the surface of a pond when we drop a pebble into it. Rossby waves disperse the energy of meridional quasi-geostrophic currents. It follows from (7.12) that stationarity ($\partial/\partial t = 0$) is possible only if

$$\frac{\partial \zeta'}{\partial x} = \frac{f}{g} V_g = 0. \tag{7.72}$$

Geostrophic motion with a finite, meridional component $V_g$ is inevitably transient. Over a sufficiently long period of time, the energy of such meridional geostrophic motion is dispersed by Rossby wave radiation.

A detailed review of Rossby waves has been published by Platzman (1968). The dynamics of these waves, like that of shelf waves, is governed by the principle of potential vorticity conservation (7.67). In shelf waves, changes in $\eta$ are caused by vortex stretching associated with changes in $D$. In planetary waves, these changes are caused by meridional changes of $f$. If potential vorticity $(\eta + f)/D$ is to be conserved, then a vertical fluid column must acquire negative or anticyclonic relative vorticity when it is displaced poleward where $f$ is larger. It acquires cyclonic relative vorticity when displaced toward the equator.

### 7.6.1  *Free planetary waves*

With $\partial f/\partial y = \beta \neq 0$, one cannot use the derivation that led from (4.1) to (4.4). The dispersion relation, (4.6) is not valid in this case. The dispersion of Rossby waves is specified instead by (7.16). That equation also covers the baroclinic case, if $a_0$ is replaced by appropriate values of $a_j$. Phase and group velocities in the eastward ($x$) and northward ($y$) directions, as obtained from (7.16), are given by

$$\frac{\omega}{l} = -\beta(l^2 + m^2 + a_j^2)^{-1}, \qquad \frac{\omega}{m} = -\beta l m^{-1}(l^2 + m^2 + a_j^2)^{-1}$$

$$\frac{\partial \omega}{\partial l} = \beta(l^2 - m^2 - a_j^2)(l^2 + m^2 + a_j^2)^{-2}, \qquad \frac{\partial \omega}{\partial m} = 2\beta l m(l^2 + m^2 + a_j^2)^{-2}$$

$$(7.73)$$

The zonal phase propagation $\omega/l$ is always negative; the waves can only propagate toward the west. Their phase speed decreases monotonically with decreasing wavelength.

The westward propagation of planetary waves can be explained by a simple kinematic argument which is analogous to the explanation of shelf waves by Fig. 7.9. Cyclonic troughs advect potential or absolute vorticity toward the equator on their western side. As the fluid moves toward the equator it experiences a decrease in the planetary vorticity $f$. The associated increase of relative vorticity accelerates the flow cyclonically. This shifts the locus of maximum cyclonic vorticity westward. Anticyclonic ridges advect absolute vorticity poleward in their western half. This causes a decrease in relative vorticity and hence anticyclonic acceleration. The center of negative, anticyclonic vorticity, therefore, is also shifted westward. The whole perturbation pattern is propagated toward the west in this way. The phase can propagate northward or southward, depending on the sign of the product $lm$.

Frequencies of some barotropic and baroclinic Rossby waves, as well as those of hypothetical gravity waves with the same length, are shown in Fig. 7.10. In constructing this diagram, it was assumed that $a_0$ and $a_1$ could be considered constant and that $l = m = k\sqrt{2}$. Horizontal variations in ocean depth and hence in $a_0$ can give rise to hybrid planetary-topographic or planetary-shelf waves. Rossby waves shorter than 50 km are theoretically possible in a linear, inviscid model, but the associated motion would be so sluggish that it could not be observed. The frequency $\omega$ has a maximum when its derivatives (i.e., the components of the group velocity) are both

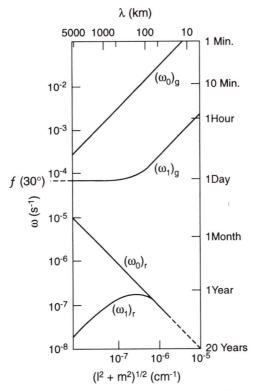

**Fig. 7.10.** Frequencies of various wave forms as a function of wavenumber ($l = m$) or wavelength: $(\omega_0)_g$ long surface gravity waves; $(\omega_1)_g$ internal inertio-gravity waves; $(\omega_0)_r$ barotropic Rossby waves; and $(\omega_1)_r$ baroclinic Rossby waves. After Veronis and Stommel (1956).

zero. It occurs when $l = a_j$ and $m = 0$. The highest possible frequency for wave mode $j$, therefore, is

$$\omega_{max} = \frac{\beta}{2} a_j. \qquad (7.74)$$

This maximum can be seen in the curve for baroclinic Rossby waves in Fig. 7.10.

Rossby waves transport energy westward ($\partial\omega/\partial l < 0$) or eastward ($\partial\omega/\partial l > 0$) depending on whether $l^2$ is smaller or larger than $(m^2 + a_j^2)$. Maxima or minima of the group velocity are reached when the dispersion $\partial^2\omega/\partial l^2 = 0$. This occurs when $l^2 = 3(m^2 + a_j^2)$ or when $l^2 = 0$. The maximum westward propagation is approached asympotically as $l^2/a_j^2 \Rightarrow 0$. Waves with a zonal wavelength much in excess of the deformation radius are nondispersive and have westward zonal group and phase velocities that are independent of the zonal wavenumber. The value of both extremes is specified by

$$\left(\frac{\partial\omega}{\partial l}\right)_{max} = \frac{1}{8}\frac{\beta}{(m^2 + a_j^2)} \quad \text{(fastest eastward)}$$

$$\left(\frac{\omega}{l}\right)_{min} = \left(\frac{\partial\omega}{\partial l}\right)_{min} = -\frac{\beta}{m^2 + a_j^2} \quad \text{(fastest westward)} \qquad (7.75)$$

Energy and information can be transported westward eight times faster than eastward. Maximum speeds decrease sharply with increasing mode number and latitude, as shown for example by the numbers in Table 7.1.

In the absence of external forcing ($\zeta' = \zeta$), the surface deformations

**Table 7.1.** Rossby Radii ($1/a_j$) and Westward Velocities ($\omega/l$) of Nondispersive Rossby Waves ($l^2 \ll a_j^2$, $m^2 \ll a_j^2$) at Two Different Latitudes

| $\phi$ | Barotropic Mode ($D_0 = 4000$ m) | | First Baroclinic Mode ($D_1 = 1$ m) | |
|---|---|---|---|---|
| | $1/a_0$ | Speed | $1/a_1$ | Speed |
| 10° | 7820 km | 11910 km day$^{-1}$ | 124 km | 2.98 km day$^{-1}$ |
| 50° | 1773 | 612 | 28 | 0.15 |

produced by horizontally propagating, barotropic Rossby waves can be specified again by expressions of the form

$$\zeta = A \cos (lx + my - \omega t).$$

The contributions of these waves to the geostrophic velocity field are

$$U_g = -\frac{g}{f}\frac{\partial \zeta}{\partial y} = \frac{mgA}{f} \cos (lx + my - \omega t)$$

$$V_g = \frac{g}{f}\frac{\partial \zeta}{\partial x} = \frac{lgA}{f} \cos (lx + my - \omega t)$$

(7.76)

To assess the dispersal of a meridional geostrophic current, one represents its configuration at some initial time $t = 0$, by a sum of terms of the form (7.76). Each of these terms can then be considered a Rossby wave that carries energy away from the current at a rate indicated by (7.73).

The actual velocity associated with a free Rossby wave is very nearly but not quite in balance with the pressure gradient, hence the term *quasi-geostrophic*. To obtain an expression for the departure from geostrophic balance, one introduces $f = f_0 + y\beta$ into the equation of motion (7.7). After some rearrangement and with consideration of (7.76), this yields

$$U - U_g = -\frac{\beta y U + \partial V / \partial t}{f_0} \approx -\frac{\beta y U_g + \partial V_g / \partial t}{f_0}$$

$$V - V_g = -\frac{\beta y V - \partial U / \partial t}{f_0} \approx -\frac{\beta y V_g - \partial U_g / \partial t}{f_0}$$

(7.77)

The ageostrophic part of the velocity field is very much smaller than the geostrophic part, but it is an essential feature of the wave because it determines the alternate convergences and divergences that cause the westward propagation. The $\beta$-driven part, which acts parallel to the isobars, is due to the fact that the geostrophic velocity that balances a given pressure gradient becomes larger at lower latitudes. The second term on the right-hand side of (7.77) represents an isallobaric movement in the direction of decreasing pressure.

The phase and group velocities of baroclinic Rossby waves can have a vertical component, provided

$$\omega_j = \frac{-\beta l}{k^2 + a_j^2} < N.$$

Following (7.53), the vertical wavenumber $n_j = N/c_j$ when $N$ is constant. The cosine term in (7.76) has to be replaced in this case by $\cos (lx + my + Nz/c_j - \omega t)$.

### 7.6.2   Forced planetary waves

Discussions of forced planetary waves can be found in review papers by Rhines (1977) and Dickinson (1978). Planetary waves can be forced only by an input of relative vorticity. In the ocean this may take place through a wind stress curl. If bottom friction is neglected, the relevant form of the governing equation (7.12) is

$$c_0^2 \left[ \left( \frac{\partial^2}{\partial x^2} + \frac{\partial^2}{\partial y^2} - a_0^2 \right) \frac{\partial}{\partial t} + \beta \frac{\partial}{\partial x} \right] \zeta' = \frac{\nabla \times \tau}{\rho}. \tag{7.78}$$

When dealing with Rossby waves, which depend on meridional changes of $f$, it is often more convenient to use the meridional velocity $V$ instead of $\zeta$ or $\zeta'$ as the dependent variable. An equation for $V$ can be obtained by the same algorithmic process that led to (7.12). It has the same structure as (7.78), but a different dimension. For a flat ocean bottom, one gets

$$D \left[ \left( \frac{\partial^2}{\partial x^2} + \frac{\partial^2}{\partial y^2} - a_0^2 \right) \frac{\partial}{\partial t} + \beta \frac{\partial}{\partial x} \right] V = \frac{\partial}{\partial x} \frac{\nabla \times \tau}{\rho}. \tag{7.79}$$

The preceding equations can also be applied to baroclinic Rossby waves if $c_0$, $a_0$, $D$ are replaced by $c_j$, $a_j$, $D_j$. Numerical solutions of (7.79) have been discussed by Anderson and Gill (1975) and by Anderson et al. (1979). An analytical approach by Longuet–Higgins (1965), was based on the quasi-geostrophic character $(V \approx V_g)$ of the Rossby wave velocity field. It involved introduction of the stream function

$$V_g = \frac{\partial \psi}{\partial x} \qquad U_g = -\frac{\partial \psi}{\partial y}.$$

Integration of (7.79) with respect to $x$ yields the following expression plus an arbitrary function of $y$

$$D\left[\left(\frac{\partial^2}{\partial x^2}+\frac{\partial^2}{\partial y^2}-a_0^2\right)\frac{\partial}{\partial t}+\beta\frac{\partial}{\partial x}\right]\psi=\frac{\nabla\times\tau}{\rho}. \tag{7.80}$$

Wind stress changes are very fast compared to the period of planetary waves. This allowed Longuet–Higgins to represent them as a delta function in time. He then solved (7.80) for several hypothetical configurations of the forcing wind stress $\tau$. An example of the resulting computations is shown in Fig. 7.11. It represents the calculated wave pattern associated with the dissolution of a geostrophically balanced surface trough, which was produced by a moving cyclone as discussed in Section 7.4.3. In the present case, the cyclone path was supposed to be a straight

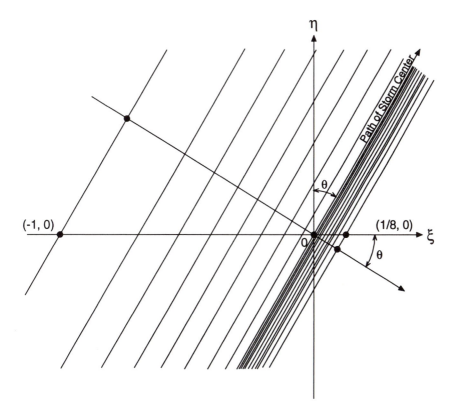

**Fig. 7.11.** The planetary wave pattern produced by a fast-moving storm making an angle $\theta$ with the northward direction. The outer edge of the wave pattern is determined by $\xi=a^2x/\beta t=-1$ in the west and by $\xi=1/8$ in the east. After Longuet–Higgins (1965).

line, making an angle $\theta$ with the meridians. Wave crests then run parallel to the storm track. In accordance with (7.75), the whole perturbation zone widens linearly with time—eight times faster in the west than in the east. The actual width of this zone is an inverse function of the angle $\theta$. For a zonally moving perturbation ($\theta = \pi/2$), the oceanic response assumes the form of a steady zonal current that diminishes exponentially away from the storm track. Strictly zonal currents cannot be dispersed by Rossby waves.

When the forcing and the resulting motion are both steady ($\partial/\partial t \equiv 0$), the integration of (7.79) with respect to $x$ yields the classic *Sverdrup relation*

$$\rho\beta DV \approx \rho\beta DV_g = \nabla \times \tau. \tag{7.81}$$

It states that under steady-state conditions, an anticyclonic wind stress torque can be balanced locally by barotropic advection of cyclonic planetary vorticity from higher latitudes and vice versa. To obtain a more precise version of this relation, one can replace $\rho DV$ on the left-hand side of (7.81) by the vertical integral of the specific momentum $\rho V$ over the column depth $D$.

The atmosphere exerts a mean anticyclonic frictional torque upon the oceans between the strongest trades and the strongest west winds in higher latitudes. It follows from (7.81) that this should be associated with a Sverdrup transport toward the equator. A poleward flow should be expected in regions of cyclonic wind stress curl, on the polar side of the westerly belts, and on the equatorial side of the strongest easterly trades.

A quantitative verification of (7.81), by separate assessments of the wind stress curl and of the meridional oceanic mass transport, has been a popular goal of oceanographic studies. Different methods to derive this transport from oceanic observations have been reviewed by Killworth and Bigg (1988). To study the development of Sverdrup transports, several authors have used a model that involves the sudden application of a constant zonal wind stress to an ocean previously at rest. In one simulation of this type, by Anderson and Killworth (1977), the stress was initially found to drive only a meridional Ekman transport, but after some time a baroclinic Rossby wave moves westward from the eastern boundary with the velocity given by (7.75). As this wave crosses the ocean it leaves a steady Sverdrup transport behind.

The zonally integrated mean meridional mass transport across a closed ocean basin must be zero. It follows that the Sverdrup relation (7.81), cannot be satisfied all along a parallel circle. There must be a poleward return flow in subtropical latitudes, where an anticyclonic wind stress forces a Sverdrup transport toward the equator. The opposite holds in zones of cyclonic wind stress torques. The two sides of (7.81) tend to have opposite signs in these return flows. The balance described by that equation is therefore not applicable to them. It was first noted in a classic paper by Stommel (1948) that the return flows are concentrated in narrow western boundary currents such as the Gulf Stream or the Kuroshio. These currents are fast enough for other forces, including bottom friction (Stommel, 1948), lateral friction (Munk, 1950), and inertial forces (Morgan, 1956), to come into play. Detailed discussions of this topic can be found in most textbooks of physical oceanography. The monographs or reviews, compiled by Stommel (1965), Kamenkovich (1977), and Veronis (1981), treat this matter in greater depth.

The Sverdrup relation deals with a steady state, but the driving wind stress curl obviously varies with time. Pedlosky (1965) found that an oscillating wind stress, even if its mean is zero, would produce a fluctuating response and could also drive a steady circulation in a bounded ocean. It does so by the excitation of long Rossby waves, with a westward group velocity. When a group of these waves reaches the western rim of an ocean basin, it is reflected as a short-wave packet, which transports energy much more slowly and is more prone to attenuation by friction. These processes contribute to the concentration of energy in the western boundary currents. Rectification of the momentum flux can be explained by an inertial mechanism. The Rossby wave velocity component normal to the coast $U_g = -\partial\psi/\partial y$ is finite on the ocean side of the boundary current. Along the coast $\partial\psi/\partial y = 0$. Lighthill (1969) integrated the homogeneous part of (7.80) with this boundary condition and showed that this leads to a rectified flow, parallel to the coast.

The rate, amplitude and location of the oceanic response to long-term wind stress changes in midlatitudes must affect the atmosphere in turn. This subject is important for our understanding of climate change. It has been discussed by many authors, including Charney and Flierl (1981), but it is probably fair to say that no definitive understanding has been reached so far.

## 7.7  Equatorial currents and perturbations

In contrast to conditions in higher latitudes, the time-scale of widespread oceanic changes near the equator can often match corresponding transients in the atmosphere. As a consequence, ocean waves of various kinds that travel along the equator play an important role in the operation of the joint atmosphere–ocean system. A great deal of ongoing research deals with this topic. Some current ideas about equatorial ocean–atmosphere feedback processes will be reviewed in Chapter 8. The basic concepts of equatorial dynamics have been discussed very clearly and in considerable detail by Gill (1982). The present section provides a brief summary of a subject with its own, rather considerable, literature.

Throughout this section it is assumed that the Coriolis parameter can be represented by the equatorial $\beta$-plant approximation

$$f \approx \beta y \quad \text{with} \quad \beta = \left(\frac{\partial f}{\partial y}\right)_{\phi=0} = 2.3 \; 10^{-11} \, \text{m}^{-1} \text{s}^{-1}. \tag{7.82}$$

The inverse Rossby radius $a_j = \beta y / c_j$ vanishes along the equator ($y = 0$). For analysis and computations, it is convenient to introduce an inverse *equatorial Rossby radius* $a_{ej}$ which is independent of latitude and specified usually by

$$a_{ej}^2 = \frac{2\beta}{c_j}. \tag{7.83}$$

In the equatorial ocean, the barotropic deformation radius is $1/a_{e0} \approx 2000$ km. The lowest order baroclinic radius $1/a_{e1} \approx 200$ km.

### 7.7.1  Balanced equatorial currents

The average wind stress along the equator is mainly zonal outside some rather limited monsoon regions. Currents are also predominantly zonal. As a first approximation, one can therefore construct a two-layer model of a steady-state ($\partial/\partial t = 0$) equatorial ocean circulation by setting $\tau_y = 0$ and $V = 0$. With neglect of the baroclinic contribution to surface elevation changes ($\nabla \zeta_1 \ll \nabla \zeta_0 \simeq \nabla \zeta$) and with $f = \beta y$, the barotropic equations of motion (7.22) assume the form

$$\frac{\partial \zeta}{\partial x} = \frac{\tau_x}{\rho g D} = \frac{\tau_x}{\rho c_0^2}$$

$$\beta y U_0 = -g \frac{\partial \zeta}{\partial y} \tag{7.84}$$

One can develop the meridional slopes of the surface and of the interface as a Taylor series of the form

$$\frac{\partial \zeta}{\partial y} = y \frac{\partial^2 \zeta}{\partial y^2} + \frac{1}{2} y^2 \frac{\partial^3 \zeta}{\partial y^3} + \cdots$$

Introduction into the second equation (7.84) yields the following first-order, quasi-geostrophic approximations for the zonal velocity on or near the equator

$$(\beta U_0)_{y\approx 0} \approx -g\frac{\partial^2 \zeta}{\partial^2 y}. \tag{7.85}$$

For the baroclinic component of motion one gets similarly

$$\frac{\partial h}{\partial x} = -\frac{\tau_x}{\rho g'\bar{h}} = -\frac{\tau_x}{\rho c_1^2} \tag{7.86}$$

and

$$(\beta\,\Delta U)_{y\approx 0} \approx g'\frac{\partial^2 h}{\partial^2 y}. \tag{7.87}$$

The first equation (7.84) indicates an approximate balance between the stress and the surface slope, in the absence of transients along the equator. Over most of the open equatorial oceans the direction of the mean wind stress is toward the west ($\tau_x < 0$). It follows that the sea surface must slope downward toward the east. The internal interface is shown by (7.86) to slant much more steeply upward toward the east. The ratio of the two slopes is $-(\partial h/\partial x)/(\partial \zeta/\partial x) = gD/g'\bar{h} \gg 1$.

Off the equator, the westward wind stress drives a poleward Ekman transport in both hemispheres. The resulting removal of water from the equatorial surface layer causes the formation of a trough ($\partial^2 \zeta/\partial y^2 > 0$) in the surface elevation and a ridge ($\partial^2 h/\partial y^2 < 0$) on the internal interface. It follows from (7.85) and (7.87) that the velocity is negative or westward in the layer above the interface and eastward in the layer below. On a bounded frictionless ocean, the upper westward flow would remove water from the eastern part of the basin and pile it up on the western boundary. The resulting surface rise in the west and fall in the east, would produce a continuously steepening, west–east surface slope and a corresponding pressure gradient. A steady-state current configuration can be established under these conditions, if the westward acceleration produced by the stress $\tau_x$ is balanced by internal friction or entrainment across the internal interface.

In fact, a quasi-steady equatorial current system is a characteristic feature of the Pacific and Atlantic. The lower, eastward flow tends to be concentrated in a thin layer of about a 200-m thickness, immediately below the shallow, wind-driven surface drift toward the west. This concentrated eastward flow is known as the *Equatorial Undercurrent*. Its narrow vertical extent indicates that higher modes $(j>1)$ cannot be neglected in a description of the phenomenon.

The Undercurrent seems to be fed mainly by inflows from north and south along the western boundary. The cyclonic wind stress curl, on the equatorial side of the trade wind maxima, causes a poleward Sverdrup transport (7.66) that is compensated by western boundary currents toward the equator. Off the equator, the poleward Sverdrup transport can be balanced geostrophically by the piling up of water along the western boundary and the resulting west–east surface slope. At the equator, this balance disappears and the existing pressure gradient drives the water toward the east. A steady state can then be maintained by friction (i.e., by mixing of the Undercurrent water with the opposing surface drift and with laterally adjacent westward currents).

This steady-state circulation has been modelled by McCreary (1981). He stipulated an eddy viscosity that varied with depth. The resulting friction increases with the mode number, or more precisely with $c_j^{-2}$. McCreary showed that the lowest modes were characterized by an approximate balance between the wind stress and the west–east pressure gradient. The main contribution to the steady Undercurrent came from higher modes, in which frictional damping was just strong enough to mask the zonal propagation of perturbations from the eastern and western boundaries.

Evidence for the mean west–east slope of the equatorial sea surface in the Pacific has been compiled by Lemasson and Piton (1968). They established a sea surface drop of about 0.5 m between the Gilbert (175° E) and Galapagos (90° W) Islands by measuring the thickness of the layer between the air–sea interface and the 700 db isobaric surface (1.0 db $=$ $10^4$ Pa $\approx 1$ m of water). The opposing upward slope of the thermocline isopycnals below forces the Undercurrent to rise as it flows toward South America. In this way, cold water is brought close to the surface in the eastern part of the ocean basin. This is probably the main reason for the large average sea surface temperature difference—more than 8°C— between the western and eastern parts of the basin. An additional cause for this difference is the advection of cold water and upwelling along the coast of South America, as discussed in subsection 7.5.1. The east–west difference in the mixed-layer thickness is also affected by differences in the buoyancy and fresh water fluxes across the sea surface.

The zonal pressure gradient, which is caused by the west–east sea surface slope, extends down to a depth of several hundred meters. It is balanced by a meridional geostrophic flow toward the equator in both hemispheres. In the surface layers this flow is opposed by the wind-driven Ekman drift away from the equator. These processes together engender a meridional circulation, which is very similar to the coastal upwelling circulation discussed in subsection 7.5.1. Yoshida (1959) used this analogy to derive an equatorial velocity field that is essentially of the same form as the monotonic parts of (7.63) and (7.64). With the coast replaced by an equatorial vertical plane, it indicates time-independent upwelling within a zone that extends to a meridional distance of order $\pm a_{e1}$ from the equator. An associated constant deepening of the equatorial surface trough and a westward surface velocity that increases linearly with time, are in keeping with the preceding discussion and with that of upwelling along a straight coast.

The actual oceans are obviously much more complicated than any conceptual ocean model with straight meridional boundaries and meridionally symmetric forcing by a zonal wind stress. In particular, the western boundary of the equatorial Pacific is much too irregular and complex topographically to be approximated realistically by a straight meridional wall. The wind stress is hardly ever strictly zonal nor uniform meridionally. Over the Atlantic and Pacific, the atmospheric *Intertropical Convergence Zone* (ITCZ) is about 8° north of the equator much of the time (see subsection 8.3.2). The associated oceanic Ekman divergence in that zone causes the formation of a secondary sea surface trough. The northern slope of this trough is balanced geostrophically by the westward-flowing *North-Equatorial Currents* of the Pacific and the Atlantic. Its southern slope, which extends approximately through the 5–10° belt, is balanced mainly by the quasi-geostrophic *Equatorial Countercurrent.* This surface current flows eastward like the Equatorial Undercurrent to which it can be connected, although the two currents always have separate cores. Immediately south of the Countercurrent, the westward surface drift along the equator can be considered a part of the *South-Equatorial Currents* that cover a broad latitude belt in all the oceans.

In the Pacific, the secondary surface trough under the ITCZ tends to have a greater depth or amplitude than the parallel equatorial trough. The associated Countercurrent moves further north and is more intense in summer than in winter. Its volume transport of about 25 Sverdrup ($25 \times 10^6 \, \mathrm{m^3 \, s^{-1}}$) toward the east is comparable to the transport of the Gulf Stream through the Florida Straights (Sverdrup et al., 1942). The surface speed in its core can reach $1.0 \, \mathrm{m \, s^{-1}}$. In addition to its geostrophic part, it is accelerated eastward significantly by the west–east surface slope (Montgomery and Palmén, 1940). The Atlantic Countercurrent is weaker and more variable. Model studies suggest that this variability is caused mainly by annual variations of the meridional wind component. The ensuing changes of the zonal Ekman drift can weaken or strengthen the current in different seasons. Monsoonal circulations modulate the strength of the trades and the location of the ITCZ. The resulting effects are more important in the Atlantic than in the Pacific. They become dominant in the Indian Ocean.

### 7.7.2 Equatorial perturbations

Wind stress fluctuations must disturb the idealized steady current pattern. References to many relevant studies can be found in a review by McCreary and Anderson (1991). A brief discussion of equatorial perturbations can be based on the unforced form of (7.12). With $\zeta'$ replaced by the meridional perturbation velocity $v$ as in (7.79) and with $f = \beta y$, the left-hand side of (7.12) is transformed into

$$\left\{ \frac{-\partial^3}{\partial t^3} + \left[ c_j^2 \left( \frac{\partial^2}{\partial x^2} + \frac{\partial^2}{\partial y^2} \right) - (\beta y)^2 \right] \frac{\partial}{\partial t} + c_j^2 \beta \frac{\partial}{\partial x} \right\} v_j = 0. \tag{7.88}$$

Introduction of

$$v_j = V(y)\chi_j(z) \cos (lx - \omega_j t) \tag{7.89}$$

into (7.88) yields, after division by $[\omega_j \chi_j c_j^2 \sin (lx - \omega_j t)]$,

$$\left\{ \frac{\partial^2}{\partial y^2} + \left[ \frac{\omega_j^2 - \beta^2 y^2}{c_j^2} - l^2 - \frac{l\beta}{\omega_j} \right] \right\} V(y) \equiv \left\{ \frac{\partial^2}{\partial y^2} + m(y)^2 \right\} V(y) = 0. \tag{7.90}$$

where $m(y)^2$ is used as an abbreviation for the term in square brackets. This term can be positive close to the equator where $y$ is relatively small. The perturbations there have the character of waves that can propagate poleward with a meridional wavenumber $m(y)$. The parameter $\beta y = f$ becomes larger with increasing latitude, causing $m(y)$ to decrease, until $m(y_c)^2 = 0$ at a critical distance $y_c$ from the equator. This turning latitude is specified by

$$(\beta y_c)^2 = \omega_j^2 - \left( l^2 - \frac{l\beta}{\omega_j} \right) c_j^2. \tag{7.91}$$

Equatorial waves are totally reflected at their critical latitude. The solution of (7.90) becomes exponential beyond this latitude and waves cannot propagate into that region. They are trapped within an equatorial wave-guide of half width $y_c$. The causes for this meridional constriction were discussed in subsection 7.3.1.

Matsuno (1966) showed that a solution of (7.90), which lets $V(y) \Rightarrow 0$

with increasing latitude, can be obtained only if $y_c$ and hence $\omega$ satisfy the quantum condition

$$(\beta y_{cr})^2 = \omega_{jr}^2 - \left(l^2 - \frac{l\beta}{\omega_{jr}}\right)c_j^2 = (2r+1)\beta c_j \qquad (r = 1, 2, 3, \ldots) \qquad (7.92)$$

With $\omega_{jr}$ specified by (7.92), the solution of (7.90) can be expressed by a sum of terms that have the form

$$V_j(y) = V_{jr}(y) = 2^{-r/2}e^{-1/4(ya_{ej})^2}H_r(2^{-1/2}ya_{ej}), \qquad (7.93)$$

where $H_r$ is the Hermite polynomial of order $r$. These polynomials are listed in mathematical tables. The exponential in (7.93) indicates that the amplitude of the perturbations becomes small at a distance $2/a_{ej}$ from the equator. For the lowest baroclinic mode that is about 400 km.

The dispersion relation (7.92) can be simplified by the same scaling arguments that were used in Subsection 7.1.3. This again allows us to distinguish between different wave types

$$\omega_{jr}^2 \approx (lc_j)^2 + (2r+1)\beta c_j = c_j^2[l^2 + (r+0.5)a_{ej}^2] \qquad (\text{if } \omega^3 \gg l\beta c_j^2) \qquad (7.94)$$

This frequency relation has the same structure as does (7.55). It specifies perturbations that behave like plane gravity waves in the immediate vicinity of the equator. As these waves propagate poleward, $f$ and hence $a_j^2$ increase. This changes them into inertio-gravity waves with orbits that are no longer confined to vertical planes. The group velocity turns more zonal and the phase velocity vector more meridional with increasing latitude, until the waves are turned back again toward the equator at the critical latitude $y_{cr}$, which was defined by (7.92).

By the same reasoning, the dispersion relation (7.92) assumes the form

$$\omega_{jr} \approx \frac{-\beta l}{l^2 + (2r+1)\beta/c_j} = \frac{-\beta l}{l^2 + (r+0.5)a_{ej}^2} \qquad (\text{if } \omega^3 \ll l\beta c_j^2) \qquad (7.95)$$

Waves that are governed by this frequency relation are known as *equatorially trapped Rossby waves*. They can only propagate westward ($\omega_{jr} < 0$) and they become nondispersive if $l_2 \ll (2r+1)\beta/c_j$. The group and phase velocities are then given by

$$\frac{\omega_{jr}}{l} = \frac{\partial \omega_{fr}}{\partial l} = \frac{-c_j}{2r+1}. \qquad (7.96)$$

For $c_1 \approx 1\,\mathrm{m\,s^{-1}}$ and $r = 1$, this would indicate a signal propagation of about $30\,\mathrm{km\,day^{-1}}$ from east to west along the equator. The fast propagation speed of equatorial Rossby waves accounts partly for the relatively rapid adjustment of equatorial ocean currents to changing meteorological conditions. In the atmosphere, these waves tend to be associated with westward-moving cloud clusters in or near the ITCZ.

Hybrid or mixed forms with frequencies between those indicated by (7.94) and (7.95) are obviously possible. Of some special interest are the *mixed equatorial planetary-gravity waves,* which can be specified by setting $r = 0$ in (7.92). These waves transport energy eastward, but the phase can propagate either west or east, depending on the sign of $l$.

Equatorial Rossby waves are characterized by vortices north and south of the equator, which have opposite signs and hence form mirror images of each other. This is indicated by (7.93) if one considers that $H_r(y) = -H_r(-y)$. In *equatorial Kelvin waves* the meridional perturbation pressure gradients in the two hemispheres are equal and opposite. These gradients can be balanced by the same zonal, geostrophic velocity on each side. Their dynamic is essentially the same as that of coastal Kelvin waves with the coastal boundary replaced by the equatorial plane. To visualize equatorial Kelvin waves one would have to complement Fig. 7.8 by a mirror image on the other side of the boundary wall. This requires pressure maxima and minima on the equator itself, in contrast to equatorial Rossby waves that involve pressure extremes in off-equatorial latitudes.

Perturbation velocities in a two-layer model of equatorial Kelvin waves are the same as those indicated by the periodic parts of (7.63) and (7.64), with $y = 0$ on the equator and the factor $\exp(-a_j y)$ replaced by $\exp(-a_{ej}^2 y^2)$. All Kelvin waves are gravity-driven and nondispersive with phase and group velocities specified by (7.66). Equatorial Kelvin waves can only propagate eastward along the equator. Kelvin waves, therefore, can circle an ocean basin; moving toward the equator along a western boundary, eastward along the equator, and poleward again along the eastern boundary. Energy can reach or leave the equatorial ocean in this way.

In the central open ocean, energy cannot be radiated easily into higher latitudes. It was seen in subsection 7.7.1 that a constant zonal wind stress tends to produce a monotonically increasing slope of the sea surface and the internal isopycnic surfaces. When this slope can no longer be balanced by the opposing wind stress, the gravitational instability is released by a Kelvin wave, which moves water back from west to east. This development is favored by temporary surges in the off-equatorial trades, or alternatively by a weakening or reversal of the eastwinds along the equator itself.

Either change increases the cyclonic wind stress curl in the zone adjacent to the equator. This enhances the poleward Sverdrup flow. It also generates long Rossby waves, which carry energy and cyclonic vorticity toward the west. When these waves reach the basin boundary, they strengthen the equator-ward transport in the western boundary currents. The resulting confluence at the western end of the equatorial zone raises the sea surface height and lowers the thermocline depth there. The associated steepening of the slope of both the sea surface and the thermocline, increases the gravitational instability, making its release by Kelvin waves more probable. These waves carry then warm surface waters back from west to east. Acceleration in the opposite direction below weakens or suppresses the eastward transport of water in the Undercurrent.

Oceanic perturbations are generated by these vorticity inputs, as well as by temporary changes in the zonal gradient of the wind stress, the air pressure, or the surface fluxes of fresh water and buoyancy. In the terminology of subsection 7.1.1, all these changes could be represented by terms of the form $\partial^2 \zeta_F / \partial x \partial t$. The resulting local sea surface deformations and local current changes are propagated zonally through the equatorial wave guide, both westward and eastward, until they are dissipated or reflected in different forms at the meridional boundaries. Equatorial Kelvin waves can transmit signals from west to east across the pacific in about 100–120 days. transmission in the opposite direction by the fastest ($r = 1$) equatorial Rossby wave takes about 1 year. These time scales permit feedback interactions with the annual cycle in the atmosphere, which are the subject of discussion in subsection 8.4.2.

# 8

## LARGE-SCALE FORCING BY SEA SURFACE BUOYANCY FLUXES

This chapter deals with convective fluxes of sensible heat, moisture, and salinity that originate at the sea surface. In Section 8.1 we consider the relative influence of oceanic and atmospheric variability upon these fluxes. The general character of deep convection and its occurrence in the polar oceans is discussed in Section 8.2. The case of deep convection over the ocean in the tropical atmosphere, which is somewhat more complicated because of compressibility and cloud formation, is discussed in Section 8.3. Finally, in Section 8.4, we consider some of the long-term ocean–atmosphere feedback processes.

### 8.1 The predominant direction and variability of air–sea interactions

Kinetic energy in the atmosphere–ocean system is derived mainly from an upward flux of buoyancy. The resulting redistribution of mass reduces available potential energy APE and lowers the centre of gravity. In turn, APE is generated, primarily by non-adiabatic processes: unequal absorption and emission of radiation; local release of latent heat in the atmosphere; local salinity changes in the ocean; and unequal heat conduction from the boundaries.

The total mass of the oceans is about 280 times that of the atmosphere; their heat capacity is nearly 1200 times larger. Oceanic response times to external forcing are correspondingly slower. Although the annual irradiation cycle affects only a small part of the water mass, the thermal inertia is strong enough to prevent large or fast temperature variations. It is well known that this has a dominant influence on the whole terrestrial climate. This influence is particularly strong in the marine temperate regions. Figure 5.10 showed that even the daily temperature changes of the surface waters are smaller than those in the air.

By virtue of their mechanical and thermal inertia, the oceans tend to play the role of a flywheel in the air–sea system. The atmosphere is the more volatile and more variable partner. It supplies mechanical energy to the oceans at a rate that has a very skewed distribution in space and time because the work of the wind stress is proportional to the third power of the windspeed. This creates a strong bias in favour of restricted stormy areas. The heat and moisture flux tends to change linearly with the windspeed, but these fluxes are affected also by the temperature and vapour pressure differences between the water surface and the air. Short-term changes in these differences are caused mainly by the advection of air masses with different properties and not by local changes in water temperature. This is most noticeable in extratropical latitudes, where the upward flux of latent heat from the sea surface is often more than ten times larger in the cold rear of travelling cyclones than in the warm air ahead of the centre.

The mainly atmospheric origin of short-term, local heat flux variations was demonstrated by Kraus and Morrison (1966), with a statistical analysis of atlantic weather ship records. Figure 8.1, taken from their paper, compares the average daily variability of the sea–air temperature difference within calendar months with the variability due to the difference between the same months in different years. The data were collected over a period of 12 years on weather stations A (63°N, 33°W), D (44°N, 41°W), and E (35°N, 48°W). The variability is specified here by the variance of the temperature difference $V(T_s - T_a)$, where $T_s$ is the sea temperature and $T_a$ the air temperature. The variance is necessarily positive, and it is represented by the total height of the column over the zero level in Fig. 8.1. The diagrams on the left-hand side show variance estimates within months; those on the right, estimates of the variance between the means of the same months in different years. The variance can be split into two terms

$$V(T_s - T_a) = CV\{T_s(T_s - T_a)\} - CV\{T_a(T_s - T_a)\}. \tag{8.1}$$

The first term on the right-hand side, representing the covariance of $T_s$ with the temperature difference $T_s - T_a$, is a measure of the effect which temperature variations have on the sea–air temperature difference. This term is represented by the shaded part of the columns in the figure. Similarly, the effect of air temperature variations is shown by the blank parts of the columns. The figure demonstrates that, over monthly periods, the fluctuations of the sea temperature tend to contribute very little to the variance of the sea–air temperature difference; most of it can be attributed to fluctuations in the atmosphere. The day to day variability of vertical

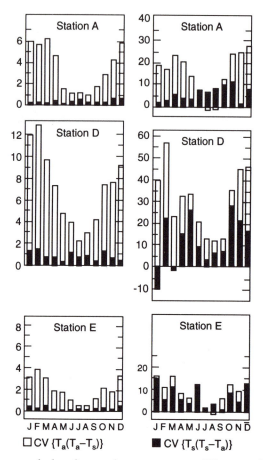

**Fig. 8.1.** Variance analysis of sea–air temperature difference (K²) at weather stations A, D, and E. Figures on left are estimates of variance 'within months'; figures on right are estimates of variance between the same calendar month in different years. The height of the shaded column represents covariance of sea temperature with sea–air temperature difference. Blank columns represent covariance of air temperatures with sea–air temperature difference. After Kraus and Morrison (1966).

transfers across the interface in temperate latitudes is largely determined by the passage of individual synoptic atmospheric disturbances. The ocean cannot cool or heat much during the lifetime of these perturbations. Sea-temperature variations become more important over longer periods. This can be seen on the right-hand diagrams of Fig. 8.1, though the large difference between different calendar months suggests that the record period of 12 years may be somewhat short for definite quantitative conclusions.

Figure 8.1 suggests also that the effect of sea-temperature variations on the sea-air temperature difference becomes relatively more important in summer. This is due to the absence of large horizontal temperature contrasts in the atmosphere during that season. Different winds do not advect air masses with greatly differing temperatures. In contrast, the sea-surface temperature variations are relatively large in summer because that is the season when the surface mixed-layer has the smallest depth.

The same argument can be used to explain the comparative increase of the oceanic influence on flux variability with decreasing latitude. Horizontal temperature gradients in the tropical marine atmosphere are weak. On the other hand, the relatively high translation speed of Rossby waves and other perturbations in low latitudes tends to make the sea-surface temperature there more variable, both in space and time. The effect is enhanced by the approximately exponential increase of the saturation vapour pressure with temperature. This causes sea surface temperature fluctuations of a given amplitude to have a much larger effect on the vertical flux of latent heat in the tropics than it does in colder regions.

## 8.2 Deep convection

### 8.2.1 *The general character and organization of deep convection*

Absorption of solar energy in the upper ocean and emission of infrared radiation from the middle and upper troposphere produce together a predominant upward flux of buoyancy through the air–sea interface. Most of this flux is contained in the mixed layers of the two media. Its role in the generation of coherent boundary layer structures and in mixed-layer deepening was discussed in Chapter 6. In this section we are concerned with convective processes, which include vertical exchanges through the whole depth of the ocean or the troposphere. In contrast to conditions in PBL, these processes tend to be driven primarily by an organized upward buoyancy flux, with shear playing a locally negligible or subsidiary role. Deep convection determines the characteristics of ocean bottom waters, as well as that of air in the upper layers of the tropical and subtropical atmosphere.

Deep convection is a highly intermittent process, which is restricted to relatively small areas. This can be explained qualitatively by the difference between the vertical buoyancy transports produced by convection and diffusion. The former is much faster than the latter. To explain the narrow regional confinement of deep-water formation, Stommel (1962) investigated the buoyancy-driven motion of fluid in an elongated, narrow tank.

The upper surface of the fluid could be cooled, strongly at one end and rather weakly at the other end. With an initially uniform temperature, cold and dense fluid is convected downward rapidly from the most intensely cooled part of the surface. Spreading of this dense fluid near the bottom generates a stable stratification throughout the basin, including the more mildly cooled regions. This tends to restrict convection everywhere to a relatively shallow surface mixed layer, except for the most strongly cooled region where cold water sinks to the bottom. The gradual filling of the basin with dense fluid from this deep-convecting region is counteracted by entrainment and small-scale turbulence, which heats the upwelling water at a relatively shallow depth over most of the basin area. A similar process results from a heating gradient along the bottom of the tank. The potentially most buoyant fluid rises from the bottom to the surface where it spreads out, causing a wide region to become statistically stable. This is later illustrated schematically for the tropical atmosphere by Fig. 8.3.

Convection in geophysical fluids necessarily involves a hierarchy of structures and length-scales. Over a warm ocean surface, small-scale turbulence, longitudinal rolls, and other boundary layer structures tend to increase buoyancy within the confined thickness of the marine PBL. Similar processes convert an upward surface buoyancy flux into a mean density increase of the ocean mixed layer. Together with the concomitant thickening of the PBL by entrainment, they cause a gradual increase of convective instability at its outer boundary. The buoyancy anomaly is advected by low-level winds into areas of convergence over the warmest sea surface areas. Within the ocean, dense surface water is similarly advected toward the most strongly cooled regions. The accumulated convective instability can be released there by random fluctuations, by the effect of boundary irregularities, or by synoptic-scale perturbations that cause additional convergent Ekman transports. The resulting deep convection can take the form of relatively large fluid parcels or of more persistent plumes, which 'cream off' the buoyancy anomaly and transport it vertically away from the boundary layers. These convective elements tend to become organized in turn into larger rotating structures.

### 8.2.2  *Laboratory experiments and dimensional analysis*

Rotation played no role in Stommel's (1962) study. Convection in a rotating tank was studied by Saunders (1973). Recent experiments by Fernando et al. (1991) involved a rotating vessel, which was heated from below. By contrast, Maxworthy and Narimousa (1993) produced convective instability by showering salt water on a revolving tank filled with fresh water. This caused first the formation of a saline, turbulent surface layer.

When this layer had gained some thickness, the instability was released by fingerlike, convective plumes. After these had spread out along the bottom, anticyclonic vortices with a larger diameter formed. At the same time, the converging fluid near the surface above began to rotate cyclonically.

In their analysis, Maxworthy and Narimousa denoted the thickness of the turbulent, but initially nonconvective boundary layer by $h_f$ and fluid surface buoyancy flux by $B$. They also stipulated an initial near-surface velocity scale of order

$$w_{*f} \propto (Bh_f)^{1/3} \tag{8.2}$$

with a corresponding Rossby number

$$Ro_f \propto \frac{w_{*f}}{fh_f} \propto f^{-1}h_f^{-2/3}B^{1/3} \tag{8.3}$$

As the saline front advanced downward, $h_f$ increases and $Ro_f$ becomes small until it approaches unity when rotational effects become significant. At that depth

$$h_f = z_c = \text{const.} \times B^{1/2}f^{-3/2}. \tag{8.4}$$

Fluid elements that sink below $z_c$ acquire cyclonic relative vorticity by vertical stretching. The associated density anomaly corresponds to a reduced gravity of order

$$g' \propto \left(\frac{B^2}{z_c}\right)^{1/3} = \text{const.} \times (Bf)^{1/2}. \tag{8.5}$$

Studies by Fernando et al. (1991) suggest values of 12.7 and 0.43 for the nondimensional proportionality constants in (8.4) and (8.5). These numbers would yield a critical depth $z_c \approx 800\,\text{m}$, if one assumes a surface buoyancy flux $B \approx 10^{-8}\,\text{m}^2\,\text{s}^{-3}$ and $f \approx 1.2 \times 10^{-4}\,\text{s}^{-1}$ in the polar ocean. The corresponding negative buoyancy would then be of order $g' \approx 0.5 \times 10^{-6}\,\text{m}\,\text{s}^{-2}$. Plume diameters could be expected to be of same order as $z_c$.

This argument is applicable only if $z_c$ is smaller than the total fluid depth $D$. In the ocean, with $D \le 4000\,\text{m}$, it follows from (8.4) that this inequality cannot be satisfied if $B > 1.2 \times 10^{-7}\,\text{m}^2\,\text{s}^{-3}$, which corresponds to an upward heat flux of only about $10 \, \text{W m}^{-2}$ at a water temperature of $0°C$. Surface cooling of the polar ocean will frequently be much more intensive. Similarly, in the equatorial atmosphere, values of $z_c$ as derived from (8.4) tend to exceed the height of the troposphere because $f \rightarrow 0$. The rotation of air in waterspouts and tornadoes is probably not influenced significantly by the Coriolis force, but is caused primarily by convergence in the presence of horizontal wind shear and by the tilting of horizontal vortex lines into the vertical direction.

Dimensional reasoning suggests that the radius of nonrotating plumes might be of order

$$r \propto \left(\frac{K^3}{B}\right)^{1/4}, \qquad (8.6)$$

where $K$ is an eddy coefficient that characterizes lateral mixing by small-scale turbulence at the periphery of the plume. Dimensional considerations also suggest that in this case

$$g' \propto \left(\frac{B^3}{K}\right)^{1/4}. \qquad (8.7)$$

The proportionality factors in (8.6) and (8.7) are uncertain. In physical terms, both (8.5) and (8.7) indicate that the density anomaly in the plumes increases with $B$. The last equation shows also that $g'$ is reduced by lateral mixing, as could be expected. If $K$ were relatively large, then only plumes with a sufficiently large diameter could survive and this is indicated by (8.6).

After convective plumes or fluid parcels have reached the bottom at depth $D$, the density anomaly is spread outward by internal waves that propagate with a velocity of order $c \approx (g'D)^{1/2}$. The local density anomaly is ultimately balanced by a quasi-geostrophic anticyclonic vortex with a diameter that corresponds to the local deformation radius $R_d = c/f$. A corresponding cyclonic vortex is formed above.

### 8.2.3  *Deep convection and bottom-water formation in the oceans*

The oceans are stably stratified. The water with potentially the highest density is found close to the bottom. In the present terrestrial climate, they are produced by the sinking of very cold surface waters down to great depth. Figure 2.1 and Table 2.1 indicated that small temperature fluctuations have relatively little effect on the density of sea water near $0°C$. The relative influence of salinity variations on sea water density becomes

arly in the eastern Mediterranean, where the large evaporation–
ation difference engenders a salinity of 38.65 ppt. With a tempera-
+13.7°C these waters have a specific volume anomaly of
ton$^{-1}$. However, the Strait of Gibraltar has a sill depth of only
nd the Greenland–Scotland ridge rises to a depth of 600–800 m.
editerranean and Arctic bottom waters mix with more buoyant
s they escape across these ridges and flow down the slope on the
de. This makes them less dense than the Antarctic Bottom Water
/). The NADW, which is formed by a combination of water from
rflow across the Greenland–Scotland ridge with water produced by
ion in the Labrador Sea, spreads out all across the world ocean
ve the AABW. Mediterranean water is actually still denser than
BW after its exit from the Gibraltar Strait, but the relatively small
ge volume mixes with the environment. As a result, it cannot sink
elow 1200 m and remains confined mainly in the North Atlantic.
p water production affects the meridional distribution of $CO_2$ in the
here. The solubility of this gas decreases with increasing tempera-
s shown, for example, in Table 2.3. Following anthropogenic
n of $CO_2$ into the atmosphere, a new state of near, but not perfect,
ium can be established rather rapidly with the oceanic mixed layer
cal and temperate latitudes. In high latitudes, deep convection in
an will continue to act as a sink, resulting in an increased
nal gradient and poleward transport of atmospheric $CO_2$. This
affairs can be expected to last until the deep water has come into
ium with the atmospheric $CO_2$ concentration. During this period,
ar oceans will continue to act as cold spots that drain both $CO_2$ and
m the atmosphere. Bretherton (1982) estimated that this process is
o inhibit a rapid rise in atmospheric temperatures for at least 30
rhaps several hundred years. A better quantitative assessment of
ay time would require consideration of the effect of surface heating
p convection. This could be obtained only in the framework of
l ocean–atmosphere models, which could provide prediction of
, run-off, and cloud distribution.

## he tropical atmosphere

### Deep convection in the presence of clouds and precipitation

ix of sensible and latent heat from the oceans is largest over the
l half of the globe. It provides the major part of the energy that
the global atmospheric and hydrological circulations. Heat is

correspondingly larger at low temperatures. To s
surface waters have to be exposed to intense surf
have to be preconditioned by a relatively high salini
are too fresh, they freeze without sinking, however
be.

The best-documented descriptions of deep con
by the MEDOC group in the Mediterranean
summarized by Killworth (1976) and by Gasca
processes in the Labrador Sea have been analyzed
(1983). These studies confirm that deep convec
process, which tends to involve transient, intens
saline water. The process is commonly associated
cold winds. The hierarchical arrangement of differe
during convection is similar to that observed in the
Active convective plumes or 'chimneys' were re
Clarke (1983) to persist for a few hours only and t
than one kilometer. In the Labrador Sea these chi
reach the bottom, but they communicate with th
Water (NADW), about 2000 m below. Several of 1
are embedded in mesoscale rotating eddies wi
kilometers, corresponding to the local, internal
These eddies in turn drift within a larger area
circulation.

The production of deep water always occurs
atmosphere. It can involve free convection in the
currents that sink along the slope of the contine
1979). Water in freely convecting plumes can arri
with relatively little dilution. On the other hand, de
a rough surface. They are always very turbulent a
good deal of water from the overlying strata. Th
limit the depth to which these waters can sink and
out. This depth depends upon the density an
transport of the current, as well as upon the chara
water that is produced by free convection and by
Petersen (1978) computed the different depths
density currents within the same basin could sir
model can be found also in Kraus et al. (1978).

The bottom waters of the world ocean orginate
of Antarctica. With a potential temperature of
about 34.66 ppt, the specific volume of Weddell
from that of standard sea water ($T = 0°$, $s = 35.0$
Much denser waters can be found at depth in

advected poleward by the atmosphere until, ultimately, much of it is lost into space in the form of radiation. Evaporation from the tropical oceans feeds the monsoon rains and most of the great rivers that drain the continents. All these events are affected significantly by sea surface temperature variations and by the resulting ocean–atmosphere interactions.

The vertical distribution of moisture tends to reduce the hydrostatic stability of the tropical atmosphere. This stability can be specified in a number of ways. It is basically determined by the vertical gradient of moist air entropy as formulated by (2.30). It follows from (2.31), that this entropy can be specified also with good approximation by the logarithm of the equivalent potential temperature $\Theta_e$. As an alternative, some investigators have used the *moist-static energy*, which is defined by

$$\sigma = gz + c_p T + Lr,$$

where $r$ is the mixing ratio and $L$ is latent heat. A careful comparison of $\Theta_e$ and $\sigma$ by Madden and Robitaille (1970) shows $\Theta_e$ to be conserved exactly during pseudo-adiabatic changes, while $\sigma$ is conserved only approximately. The difference is not significant for practical purposes.

A typical vertical distribution of $\Theta_e$ is reproduced in Fig. 8.2. Its decrease from the surface to a minimum in the lower troposphere is characteristic for the marine tropics. The level of this minimum coincides approximately with the top of the cloud-topped mixed layer, discussed in Section 6.4.4. It also coincides with the level of maximum radiative cooling. Above that level, the cooling tends to decrease with height, though the radiation balance remains negative on the average throughout the troposphere. The high values of $\Theta_e$ just below the tropopause must therefore be associated with a convective supply of heat from lower levels and ultimately from the sea surface.

The concept of convective instability was specified in Section 2.2. When an air layer with upward decreasing $\Theta_e$ is lifted until it becomes saturated throughout, it attains a lapse rate in excess of the moist adiabatic, which makes it absolutely unstable and causes it to overturn. Condensation can also result from local layer thickening, caused by internal gravity waves in the inversion and the stable region above. Under suitable conditions, this process may help to explain the occurrence, spacing, and quasi-regular arrangement of trade-wind cumuli.

When air is lifted pseudo-adiabatically, without mixing with the surrounding air, its equivalent potential temperature $\Theta_e$ remains constant.

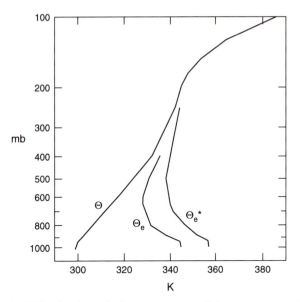

**Fig. 8.2.** Vertical distribution of the mean potential temperature $\Theta$ and of the equivalent-potential temperature $\Theta_e$ in the Caribbean during the hurricane season of July–October. $\Theta_e^*$ is the equivalent-potential temperature of saturated air at the same temperature and pressure. After Jordan (1958).

Therefore, the process is represented by a straight vertical line in Fig. 8.2. The curve marked $\Theta_e^*$ represents an equivalent potential temperature, which the environmental air would possess if it received enough additional moisture to become saturated without change of pressure or temperature. As indicated by Ooyama (1969), the lifted air is always more buoyant than the ambient air above the level where the vertical line from its place of origin intersects the $\Theta_e^*$ curve. Above this level, the displaced parcel is accelerated upwards. In the absence of mixing with the environment, one can expect air to ascend to a level where its equivalent-potential temperature becomes again approximately equal to that of the ambient air. This tends to occur high in the tropical troposphere, usually above the 200-mb level (i.e., above a 12-km height). As pointed out by Betts (1982), the actual rate of ascent to that level will be affected significantly by the greater density of the air–water mixture in clouds, as compared to saturated air that contains no condensed water.

Malkus (1962) suggested that no ascent of surface air to the tropo-pause, without significant dilution by mixing with environmental air, can occur in towering tropical cumulus clouds, the 'hot towers'. The diameter of these structures would have to be larger than the one indicated by (8.6). Air that is being pumped into the upper tropical troposphere within these towers can then retain the characteristic equivalent potential temperature of the surface mixed-layer over the tropical oceans. After it has spread out below the tropopause, it may sink again at some distance from where it has risen. The sinking air is cooled by radiation until some of it reaches the top of the PBL, where its heat content is a gain augmented by turbulent transports from the sea, until it is drawn off once more by another hot tower.

The observed vertical distribution of $\Theta_e$ in the undisturbed tropical atmosphere can be explained qualitatively by the subsidence and radiative cooling of air with an initially very high equivalent potential temperature. At a relatively low level, the subsiding air reaches a $\Theta_e$ minimum, through which it is entrained into a shallow mixed layer above the sea surface. Below the tropopause, the equivalent potential temperature of the subsiding air is about the same as that of air in contact with the warmest sea surface region. In this way the sea surface temperature determines the value of $\Theta_e$ in both the boundary layer and below the tropopause. By itself, this model does not explain the source of the threshold energy that is needed to lift the surface air to a level, where it becomes more buoyant than its environment. The supply of this threshold energy is the topic of Section 8.3.2.

### 8.3.2  The InterTropical Convergence Zone (ITCZ) and the Hadley circulation

A schematic meridional cross-section of the marine tropical atmosphere is shown in Fig. 8.3. Following Emanuel (1988), regions with different regimes are numbered 1–4. Region 1, the locus of deep convection, is the ITCZ. Reasons for the narrow meridional width of that zone were discussed in subsection 8.2.1. Although the core of the hot cumulus towers may contain undiluted air with near-surface values of $\Theta_e$, intensive mixing is likely to occur along their periphery. The equivalent potential tempera-ture of the air that sinks between the towers is therefore not much lower than that of the rising air. Just below the tropopause, very cold air $(T < -60°C)$, which cannot contain much moisture, flows from region 1 into region 2. As this air flows poleward and subsides, it loses heat, and hence potential temperature, by radiation, although it gains sensible heat and becomes warmer by compression. The trade inversion constitutes the top of the cloud-capped surface mixed layer. It is generally found at 1–3 km altitude. The subsiding air is entrained through this inversion from region 2 into region 3, as was described in Section 6.4. Its moisture content is increased in region 3 by evaporation from generally nonpre-cipitating trade-wind cumulus clouds. Finally, it enters the subcloud layer, region 4, which is typically only about 500 m deep. Under undisturbed

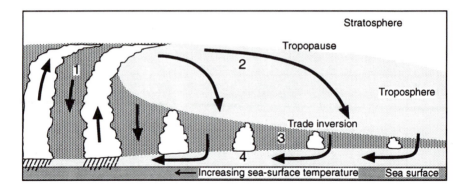

**Fig. 8.3.** Schematic meridional cross-section of the marine tropical atmosphere. The circulation is driven by a sea surface temperature increase of about 7°C from the left side to the right side of the diagram. Air ascends in the deep-convective region 1. It slowly subsides in region 2, losing energy by infrared radiation. In region 3 it is moistened by evaporation from nonprecipitating trade-wind cumuli. It returns toward low latitudes in the boundary layer, region 4, where it receives sensible and latent heat from the ocean. After Emanuel (1988).

conditions, the relative humidity within this layer is generally about 75–80 per cent. As the air flows toward the equator over increasingly warmer water, its vapour-mixing ratio, and hence its equivalent potential temperature $\Theta_e$, increases again, until it ascends once more into the ITCZ. Figure 8.3 is a two-dimensional section of the three-dimensional *Hadley circulation,* which also involves zonal winds in region 2 and the trades which blow mainly from east to west in regions 3 and 4. The Hadley circulation is driven primarily by the density difference between regions 1 and 2. It can be expected that it will be speeded up whenever an excess of heat is released in the ITCZ. The time needed for this acceleration to effect the trade winds in other latitudes is equal to the travel time of meridionally propagating, internal gravity waves. Ultimately it will tend to produce a surge in the low-level wind, causing an instantaneous increase of the local heat and vapour flux from the sea. It then may take 2 or 3 weeks for the resulting excess moisture to reach once more the convective area of the ITCZ. The whole sequence of processes can give rise to oscillations with preferred periods that correspond to the average lag time between evaporation in the subtropics and precipitation in the ITCZ (Kraus, 1959).

Average values of $\Theta_e$ just below the tropical tropopause are of order 360 K. Saturated surface air at a pressure of 1000 mb can have an equivalent potential temperature of 360 K only if its actual temperature is about 27°C. Cooler or dryer air cannot penetrate to the tropopause. A threshold sea surface temperature of 27°C seems to be a necessary condition for deep convection in the ITCZ. In general, the air in immediate contact with the surface is not saturated. As noticed by Riehl (1963), the system begins to become locally unstable under these conditions only when air from the surface mixed layer has been lifted well above the condensation level. Energy for the initial lifting has to be supplied by the large-scale flow, mainly through Ekman transport convergence. Deep cumulus clouds or hot towers can develop only after this stability threshold has been overcome.

The relatively large specific volume of the warm, moist air within the hot towers causes the isobaric surfaces above to bulge upward. The amplitude of this local bulge increases with height. The resulting pressure gradients cannot be balanced geostrophically by the weak Coriolis force of low latitudes. In other words, the lifetime of cumulus towers is very short compared to the half-pendulum day which characterizes the time scale needed for geostrophic adjustment. A stratified fluid reacts to such an unbalanced, impulsive perturbation with the emission of internal gravity waves. These can transport both energy and mass away from the perturbed region in the upper troposphere. The horizontal mass flow divergence causes a reduction of pressure below. The resulting inward-directed pressure gradient near the surface favours the additional convergence of moist boundary layer air into the perturbed region where it can be used to fuel additional cumulus development. This cycle of events was first analyzed by Charney and Eliassen (1964), who called it *Convective Instability of the Second Kind* (CISK). It represents a self-amplifying process, which will continue until suppressed by friction or by an exhaustion of the supply of moist, warm boundary layer air.

The ITCZ is particularly persistent and sharply defined over the North Pacific and Atlantic between 5 and 10°N. It occasionally appears over the southeastern Pacific between 5 and 10°S. It is rarely, if ever, found at the equator itself, except in the vicinity of the continents. These peculiarities are probably the result of ocean surface conditions. Various mechanisms have been proposed to account for the existence of a conspicuously cloud-free 'equatorial dry zone'. Bjerknes et al. (1969) suggested that the equatorial upwelling, discussed in Section 7.7.1, brings cool water to the surface and that this inhibits convection above. The same conclusion was drawn by Pike (1971) from the numerical integration of a coupled atmosphere–ocean model. On the other hand, Charney (1966) proposed a model that does not require an equatorial minimum in the sea surface temperature. Location of the zone of maximum convection depends in Charney's model on a decrease with latitude of convective instability and an increase with latitude of Ekman transport convergence. Together, these two processes produce maximum growth rates, and hence ITCZ development, several degrees away from the equator.

There is some weakness in both these explanations. The equatorial cold-water belt does not exist in the western Pacific, but the ITCZ continues to be located at a distance of several hundred miles from the equator even there. Charney's conditions for instability are also absent at many ITCZ locations. It is conceivable that these inconsistencies are due to the assumption of stationarity upon which both hypotheses are implicitly based. Analysis of satellite photographs has shown that the traditional image of the ITCZ, as a more or less stationary and continuous line of clouds, is not true. It consists rather of a number of cloud clusters, which travel from east to west and are separated by large expanses of relatively clear skies. This led Holton et al. (1971) to a view that considers the ITCZ as the locus of cloud clusters that are associated with zonally propagating, wavelike perturbations.

The propagation and the velocity field associated with these perturbations are governed by the relations discussed in subsection 7.7.2, which is applicable to both oceanic and atmospheric, equatorial waves. As indicated by (6.9), the Ekman layer becomes infinitely deep at a latitude where the perturbation frequency $\omega = f$. In the case of a continuous frequency spectrum, this singularity is spread out over a zonal belt. Holton and his colleagues (loc. cit.) found that it is associated with a maximum of horizontal convergence in the PBL. The air above is pushed upward, which favours the formation of cloud clusters. The convergence is further enhanced by the CISK process (i.e., by the work of buoyancy in the hot towers), which the clusters contain. Clusters are observed to follow each other at intervals of about 4–5 days, i.e., with an angular frequency $\omega \approx 1.5 - 2 \times 10^{-5}\,\mathrm{s}^{-1}$. The corresponding critical latitude range is 6–7°, which is close to the most common ITCZ position.

The frequency and wavenumber of ITCZ perturbations are necessarily also affected by the conditions at the interface. It may be argued that high sea surface temperatures and a good moisture supply at a particular latitude will favour the release of latent heat in perturbations that travel westward with a frequency that corresponds to the value of $f$ at that latitude. Explanations of the ITCZ position as the locus of maximum sea-surface temperatures or of perturbations with frequencies $\omega = f$ are, therefore, not mutually exclusive. If the preferred latitude and frequency of the tropical perturbations depend on sea-surface temperature, the opposite holds also true: the distribution of water masses near the surface and their temperature reflects in turn the atmospheric circulation above. Conditions in both media are variable. Once perturbations have formed, however, they will tend to persist at the same latitude, with their energy maintained by the inherent convective instability of the tropical atmosphere–ocean system.

### 8.3.3  *Hurricanes*

The hurricane literature is extensive. A monograph on the topic has been published by Anthes (1982). Historical and climatological data have been compiled by Dunn and Miller (1964) and by Alaka (1968). Current hurricane theory has been discussed by Emanuel (1987), Rotunno and Emanuel (1987), and, in a more condensed form, by Emanuel (1988).

The word 'hurricane' was derived from the name of the evil god Huracan of the Caribs. It was applied originally only to storms in the western tropical Atlantic. Similar storms are called typhoons in the western Pacific area. They are simply known as cyclones in Australia and on the Indian subcontinent. For simplicity's sake, we shall use the word 'hurricane' for all vigorous, revolving tropical or subtropical storms, wherever they may occur. These storms can cause storm surges that kill people by the thousands or hundred of thousands, as mentioned in subsection 7.5.4. When hurricane Andrew crossed the southern tip of Florida in August 1992, its winds alone caused property damage estimated at some $20 billion.

All hurricanes conform to the same universal pattern, but no two hurricanes are the same. This justifies their having individual names. As a first approximation, hurricane dynamics can be considered an axisymmetrical analogue of ITCZ dynamics. The main differences are caused by the somewhat higher latitude, which makes the Coriolis force more important, and by the much lower central surface pressure, which has been observed in some hurricanes to drop below 870 mb.

The ascent of undisturbed surface air through the troposphere cannot produce a buoyancy distribution that would cause the very low pressures found in the centre of active hurricanes. It appears that the development of such storms involves yet another amplification process that further enhances the equivalent potential energy of the ascending air. The air in the marine boundary layer expands as it spirals inward to progressively lower pressures near the storm's centre. However, sensible heat supply from the warm ocean surface below causes its temperature to remain practically constant. The air therefore begins its ascent near the storm centre with a higher specific entropy, $S$, or equivalent potential temperature, $\Theta_e$, than is found in the undisturbed tropical marine PBL. For constant sea surface temperature, $T_0$, it follows from (2.30) or (2.31) that

$$c_p^{-1} \frac{\partial S}{\partial p} \approx \frac{1}{\Theta_e} \frac{\partial \Theta_e}{\partial p} = -c_p^{-1} \left( \frac{1}{p-e} \frac{Lr_*}{T_0} + \frac{R}{p} \right). \tag{8.8}$$

The first term in brackets on the far right-hand side of (8.8) represents the potential temperature increase, caused by a pressure decrease at constant temperature $T_0$. It follows from definition (2.23) that the saturation mixing ratio, $r_*$, must also increase with decreasing pressure, if the saturation vapour pressure $e_s(T_0)$ remains invariant. This augmentation of $r_*$ is represented by the last term. With a pressure drop $\Delta p = -60$ mb along the trajectory and an ocean surface temperature $T_0 = 300$ K, (8.8) indicates an increase in $\Theta_e$ of about 6°C. The importance of this additional energy supply to air in a hurricane was first stressed by Malkus and Riehl (1960).

Transients of the temperature and pressure distributions are likely to be small during the period when a fluid parcel spirals in towards the storm centre. Using (5.77), the sensible heat balance of a well-mixed hurricane inflow layer can therefore be described by

$$\rho c_p \left[ \frac{\partial(\Theta h)}{\partial t} + \frac{\partial(rhU_r\Theta)}{r\,dr} + C_Q U(\Theta_0 - \Theta) \right] \approx 0, \tag{8.9}$$

where $h$ is the height of the inflow layer, $r$ the distance from the storm centre, $U_r$ the radial component of the near-surface velocity $U$, and $\Theta$, $\Theta_0$ are the potential temperatures of the mixed layer and of the sea surface. The third term in (8.9) indicates the sensible heat transfer from the sea surface. Under steady-state conditions ($\partial/\partial t = 0$), an estimate of the inflow angle $\sin \alpha_0 = U_r/U$ can be derived from (8.9) as a function of the temperature and pressure distributions.

The radial inward increase of the vapour mixing ratio in the inflow layer can be described by an equation that has the same form as (8.9). Assuming the coefficients $C_Q$ and $C_E$ in (5.77) and (5.78) to be equal, one can express both changes in terms of $\Theta_e$.

$$\frac{\partial(\Theta_e h)}{\partial t} + \frac{\partial(rhU_r\Theta_e)}{r\,\partial r} + C_Q U(\Theta_{e0} - \Theta_e) \approx 0, \tag{8.10}$$

where $\Theta_e$ and $\Theta_{e0}$ are equivalent-potential temperatures of air in the mixed layer and of air in immediate contact with the sea surface.

When the storm crosses a coastline, $\Theta_{e0}$ is reduced immediately. The associated rise of central pressure and reduction of storm intensity are demonstrated by Fig. 8.4, which is based on Ooyama's (1969) calculations. Similar, but more gradual effects are produced when the storm moves over a colder sea surface. Stratification does not seem to affect $C_Q$ in (8.10) significantly because the strong hurricane winds automatically maintain a

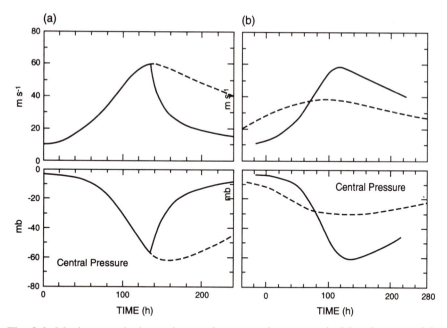

**Fig. 8.4.** Maximum velocity and central pressure in a numerical hurricane model. (a) The effect of landfall. The full line represents conditions with the energy flux from the sea surface cut off at $t = 134$h of simulated hurricane development time. (b) Effect of sea surface temperature. The full line represents developments with $T_s = 27.5°C$, dashed line with $T_s = 25.6°C$. The time of origin of the latter case is $-100$ h (off the diagram); initial conditions were identical. After Ooyama (1969).

forced convection regime. A study by Leslie and Smith (1970), who assumed that $C_Q = C_D \sim (u_*/G)^2$, suggests that the boundary layer inflow and the induced vertical circulation depend critically on radial variations of this parameter. The symbol $G$ stands here for the *gradient wind.* As discussed in Section 1.5 it involves a balance between pressure, Coriolis, and inertial forces. For circular motion it is defined by

$$\frac{G^2}{r} + fG = \rho^{-1} \frac{\partial p}{\partial r} \equiv fU_g,$$

which makes $G$ smaller than $U_g$ in a cyclonically revolving storm $(r > 0)$. The ratio $u_*/G$ increases with the surface roughness length $z_0$, as is indicated by (5.22) and (5.76). When a hurricane moves inland, the increase in surface roughness contributes to the attenuating effects of a reduced latent heat supply from the surface.

The frictionally induced boundary-layer inflow $hU_r$ decreases sharply as the air approaches the eye-wall. The resulting convergence of the boundary-layer flow is responsible for the continuing convective activity in the inner rain area of the storm. On the other hand, the boundary-layer flow is also nearly zero at a large distance from the storm centre where the winds are weak. It follows that the boundary-layer inflow must have a maximum at some distance beyond the eye-wall, and that this distance must depend critically on horizontal variations in turbulent mixing. Outside this belt of maximum inflow, the Ekman layer flow is divergent. It sucks air downward from the free atmosphere into the PBL. This tends to decrease the equivalent-potential temperature of the boundary-layer air by mixing with drier air from above. The convective activity in the storm area can be maintained only if the supply of heat and moisture from the ocean is sufficiently rapid to compensate for this dilution of the boundary-layer air.

The central pressure in a steady-state hurricane can be expected to be a function of the sea surface temperature. To estimate a lower bound for this pressure, Emanuel (1988) treated the phenomenon as a Carnot cycle, which converts thermal into mechanical energy and has an efficiency

$$\epsilon \equiv \frac{T_0 - T_1}{T_0},$$

where $T_0$ is the temperature of the heat source and $T_1$ that of the sink or cold source. In the present case $T_0$, the sea surface temperature, is approximately 300 K; $T_1$ was considered by Emanuel to be about 200 K. This would give the hurricane heat engine an efficiency of 33 per cent, which seems rather high.

The amount of heat that is received by an air parcel as it spirals in toward the eye-wall is $\Delta Q = T_0(S_c - S_a)$ where $S$ is the entropy and the subscripts $c$ and $a$ denote boundary-layer conditions near the centre and outside the periphery of the storm. Carnot's theorem indicates that a complete circuit of a unit mass of air through the hurricane system (i.e., rise in the eye-wall), sinking in the environment and return to the eye-wall in the surface boundary layer, generates an amount of mechanical energy that is given by

$$E = \epsilon \Delta Q = \epsilon T_0(S_c - S_a) \approx \epsilon c_p T_0 \ln \frac{\Theta_{ea}}{\Theta_{ec}} \qquad (8.11)$$

The generation of this mechanical energy is almost totally balanced by frictional dissipation at the ocean surface. As the air is driven against friction by the radial pressure gradient, a steady state can be maintained only if frictional dissipation is balanced in turn by the work of the pressure force. This implies that

$$\rho^{-1}\Delta p = RT_0 \ln \frac{p_a}{p_c} = -E. \tag{8.12}$$

Elimination of $E$ between the last two equations yields

$$p_c = p_a \left(\frac{\Theta_{ea}}{\Theta_{ec}}\right)^{\epsilon/R}. \tag{8.13}$$

Since $\Theta_e$ is itself a function of $p$, the relation (8.13) represents a implicit estimate of the central pressure $p_c$ as a function of conditions in the undisturbed tropical atmosphere. This estimate is a lower bound because the balance represented by (8.12) is generally not achieved completely. A comparison with observations can be found in Emanuel (1988).

Observational data suggest that hurricanes do not develop if the surface mixed layer of the underlying tropical ocean is less than 60-m deep. Shallower layers may be cooled too rapidly by the entrainment of colder water from below, as discussed in Section 7.5 (see Fig. 7.5). After development has started, the storm draws large quantities of water from the surface. Using tritium as a tracer, Ostlund (1967) found that more water is evaporated within a hurricane inflow area than is precipitated locally. Export of the remaining water mass from the storm area may be a significant factor in the global hydrological circulation.

Hurricanes must collect latent heat from the ocean to survive. They can develop only because the warm upper ocean and the unsaturated air above represent a system that is far from equilibrium. However, this development can be initiated only by convergence of air into a pre-existing, near-surface pressure minimum. In general, such minima come into existence through wavelike perturbations of the general tropical circulation, but they can also be produced by inertial instabilities at frontal boundaries. This last process seems to play a significant role in the development of revolving storms in the Arctic. The intensity of these storms can approach that of tropical hurricanes. Like the latter, they draw their energy from a large entropy or equivalent potential temperature difference between the sea surface and the overlying air. However, in contrast to tropical storms, the instability is caused more by differences in sensible heat than by differences in moisture content. Large, highly unstable, vertical temperature gradients can occur in air that streams out over the open ocean from an ice-covered surface. A discussion of revolving storms in the Arctic, together with a list of relevant references, can be found in a review by Businger (1991).

## 8.4   Some low-frequency ocean–atmosphere feedback processes

Climate records, such as sea surface temperature, surface pressure, precipitation, cloudiness, and so on, are measured at a large number of locations, usually at discrete time intervals. Some of these variables, such as satellite pictures or river discharges, involve a measure of spatial and/or temporal integration. Investigators have found numerous signals suggesting teleconnections (i.e., covariances or lag correlations) between different data series at different places. Some of these relationships may be real and have global climatological implications. The problem is that the system is noisy and that the time covered by pertinent, quantitative observational records is relatively short. This makes it difficult to demonstrate the existence of stochastic connections with a high degree of confidence. For the same reason, it is also difficult to find convincing statistical support for assumed, conceptual relationships.

We shall restrict ourself here to a description of various conceptual models of coupled ocean–atmosphere fluctuations. Models of this type illustrate current physical ideas, but they also distort reality through gross simplifications, or through reductions of the actual geographical variability. Numerical simulations with general circulation models (GCM) can be more realistic, but by trying to represent the complexities of the real world they also become inevitably less transparent physically. A full account, of the many GCM studies that dealt—one way or other—with fluctuations of the ocean–atmosphere system, goes beyond the scope of this book. Comprehensive lists of references can be found in the various review papers that are quoted later, separately for each of the following subsections.

The first two of these subsections deal with tropical phenomena. The presence of cold water, not far below the warm surface layer in large parts of the tropics, permits large and relatively rapid changes of the sea surface temperature. These changes can be initiated by shifts in the surface wind stress. When they do occur, they can in turn greatly influence the state of the atmosphere above. Coupled interactions between the tropical atmosphere and ocean are facilitated further by the absence of rotational constraints, which makes the upper equatorial ocean particularly sensitive to relatively small wind stress changes. The travel time of equatorial perturbations across ocean basins is also sufficiently fast to match the seasonal cycle in the atmosphere. Together these arguments suggest that observed annual and interannual fluctuations, in the upper Pacific and in the atmosphere above, result from a dynamic coupling between the two media. This topic is discussed in Section 8.4.1. A different line of reasoning, reviewed briefly in Section 8.4.2, links monthly and annual fluctuations of the Indian monsoon with sea surface temperature variations in the Arabian sea. The last subsection deals with much slower, extratropical interactions between deep water formation and the hydrological cycle in the atmosphere.

### 8.4.1    El Niño and the Southern Oscillation (ENSO)

Anomalously warm water appears off the coast of Peru and Ecuador at recurrent intervals of about 3–8 years. The phenomenon can have catastrophic consequences locally. Known as El Niño, it has been observed over several centuries and thus represents the best-documented, quasi-periodic climate fluctuation with an interannual time-scale. During his studies of the Indian monsoon, Sir Gilbert Walker, the director of the Indian weather service, noticed an oscillation of similar period in the surface air pressure difference between the Southern Pacific (Tahiti) and Indonesia (Jakarta). Walker (1928) called this the *southern oscillation*. Possible worldwide repercussions of this oscillation were investigated statistically in a series of papers by Berlage, who summarized his findings in 1966. A physical theory, connecting the southern oscillation with the El Niño through air–sea interaction processes, was first conceived by Bjerknes (1966, 1969).

Since Bjerknes published his seminal paper, interest in the topic has grown exponentially We shall limit ourselves therefore to a brief account of physical concepts that are widely accepted by the research community. A collection of relevant papers can be found in a book edited by Nihoul (1985). Comprehensive reviews of theories and relevant observations have been published subsequently by Enfield (1989) and Philander (1990). The predictability of the phenomenon has been discussed specifically by Barnett et al. (1988) and Latif et al. (1990). Results obtained in general circulation models are compared in a compilation by twenty authors, who were all involved in this type of work (Neelin et al., 1992). All of these publications contain comprehensive lists of references. The writing of this section was facilitated by McCreary and Anderson's (1991) overview of ENSO models. However, the authors warn that their summary will soon be superseded by ongoing research.

The equatorial current system was considered briefly in subsection 7.7.1. On the equator, off the coast of South America, the ocean surface temperature is much of the time about 8°C colder than its latitudinal mean. The equatorial upwelling, discussed in Sections 7.3.1 and 8.3.2, as well as advection of cold water from boundary currents along the margin of the South American continent, produce a cold water tongue along the equator. Much of the time, that tongue extends westward from South

America over a distance of about one hundred degrees of longitude, or more than 10,000 km. The cold water temperatures inhibit an upward heat flux from the ocean surface, suppressing convection. The ensuing dearth of clouds along the equator has been discussed in subsection 8.3.2.

By contrast, ocean surface temperatures are always relatively high among the islands of Indonesia and the western Pacific. Convection, and the associated release of latent heat in the air above, lifts isobaric surfaces upward in the upper troposphere. On the other hand, the absence of significant condensation heating over the eastern equatorial Pacific, causes the pressure in the upper troposphere to be relatively low. The resulting pressure gradient from west to east cannot be balanced on the equator by the Coriolis force. It therefore drives a direct zonal circulation in the equatorial plane with air rising over Indonesia, west winds in the upper troposphere, and east winds near the surface. This circulation is intensified by the CISK process described in Section 8.3.2. A second cell, over the Indian Ocean to the west, produces a circulation in the opposite sense, with a descending branch over cold waters off the Somali coast and a low-level acceleration from west to east along the equator in the lower atmosphere. Bjerknes coined the name *Walker circulations* for these two circulation cells in the equatorial atmosphere.

The Pacific Walker cell and the Hadley circulation both involve low-level east winds in low latitudes. They both convert potential energy directly into kinetic energy. However, relatively warm equatorial surface temperatures strengthen the Hadley circulation and the associated trade winds. On the other hand, cold surface waters in the eastern equatorial Pacific boost the Walker cell, and this reinforces the low level easterlies along the equator. Bjerknes (1969) argued, that the resulting increased surface wind stresses along the equator enhances upwelling of cold water in the eastern, equatorial Pacific. The equatorial surface temperature anomaly is, therefore, self-amplifying at this stage. As the tongue of cold water becomes wider and longer, the amount of energy that is fed into the meridional Hadley circulation becomes smaller. The associated weakening of the trades over the eastern Pacific will be particularly pronounced if it coincides with their seasonal decline. After some time, the ensuing reduction in the oceanic Ekman transport divergence lessens the oceanic east–west temperature gradient along the equator, and hence the energy input into the Walker circulation. A further decrease of the easterlies consequently takes place, which may ultimately cause them to be replaced by predominantly light variable winds or calm along the equator. Cessation of upwelling permits then a temperature rise of the equatorial surface waters in the eastern Pacific, which enhances the meridional temperature gradient. The resulting intensification of the Hadley circulation augments

Ekman transport divergence, causing renewed surface cooling in the eastern equatorial Pacific, until the whole cycle repeats itself. As the Walker circulation involves an east–west surface pressure gradient, the stipulated cycle could explain both the southern oscillation and the quasi-periodic appearance of warm water off the coast of South America.

Later oceanographic data show that warm episodes in the eastern Pacific involve also a large flux of heat within the ocean, from west to east, along the equator. This redistribution of heat entails a change in mixed-layer depth which can be associated with the propagation of equatorial Kelvin waves towards South America. As indicated in Section 6.4, entrainment tends to be inversely proportional to mixed-layer depth. Shallow layers are cooled rapidly by entrainment of cold water from below. Many models of the ENSO cycle use therefore the mixed-layer depth as an indicator of both surface temperature and upper-ocean heat content. This usage is reasonable for the eastern equatorial Pacific. It is less warranted in the west, where surface temperatures are less sensitive to changes in the relatively deep mixed-layer depth that commonly exists in that region.

Modern conceptual models of the ENSO cycle can be divided into two categories: dynamic ocean models with a passive atmosphere and models involving interactions with a thermodynamically active atmosphere. One example of the first approach can be found in a paper by McCreary (1983). It entails linearized equations of motion for the ocean. The model atmosphere consists simply of two patches of zonal wind stress. The eastern equatorial Pacific is assumed to be cold when the depth $h$ of the mixed-layer or pycnocline there is less than a stipulated critical depth $h_c$. A Walker circulation with westward wind stress $\tau_w$ along the mid-Pacific equator is then switched on. When the eastern equatorial Pacific is warm $(h > h_c)$, the meridional temperature gradient is large, and that implies a Hadley circulation with an extra-equatorial, westward wind stress $\tau_h$ over the eastern part of the ocean. The resulting Ekman divergence and upwelling reduce the layer thickness along the equator. This ultimately causes $h$ to become smaller than $h_c$, and the Walker circulation $(\tau_w > 0)$ is then switched on. This makes the stress curl in low latitudes off the equator negative. The ensuring deepening of the layer there causes the emission of westward-propagating Rossby waves. After reflection at the western boundary, these perturbations return as equatorial Kelvin waves and that increases the layer thickness until $h > h_c$. The cycle can then repeat itself. Its time-scale is determined by the travel time of the Rossby waves to the western boundary, plus their return time after reflection there.

A modification of McCreary's original model by McCreary and Anderson (1985) leads to solutions that adjust to one of two equilibrium

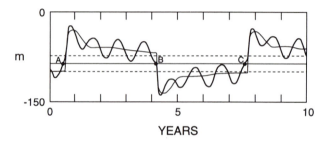

YEARS

**Fig. 8.5.** The heavy line indicates the time development of the mixed-layer depth $h$ in the eastern equatorial Pacific. The thin line represents computed values of $h$ without contribution from the externally forced annual response. The solid horizontal line marks the stipulated critical depth $h_c$ ($=-86.5$ m). The dashed lines show the two possible equilibrium values of $h$ when there is no annual forcing. Cold (warm) events are triggered at points A, C (B). Rossby waves reflected from the western boundary gradually adjust $h$ to one or the other of the two equilibrium states until the annual cycle is able to cause another switch. After McCreary and Anderson (1985).

states, as shown in Fig. 8.5. Inclusion of a background wind stress with seasonal or random variations leads to solutions that jump from being near one equilibrium state to the other one. When these jumps occur $\tau_w$ is switched on or off. The figure shows that the jumps cause the model layer depth $h$ to overshoot its equilibrium value. Relaxation toward equilibrium depth is due to the emission of Rossby waves and their reflection from the western boundary. In the natural ENSO sequence, the cold state of the eastern equatorial Pacific ($h < h_c$) lasts longer than the warm state. One can obtain similar asymmetric solutions in the models by adjusting the value of $h_c$. The equations used for the original model and for its modification can be found in McCreary and Anderson's (1991) review. That paper also includes descriptions of several later models with a passive atmosphere.

   The first models with a simplified, but thermodynamically active, atmosphere seem to have been conceived by Gill (1980, 1985). He parameterized the upper layer depth $h$ of a two-layer ocean by $h = \bar{h} + \zeta g/g'$, where $h$ is an intitial layer thickness and $\zeta$ is the sea surface height anomaly. The sea surface temperature $T_0$ is a linear function of the mixed-layer depth ($T_0 \propto h$). The ocean interacts with the bottom layer of the troposphere, which is assumed to have a constant buoyancy frequency, $N$, and a mean height, which is denoted here by $H$. Coupling between the two media is produced by the forcing functions. The atmosphere is driven by the release of latent heat, which is parameterized by a precipitation rate, $P$, that is in turn proportional to the surface temperature and hence to the ocean layer depth ($P \propto h$). The ocean is driven by a zonal wind stress, $\tau$, which is assumed proportional to a variable wind velocity, that can be derived from the atmospheric equations of motion, making $\tau$ also a function of $P$, and hence of $h$.

Gill's original scheme involved damping terms that parameterized friction along the internal interfaces in both media. These were omitted by Lau (1981) in a simpler system that was restricted to motion parallel to the equatorial plane. Without coupling between the two media, these waves would propagate with velocities $c_0 = (g'h)^{1/2}$ in the ocean and $c_a = NH$ in the atmosphere. Seeking solutions proportional to $\exp(ikx - i\omega t)$, Lau derived the implicit dispersion relation

$$(\omega^2 - c_a^2 k^2)(\omega^2 - c_0^2 k^2) - (Ac_0 k)^2 = 0. \qquad (8.14)$$

The parameter $A$, which has the dimension of inverse time, measures the strength of the coupling between the two fluids. Only real positive roots of (8.14) are of interest. Allowing for the inequality $c_a^2 \gg c_0^2$, approximate values of two relevant solutions are

$$\omega_a \approx c_a k \qquad \omega_0 \approx c_0 k \left[ 1 - \left( \frac{A}{c_a k} \right)^2 \right]^{0.5}. \qquad (8.15)$$

These equations suggest that the propagation speed $\omega_a/k$ of the atmospheric perturbation is not much affected by the coupling parameter $A$, but that the propagation of the oceanic perturbation is slower than $\omega_0$ and becomes imaginary if $(A/(c_a k))^2 > 1$. The resulting instability seems to be a feature of other coupled ENSO models. It is explained by McCreary and Anderson (1991) as being due to the basic assumption that $P \propto h$. When $h$ becomes larger, $P$ is increased. The additional release of latent heat enhances wind convergence in the lower atmosphere. In the absence of a strong Coriolis force near the equator, the convergence in the atmosphere and in the ocean have the same sign. Intensified oceanic convergence further deepens the mixed-layer, causing additional increases in the model precipitation, $P$, and so on.

The ENSO models, described earier, emphasize two different processes. In McCreary's simulation and in other similar models with a passive atmosphere, the oscillation period is determined by the slow propagation of oceanic Rossby waves across the Pacific and by their reflection from the western boundary. In the coupled models with a thermodynamically active atmosphere, the oscillation period is a function of $A$. The resulting oscillation does not depend on feedback from reflected Rossby waves and could exist on an unbounded ocean. Hybrid models, involving both processes are described in the quoted literature. However, neither observations nor results from more detailed general circulation models can tell us convincingly at this stage whether Rossby waves are in fact an essential aspect of ENSO dynamics. Other unsolved problems include the sensitivity of models to different parameterizations of convective heat sources and of the wind stress. A possible connection between ENSO and fluctuations of the Indian monsoon remains also a matter of debate.

### 8.4.2 *The Somali current and the Indian monsoon*

The seasonal reversal of the powerful Somali current along the western boundary of the Indian Ocean has been an object of much oceanographic study. It is associated with a reversing wind system that has been used by Arabian sailors for more than 2000 years to carry cargo between East Africa and southern Asia. The same winds also carry locust swarms between the two continents.

The southwest (SW) monsoon brings vital summer rains to the Indian subcontinent. It usually develops in May and is fed partly from a northward extension of the easterly trade winds over the southern Indian Ocean. The winds turn northward and cross the equator in the vicinity of the African coast. Confined by the highlands of Kenya and Ethiopia, the winds assume a jetlike structure, the 'Finlater jet', which is an atmospheric equivalent of western boundary currents in the ocean. The SW monsoon relaxes in early fall. During winter, it is replaced by the NE monsoon, which blows in the opposite direction.

The onset of the SW monsoon is accompanied or followed by the northward setting Somali current. The development has been described in detail on the basis of oceanographic observations by Leetmaa et al. (1982) and on the basis of satellite temperature observations by Brown et al. (1980). It proceeds from south to north and usually starts with the appearance of two separate gyres. During the height of summer, a very strong flow transports more than $60 \times 10^6 \, \mathrm{m}^3 \, \mathrm{s}^{-1}$ northward along the Somali coast. It is accompanied by intense upwelling and very considerable cooling of the ocean surface, offshore from Kenya to western Oman in Arabia. The existence of this transient cold-water area makes the summer fisheries in the western Arabian Sea very productive. Some time during October the winds reverse again with the onset of the NE monsoon. Upwelling ceases and warm water spreads in one or several Kelvin waves back toward the equator. This part of the cycle has attracted little research attention and is correspondingly less well known.

The phase shifts of the Somali current lag the corresponding reversal of the Indian monsoon only by about one month. Model studies by Cox

(1976, 1979), Hurlbut and Thompson (1976), and Luther et al. (1985) suggest that the upwelling and the development of a northward-flowing, longshore current in early summer are initiated by local winds (i.e., by the process discussed in subsection 7.5.1). On the other hand, a study by Lighthill (1969) indicates that the development of the current system during summer may be affected by the arrival of Rossby waves, which are excited by zonal wind stress changes all across the Indian Ocean. The appearance of cold water off the African coast enhances the west–east surface temperature gradient. The resulting additional energy input into the Indian Ocean Walker cell accelerates west winds along the equator. A first-mode equatorial Rossby wave can travel 2000 km in about 24 days. This relatively short transit time suggests that resonances and positive feedbacks may play a significant role. The relative importance of remote versus local forcing in the generation of the Somali current system has been reviewed by Knox and Anderson (1985).

Another suggested feedback process connects fluctuations in summer precipitation over India with temperature changes on the surface of the Arabian Sea (see, e.g., Gadgil et al., 1984). During the summer monsoon period, low-level winds over the northwestern Arabian Sea can remain consistently stronger over periods of months, than found anywhere else, with the possible exception of the southern ocean. The stress, exerted by these winds on the sea surface, is reduced by a high temperature difference between the warm air and the cold upwelling water. Variations in the surface temperature pattern can therefore affect the stress curl and the local upwelling in a way that is analogous to that produced by the changes in surface velocity, as discussed at the end of subsection 7.4.2. In the case of the monsoon, it tends to move the upwelling area away from the coast toward the open Arabian Sea. On the other hand, increased downwind temperature gradients toward India accelerate the winds. This tends to enhance convergence and precipitation over the continent, which in turn accelerates the wind circulation by a CISK-like process. It has been suggested that the resulting sequence of interaction between windspeed, precipitation, and the location and intensity of oceanic upwelling may account for fluctuations of the monsoon within the summer season.

### 8.4.3   *Interactions between the hydrological cycle and the thermo-haline circulation*

Most of the solar radiation is absorbed in the surface layers of the tropical and subtropical oceans. The resulting shallow warm water sphere covers more than two-thirds of the total ocean surface area, but occupies only about 15 per cent of the ocean volume. This configuration is due to the slow rate of mixing between the warm surface water and the colder water below. On average, 80–90 per cent of the absorbed solar radiation is returned locally to the atmosphere, mainly by evaporation, but also by an upward flux of sensible heat and by infrared radiation. The amount of solar energy, which is retained by the water for a significant period of time, varies annually and geographically (see, e.g., Budyko, 1974). Most of it is advected poleward by quasi-geostrophic currents along the continental boundaries. Water in these currents tends to be relatively saline because of the evaporation to which it had been subjected in its source region. When these waters reach high latitudes and are exposed locally to intense cooling, they can become dense enough to sink. They then spread out through the deeper parts of all the interconnected ocean basins, until upwelling and entrainment of cold water brings them back into the warm water sphere at low latitudes. Passage through the whole circuit is known as the *thermo-haline circulation.*

The idea of a meridional circulation, driven by high-latitude cooling and warming in the tropics, was first conceived by Benjamin Thompson, Count Rumford (1800). It is a meridionally stretched, inverted counterpart of the Hadley circulation, which was discussed in subsection 8.3.2. In fact, one could depict it by rotating Fig. 8.3 through 180°. In reality, conditions are greatly complicated by the action of the Coriolis force, by the presence of continental boundaries, and by the nonlinearity of the equation of state. The thermo-haline circulation is driven by meridional density differences. Density anomalies can be created by the flux of fresh water across the air–sea interface and by the associated change in the upper-ocean salinity. We found in subsection 8.2.3 that the influence of small salinity changes upon sea water density increases with decreasing temperature. The relative contribution of the fresh-water flux to the driving of the thermo-haline circulation, therefore, is a function of the local upper ocean temperature, which is itself a function of the circulation. This feedback loop is conducive to the existence of multiple steady states, periodic fluctuations and even reversals of the thermo-haline circulation.

During most of the earth's history, the deep ocean and the polar regions were much warmer than they are today. Ocean bottom waters have cooled monotonically from about 15°C in the early Eocene some 50 million years ago. They approached their present near-freezing values only about 10 million years ago, after considerable amounts of ice first began to accumulate on Antarctica. During the same time interval, the surface temperature difference between the tropics and the polar regions increased from about 10°C to nearly 30°C. These numbers are supported by measurements by Shakleton (1978) and other investigators. Several authors (Kraus et al., 1978; Brass et al., 1982) have tentatively explained the warmer, ice-free climate of the past by a reverse thermo-haline circulation. This involves upwelling in high latitudes of warm, highly saline bottom waters, which were formed by evaporation and convection from the surface in the zonally interconnected tropical ocean basins of that time.

In our age, the North Atlantic plays a dominant role in the production of deep water. Deep convection there is affected strongly by the presence or absence of relatively fresh water near the surface (i.e., by changes in the salinity). Such salinity anomalies can be caused by changes in the precipitation–evaporation difference, by variation of river run-off, or by oceanic and continental ice formation. There is considerable evidence that these hydrological processes have interacted with the production of NADW (see Section 8.2.3), and hence with the thermo-haline circulation, over many different time-scales. A comprehensive review of this topic and its link to climate has been published by Weaver and Hughes (1992).

A suggestion by Winton and Sarachik (1993), that the strength of the thermo-haline circulation varies significantly with its estimated overturning period of about 350 years, finds support in the analysis of Greenland ice core data by Stocker and Mysak (1992). Fluctuations with a somewhat longer period have been related by Rooth (1982) and others to changing amounts of ice on the continents. The overall time-scale of glacial and interglacial periods is probably set by insolation changes associated with cyclical variations in the earth's orbital parameters. When these variations favor melting of the continental glaciers, the accumulation of fresh melt water in the upper North Atlantic tends to shut off convection and deep water production. The resulting reduction of the oceanic heat transport toward the Arctic may cause a return to colder conditions in high latitudes. Without NADW production, the export of salinity to other oceans is shut off, causing the North Atlantic to become more saline. This tendency may be enhanced by brine rejection from a growing sea ice cover. When the salinity attains a critical level, deep convection and a thermo-haline circulation is turned on again, leading to a renewal of the melting and warming process.

The indicated sequence of processes has been used by Broecker et al. (1990) to explain climate oscillations with a time scale of about 1000 years, which have occurred at least four times since the melting of the continental ice caps begun. The best documented of these cold events is the Younger Dryas, which happened about 11,000–10,000 years ago. It seems that the change from cold to warm conditions can occur very rapidly; Dansgaard et al. (1989) suggest that the transition from the cold Younger Dryas to a much warmer climate took only about 20–50 years.

Similar oscillations may have occurred during the onset of a glacial period. The storage of water in the growing ice sheets and the reduction of

**Table 8.1.** The Heat Budget of the Norwegian Sea*

| | |
|---|---:|
| Heat loss through the sea surface | −394 |
| Heat used to melt imported ice | −67 |
| Export by ocean currents into the Arctic basin | −54 |
| Import by ocean currents from the Atlantic | 515 |

* Results are given in units of $10^{18}$ J year$^{-1}$.

run-off from the northern continents must have increased the salinity of the North Atlantic. The resulting enhancement of convection and of the thermo-haline circulation could have brought more warm water to the subpolar North Atlantic, causing temporary reductions or reversals of the trend toward a glacial climate.

The fragility of the present thermo-haline circulation can be demonstrated also in other ways. The production of NADW is closely linked to the existence of an ice-free Norwegian Sea. At a latitude between 62° and 78°N, it is the only body of permanently ice-free water as close to the pole in either hemisphere. Table 8.1, for the heat budget of the Norwegian Sea, was prepared by Kraus et al. (1978) on the basis of data by Sverdrup et al. (1942), Budyko (1963), Untersteiner (1964), and Worthington (1976).

It might be noted that only $67 \times 10^{18}$ J year$^{-1}$ are used at present to keep the Norwegian Sea open. Distributed over its area of about $2.2 \times 10^{12}$ m$^2$, this is equivalent to less than 1.0 W m$^{-2}$. If this small amount of heating is sufficient to keep the Norwegian Sea from freezing over, an equivalent heat supply would presumably also keep the Southern Ocean—which has about the same latitude—free of pack ice. The necessary amount of heat might be supplied by salty water from below with a temperature of 10°C and an upwelling velocity of only about 0.7 m year$^{-1}$.

The possibility of self-induced fluctuations and flip-flops has been demonstrated in a variety of ocean models. Most of these represent fresh water fluxes at the ocean surface as a surface boundary condition on salinity. Changes in the near-surface ocean temperature and salinity can then be parameterized as being proportional to the difference between $T, s$ and some externally imposed values $T_a, s_a$

$$\frac{\partial T}{\partial t} = k_T(t_a - T) + \psi(T) \qquad \frac{\partial s}{\partial t} = k_s(s_a - s) + \psi(s), \qquad (8.16)$$

where $\psi$ represents changes caused by internal processes. Some justification for this type of surface flux parameterization in large-scale ocean models has been given by Haney (1971). The relaxation of temperature toward $T_a$ is presumably faster than that of salinity toward $s_a$ and that implies that $k_T > k_a$. Welander (1982) showed that this inequality can cause thermo-haline oscillations even in an essentially one-dimensional system. His model involves two well-mixed layers in a vertical fluid column. Temperature and salinity can vary with time in the shallow upper layer as indicated by (8.16). They are assumed to have constant values $T_d, s_d$ in the much deeper reservoir below ($T_d < T_a, s_d < s_a$). Initial values of the upper-layer temperature and salinity are stipulated to be close to those in the lower layer ($T \approx T_d, s \approx s_d$). Because of $k_T > k_s$, rapid warming of the upper layer at first increases the stability of the system. However, the warming becomes progressively less as $T_a - T$ approaches zero, while the salinity continues to increase. This ultimately makes the water in the upper layer heavier than the water below, causing rapid mixing by convection and hence reestablishment of the initial conditions ($T \approx T_d, s \approx s_d$). The cycle can then repeat itself.

Welander's model involved stratification, but no horizontal transports. A rudimentary, two-dimensional model of the thermo-haline circulation in an unstratified system had been developed earlier by Stommel (1961). It also consisted of two reservoirs, but these were placed side by side and connected by an upper and a lower thin pipe. Fluid could therefore circulate through this configuration. The fluid in each of the two reservoirs or boxes was assumed to be well-mixed, but their temperatures and salinities differed and they were each forced by different values of $T_a$ and $s_a$. Stommel showed that this system had two different equilibria and that the fluid could circulate one way or the other way, depending on the relative values of the forcing parameters and of the time constants $k_T, k_s$. The Stommel and Welander concepts were combined later by Zhang (as quoted by Weaver and Hughes, 1992) into a four-box model that allowed for both stratification and circulation. More details about these models can be found in Weaver and Hughes, which also provides many additional references.

Stommel's box model and its later derivatives essentially simulated a thermo-haline circulation in one hemisphere, with a flow of cold, dense water through the lower pipe toward a warmer 'equatorial' box and a warm flow through the upper pipe toward a 'polar' box. In the real ocean, much of the water that is convected down into the deep and bottom water reservoirs ascends again in relatively high latitudes. The return to the surface need not occur in the same basin or hemisphere. Some NADW flows south along the western boundary, to be mixed upward again in the Antarctic region. The deep southward flow, together with a compensating northward flow at intermediate levels, may cause a northward heat transport in the South Atlantic (see, e.g., Hastenrath, 1982). Rooth (1982) and later Welander (1986) modelled this asymmetric circulation by adding a second polar box to Stommel's original two-box model. They showed that the circulation in such a three-box model can be asymmetric, reaching from pole-to-pole, even if the forcing is symmetric about the equator. Additional steady-state equilibria can be associated with superposed symmetric equator–pole circulations. A pole-to-pole circulation cell was found also by Bryan (1986) in this three-dimensional general ocean circulation model. Sufficiently large perturbations can reverse this asymmetric circulation. If that were true for the real ocean, then it would allow for periods of heat transport toward the equator in the North Atlantic.

The fixed, stipulated values of the relaxation parameters $T_a$, $s_a$ in the preceding constructs precluded consideration of the atmospheric response to different thermo-haline circulations. An attempt to simulate this response was made by Birchfield (1989). His ocean model involved side by side polar and equatorial boxes in the upper ocean, with one deep-sea box below. For the atmosphere he used a simple 'energy balance model', with features that had been described by Sellers (1969) and by Budyko (1969). Model integrations produced a cold deep ocean with sinking in high latitudes, if the atmospheric meridional moisture transport was below a critical threshold value. Above this value a double equilibrium became possible, with either a cold-fresh or a warm-saline deep ocean box. When the poleward atmospheric moisture transport was increased still further, the resulting salinity decreases in the upper polar ocean box inhibited convective mixing with the deep water. The system settled then into a warm-saline deep-water mode. The model experiments have an obvious bearing on the changes that occurred in the deep ocean between the Eocene and the Pleistocene. They were extended subsequently by Birchfield et al. (1990) to include pole-to-pole asymmetric circulations.

Operations of all quoted box models can be described by linear differential equations, which can be solved analytically to provide immediate insights into the underlying physical concepts. This shows that thermo-haline oscillations are possible, that multiple equilibria of a thermo-haline circulation can exist, and that meridionally symmetric forcing can lead to asymmetric pole-to-pole movements of sea water. These deductions are valuable, but they can obviously not convey more than a grossly oversimplified image of reality. Entrainment and mixing processes cannot be represented realistically in models that contain just a few big, widely spaced boxes. The spreading of deep water into the various ocean basins cannot be simulated either by any two-dimensional arrangement. Even less justifiable on physical grounds is the commonly used parameterization of the precipitation–evaporation difference, of river run-off and of ice formation by one single, linear relaxation formula of the form (8.16).

Manabe and Stouffer (1988) have addressed some of these difficulties in the framework of a fully coupled, general circulation model of the air–sea–ice system. They were able to integrate this system until it approached a steady-state. At present, it is difficult in models of this type to represent precipitation processes with adequate accuracy. Lack of data further hampers a sufficiently accurate specification of initial conditions. An additional constraint involves computation time. A fully coupled system encompasses many variables. To be realistic, these have to be represented with sufficient resolution, and the pertinent system of equations has to be integrated over periods that are relevant for the study of climate fluctuations. This makes comprehensive sensitivity tests very expensive on existing computers. We hope that these difficulties will be overcome in the not too distant future. It would then allow some testing of the physical hypotheses, which underlay the various conceptual models described earlier.

# 9

# REFERENCES

Agrawal, Y. C., Terray, E. A., Donelan, M. A., Hwang, P. A., Williams, A. J., III, Drennan, W. M., Kahma, K. K., and Kitaigorodski, S. A., 1992, "Enhanced dissipation of kinetic energy beneath surface waves," *Nature* **359,** 219.

Alaka, M. A., 1968, "Climatology of Atlantic tropical storms and hurricanes," *ESSA Tech. Report WB*-6.

Al-Zanaidi, M. A., and Hui, W. H., 1984, "Turbulent air flow over water waves: a numerical study," *J. Fluid Mech.* **148,** 225.

Anderson, D. L. T., Bryan, K., Gill, A. E., and Pacanowski, R. C., 1979, "The transient response of the North Atlantic—some model studies," *J. Geophys. Res.* **84,** 4795.

Anderson, D. L. T., and Gill, A. E., 1975, "Spin-up of a stratified ocean, with application to upwelling," *Deep-Sea Res.* **22,** 583.

Anderson, D. L. T., and Killworth, P. D., 1977, "Spin-up of a stratified ocean, with topography," *Deep-Sea Res.* **24,** 709.

André, J. C., and Lacarrère, P., 1985, "Mean and turbulent structures of the oceanic surface layer as determined from one-dimensinal, third-order simulations," *J. Phys. Oceanogr.* **15,** 121.

Angell, J. K., Pack, D. H., and Dickson, C. R., 1968, "A Lagrangian study of helical circulations in the planetary boundary layer," *J. Atmos. Sci.* **25,** 707.

Anthes, R. A., 1982, *Tropical Cyclones; Their Evolution, Structure and Effects, Meteor Monogr.* **41.** Boston: Am. Meteor. Soc.

Asai, T., and Nakasuji, I., 1973, "On the stability of Ekman boundary layer flow with thermally unstable stratification," *J. Meteor. Soc. Japan* **51,** 29.

Ataktürk, S. S., and Katsaros, K. B., 1987, "Intrinsic frequency spectra of short gravity-capillary waves obtained from temporal measurements of wave height on a lake," *J. Geophys. Res.* **92,** 5131.

Atlas, D., Walter, B., Chou, S., and Sheu, P. J., 1986, "The structure of the unstable marine boundary layer viewed by lidar and aircraft observations," *J. Atmos. Sci.* **43,** 1301.

Ball, F. K., 1960, "Control of inversion height by surface heating," *Quart. J. Roy. Meteor. Soc.* **86,** 483.

Banner, M. L., 1985, "Surging characteristics of spilling regions of quasi-steady breaking water waves," *Appl. Math. Prepr. AM* 86 *I.* University of N. S. W. Australia, p.1.

Barber, N. F., 1946: "Measurements of sea conditions by the motion of a floating buoy. Four theoretical notes." ARLN 2/103·40/N·2/W. Admiralty Research Laboratory, Teddington, U.K.

Barber, N. F., 1963, "The directional resolving power of an array of wave recorders," in *Ocean Wave Spectra.* New York: Prentice-Hall, p. 137.

Barnett, T. P., Graham, N., Cane, M., Zebiak, S., Dolan, S., O'Brien, J., and Legler, D., 1988, "On the prediction of the El Niño of 1986–87," *Science* **241,** 192.

Barnett, T. P., and Sutherland, A. J., 1968, "A note on an overshoot effect in wind-generated waves," *J. Geophys. Res.* **73,** 6879.

Batchelor, G. K., 1967, *An Introduction to Fluid Dynamics.* London and New York: Cambridge University Press.

Bendat, J. S., and Piersol, A. G., 1986, *Random Data: Analysis and Measurement Procedures* (second edition). New York: Wiley.

Berlage, H. P., 1966, "The southern oscillation and world weather," *Med. Verhand, K.N.M.I.* **88,** 134.

Betts, A. K., 1973, "Non-precipitating cumulus convection and its parameterization," *Quart. J. Roy. Meteor. Soc.* **99,** 178.

Betts, A. K., 1982, "Saturation point analysis of moist convective overturning," *J. Atmos. Sci.* **39,** 1484.

Betts, A. K., and Boers, R., 1990, "A cloudiness transition in a marine boundary layer," *J. Atmos. Sci.* **47,** 1480.

Birchfield, G. E., 1989, "A coupled ocean–atmosphere climate model: temperature versus salinity effects on the thermohaline circulation," *Clim. Dyn.* **4,** 57.

Birchfield, G. E., Wyant, M., and Wang, H., 1990, "A coupled ocean atmosphere box model of the Atlantic Ocean: A bimodal climate response," *J. Mar. Sys.* **1,** 197.

Bjerknes, J., 1966, "A possible response of the atmospheric Hadley circulation to equatorial anomalies of ocean temperature," *Tellus* **18,** 820.

Bjerknes, J., 1969, "Atmospheric teleconnections from the equatorial Pacific," *Mon. Wea. Rev.* **97,** 163.

Bjerknes, J., Allison, L. J., Kreins, E. R., Godshal, F. A., and Warnecke, G., 1969, "Satellite mapping of the Pacific tropical cloudiness," *Bull. Am. Meteor. Soc.* **50,** 313.

Blackadar, A. K., 1962, "The vertical distribution of wind and turbulent exchange in a neutral atmosphere," *J. Geophys. Res.* **67,** 3095.

Blackadar, A. K., and Tennekes, H., 1968, "Asymptotic similarity in neutral barotropic planetary boundary layers," *J. Atmos. Sci.* **25,** 1015.

Blanchard, D. C., 1963, "The electrification of the atmosphere by particles from bubbles in the sea," in *Progress in Oceanography* 1. Oxford: Pergamon Press, p. 72.

Blanchard, D. C., 1983, "The production, distribution and bacterial enrichment of the sea-salt aerosol," in *The Air–Sea Exchange of Gases and Particles* (P. S. Liss and W. G. N. Slinn, eds.). Dordrecht: Reidel, p. 407.

Blanchard, D. C., 1989. "Bacteria and other materials in drops from bursting bubbles," in *Climate and Health Implications of Bubble-Mediated Sea–Air Exchange* (E. C. Monahan and M. A. VanPatten, eds.). Connecticut Sea Grant College Progr. Publ. CT-SG-89-06, P. 1.

Blanchard, D. C., and Woodcock, A. H., 1957, "Bubble formation and modification in the sea and its meteorological significance," *Tellus* **9**, 145.

Bleck, R., Hanson, H. P., Hu, D., and Kraus, E. B., 1989, "Mixed layer-thermocline interaction in a three-dimensional isopycnic coordinate model," *J. Phys. Oceanogr.* **19**, 1417.

Bradley, E. F., Coppin, P. A., and Godfrey, J. S., 1991, "Measurements of sensible and latent heatflux in the Western Equatorial Pacific Ocean," *J. Geophys. Res.* **96**, 3375.

Brass, G. W., Southam, J. R., and Petersen, W. H., 1982 "Warm saline bottom water in the ancient ocean," *Nature* **296**, 620.

Bretherton, F. P., 1982, "Ocean climate modelling," *Prog. Oceanogr.* **11**, 93.

Brink, K. H., 1989, "Observation of the response of thermocline currents to a hurricane," *J. Phys. Oceanogr.* **19**, 1017.

Brocks, K., 1959, "Ein neues Gerat für störungsfreie meteörologische Messungen auf dem Meer," *Arch. f. Meteor. Geophys. Biokl.* **A2**, 227.

Broecker, W. S., Bond, G., and Klas, M., 1990, "A salt oscillator in the glacial Atlantic? I. The Concept," *Paleo Oceanogr.* **5**, 469.

Broecker, W. S., and Peng, T. H., 1982, *Tracers in the Sea.* Lamont-Doherty Geological Observatory, Palisades, NY: Eldigio Press.

Brost, R. A., Wyngaard, J. C., and Lenschow, D. H., 1982, "Marine stratocumulus layers. II. Turbulence budgets," *J. Atmos. Sci.* **39**, 818.

Brown, O. B., Bruce, J. G., and Evans, R. H., 1980, "Evolution of sea surface temperature in the Somali Basin during the southwest monsoon of 1979," *Science* **209**, 595.

Brown, R. A., 1970, "A secondary flow model for the planetary boundary layer," *J. Atmos. Sci.* **27**, 742.

Brown, R. A., 1972, "The inflection point instability problem for stratified rotating boundary layers," *J. Atmos. Sci.* **29**, 850.

Brown, R. A., 1980, "Longitudinal instabilities and secondary flows in the planetary boundary layer: A review," *Rev. of Geophys. and Space Phys.* **18**, 683.

Brown, R. A., 1982, "On two-layer models and the similarity functions for the PBL," *Bound.-Layer Meteor.* **24**, 451.

Brown, R. A., 1986, "On satellite scatterometer capabilities in air–sea interaction," *J. Geophys. Res.* **91**, 2221.

Brown, R. A., 1990, "Meteorology," in *Polar Oceanography. A. Physical Science.* New York: Academic Press.

Brown, R. A., Cardone, V. J., Guymer, T., Hawkins, J., Overland, J. E., Pierson, W. J., Peteherych, S., Wilkerson, J. C., Woiceshyn, P. M., and Wurtele, M., 1982, "Surface wind analyses for Seasat," *J. Geophys. Res.* **87**, 3355.

Brummer, B., 1985, "Structure, dynamics and energetics of boundary layer rolls from KONTUR aircraft observations," *Beitr. Phys. Atmos.* **58**, 237.

Brunt, D., and Douglas, C. K. M., 1928, "The modification of the strophic balance for changing pressure distribution, and its effect on rainfall," *Mem. Roy. Meteor. Soc.* **3**, 29.

Bryan, K., 1986, "Poleward buoyancy transport in the ocean and mesoscale eddies," *J. Phys. Oceanogr.* **16**, 927.

Bryden, H. L., 1973, "New polynomials for thermal expansion, adiabatic temperature gradient and potential temperature gradient of sea water," *Deep-Sea Res.* **20**, 401.

Budyko, M. I., 1963, *Atlas of the Heat Balance of the Earth* (in Russian). Moscow: Akad. of Sci. (see also Guide to the atlas of the heat balance of the earth. U.S. Weatherbureau, WB/T-106).

Budyko, M. I., 1969, "The effect of solar radiation variations on the climate of the earth," *Tellus* **21**, 611.

Budyko, M. I., 1974, *Climate and Life* (second edition) (D. H. Miller, ed.). New York: Academic Press.

Buriez, J. C., Bonnel, B., and Y. Fouquart, 1986, "Theoretical and experimental sensitivity study of the derivation of the solar irradiance at the earth's surface from satellite data," *Beitr. Phys. Atmos.* **59**, 263.

Busch, N. E., 1973, "On the mechanics of atmospheric turbulence," *Workshop on Micrometeorology* (D. A. Haugen, ed.). Boston: Am. Meteor. Soc., p. 1.

Businger, J. A., 1966, "Transfer of momentum and heat in the planetary boundary layer," in *Proc. Symp. Arctic Heat Budget and Atmos. Circ.* The Rand Corporation, p. 305.

Businger, J. A., 1982, "Equations and concepts," in *Atmospheric Turbulence and Air Pollution Modelling* (F. T. M. Nieuwstadt and H. van Dop, eds.). Boston: Reidel Publ. Co., p. 1.

Businger, J. A., 1988, "A note on the Businger-Dyer profiles," *Bound.-Layer Meteor.* **42**, 145.

Businger, J. A., and Delaney, A. C., 1990, "Chemical sensor resolution required for measuring surface fluxes by three common micrometeorological techniques," *J. Atmos. Chem.* **10**, 399.

Businger, J. A., and Oncley, S. P., 1990, "Flux measurement with conditional sampling," *J. Atmos. Oceanic. Tech.* **7**, 349.

Businger, J. A., Wyngaard, J. C., Izumi, Y., and Bradley, E. F., 1971, "Flux-profile relationships in the atmospheric surface layer," *J. Atmos. Sci,* **28**, 181.

Businger, S., 1991, "Arctic hurricanes," *Am. Scientist* **79**, 18.

Carslaw, H. S., and Jaeger, J. C., 1959, *Conduction of Heat in Solids* (second edition). Oxford: Clarendon Press.

Caughey, S. J., Crease, B. A., and Roach, W. T., 1982, "A field study of nocturnal strato cumulus. II. Turbulence structure and entrainment," *Quart. J. Roy. Meteor. Soc.* **108**, 125.

Champagne, F. H., Friehe, C. A., La Rue, J. C., and Wyngaard, J. C., 1977, "Flux measurements, flux estimation techniques, and fine scale turbulence measurements in the unstable surface layer over land," *J. Atmos. Sci.* **34**, 515.

Chandrasekhar, S., 1961, *Hydrodynamic and Hydromagnetic Stability.* Oxford: Oxford University Press.

Charney, J. C., 1955, "The generation of oceanic currents by wind," *J. Mar. Res.* **14**, 477.

Charney, J. G., 1966, "Some remaining problems in numerical weather predictions," in *Advances in numerical weather prediction.* Hartford, CT: The Travellers Research Center, p. 61.

Charney, J. C., and Eliassen, A., 1964, "On the growth of hurricane depression," *J. Atmos. Sci.* **21**, 68.

Charney, J. G., and Flierl, G. R., 1981, "Oceanic analogues of large-scale atmospheric motions," in *Evolution of Physical Oceanography* (B. A. Warren and C. Wunsch, eds.). Cambridge, MA: MIT Press, p. 504.

Charnock, H., 1955, "Wind stress on a water surface," *Quart. J. Roy. Meteor. Soc.* **81**, 639.

Christopherson, D. G., 1940, "Note on the vibration of membranes," *Quart. J. Math.* **11**, 63.

Church, J. A., Joyce, T. M., and Price, J. F., 1989, "Current and density observations across the wake of hurricane Gay," *J. Phys. Oceanogr.* **19**, 259.

Clarke, R. H., 1970, "Observational studies in the atmospheric boundary layer," *Quart. J. Roy. Meteorol. Soc.* **96**, 91.

Cointe, R., 1987, *A Theory of Breaking Waves.* Ph.D. Thesis, University of California, Santa Barbara.

Coleman, G. N., Ferziger, J. H., and Spalart, P. R., 1990, "A numerical study of the turbulent Ekman layer," *J. Fluid Mech.* **213**, 313.

Cox, M. D., 1970, "A mathematical model of the Indian Ocean," *Deep-Sea Res.* **17**, 47.

Cox, M. D., 1976, "Equatorially trapped waves and the generation of the Somali current," *Deep-Sea Res.* **23**, 1139.

Cox, M. D., 1979, "A numerical study of Somali current eddies," *J. Phys. Oceanogr.* **9**, 311.

Cox, C. S., and Munk, W. H., 1954, "Statistics of the sea surface derived from sun glitter," *J. Mar. Res.* **13**, 198.

Craik, A. D. D., and Leibovich, S., 1976, "A rational model for Langmuir circulation," *J. Fluid Mech.* **73**, 401.

Crawford, G. B., and D. M. Farmer, 1987, "On the spatial distribution of ocean bubbles," *J. Geophys. Res.* **92**, 8231.

Crawford, W. R., 1982, "Pacific equatorial turbulence," *J. Phys. Oceanogr.* **12**, 1137.

Crépon, M., 1974, "Genèse d'ondes internes dans un milieu à deux couches," *La Huille Blanche* **8**, 631.

Csanady, G. T., 1981, "Shelf-circulation cells," *Phil. Trans. Roy. Soc. London, Ser. A* **302**, 515.

Dacey, J. W. H., Wakeham, S. G., and Howes, B. L., 1984, "Henry's Law constant for dimethylsulfide in freshwater and seawater," *Geophys. Res. Let.* **11**, 991.

Danckwerts, P. V., 1951, "Significance of liquid-film coefficients in gas absorption," *Ind. Eng. Chem.* **43**, 1460.

Dansgaard, W., White, J. W. C., and Johnson, S. J., 1989, "The abrupt termination of the Younger Dryas climate event," *Nature* **339**, 532.

Davis, R. E., 1972, "On prediction of the turbulent flow over a wavy boundary," *J. Fluid Mech.* **52**, 287.

Davis, R. E., de Szoeke, R., Halpern, D., and Niiler, P. P., 1981, "Variability in the upper ocean during Mile. I. The heat and momentum balances," *Deep-Sea Res.* **28A**, 1427.

Davis, R. E., and Niiler, P. P., 1981, "Variability in the upper ocean during Mile. II. Modelling the mixed-layer response," *Deep-Sea Res.* **28A**, 1453.

Deacon, E. L., 1977, "Gas transfer to and across an air-water interface," *Tellus* **29**, 363.

Deacon, E. L., and Stevenson, J., 1968, "Radiation and associated observations made on Indian Ocean cruises," *Tech. Pap. Div. Metrol. C.S.I.R.O. Australia.* p.16.

Deardorff, J. W., 1970a. "A three-dimensional numerical investigation of the idealized planetary boundary layer," *Geophys. Fluid Dyn.* **1**, 377.

Deardorff, J. W., 1970b. "Preliminary results from numerical integrations of the unstable planetary boundary layer," *J. Atmos. Sci.* **27**, 1209.

Deardorff, J. W., 1970c, "Convective velocity and temperature scale for the unstable planetary boundary layer and for Rayleigh convection," *J. Atmos. Sci.* **27,** 1211.

Deardorff, J. W., 1972, "Numerical investigation of neutral and unstable planetary boundary layer," *J. Atmos. Sci.* **29,** 91.

Deardorff, J. W., 1976, "On the entrainment rate of a stratocumulus-topped mixed layer," *Quart. J. Roy. Meteor. Soc.* **102,** 563.

Deardorff, J. W., 1980, "Cloud top entrainment instability," *J. Atmos. Sci.* **37,** 131.

Deardorff, J. W., 1981, "On the distribution of mean radiative cooling at the top of a stratocumulus-capped mixed-layer," *Quart. J. Roy Meteor. Soc.* **107,** 191.

Deardorff, J. W., 1983, "A multi-limit mixed-layer entrainment formulation," *J. Phys. Oceanogr.* **13,** 988.

Deissler, R. G., and Eian, C. S., 1952, "Analtyical and experimental investigation of fully developed turbulent flow of air in a smooth tube with heat transfer with variable fluid properties," *NACA Tech. Note* 2629.

de Leeuw, G., 1987, "Near-surface particle size distribution profiles over the North Sea," *J. Geophys. Res.* **92,** 14, 631.

de Leeuw, G., 1989, "The occurrence of large salt water droplets at low elevations over the open ocean," in *Climate and Health Implications of Bubble-Mediated Sea–Air Exchange* (E. C. Monahan and M. A. VanPatten, eds.). Connecticut Sea Grant College Progr. Publ. CT-SG-89-06, p. 65.

Desjardins, R. L., 1972, *A Study of Carbon-Dioxide and Sensible Heat Fluxes Using the Eddy Correlation Technique,* Ph.D. Thesis, Cornell University.

Dickinson, R. E., 1978, "Rossby waves—long period oscillations of oceans and atmospheres," *Ann. Rev. Fluid Mech.* **10,** 159.

Dobson, F. W., Hasse, L., and Davis, R. E. (eds.), 1980, *Air–Sea Interaction; Instruments and Methods.* New York: Plenum Press.

Dobson, F. W., and Smith, S. D., 1985, "Estimation of solar radiation at sea," in *The Ocean Surface* (Y. Toba and H. Mitsuyasu, eds.). Dordrecht: Reidel, p. 525.

Dobson, F. W., and Smith, S. D., 1988, "Bulk models of solar radiation at sea," *Quart. J. Roy. Meteor. Soc.* **114,** 165.

Donelan, M. A., 1990, "Air–sea interaction," in *The Sea: Ocean Engineering Science* (Volume 9) (E. D. Goldberg et al., Eds.). New York: Wiley-Interscience, p. 239.

Donelan, M. A., Hamilton, J., and Hui, W. H., 1985, "Directional spectra of wind-generated sea waves," *Phil. Trans. Roy. Soc. London, Ser. A* **315,** 509.

Donelan, M. A., and Hui, W. H., 1990, "Mechanics of ocean surface waves," in *Surface Waves and Fluxes* (Volume I) (G. L. Geernaert and W. J. Plant, eds.). Boston: Kluwer Acad. Publ., p. 209.

Donelan, M. A., and Pierson, W. J., 1987, "Radar scattering and equilibrium ranges in wind-generated waves—with application to scatterometry," *J. Geophys. Res.* **92,** 4971.

Doronin, Yu. F., and Khelsin, D. E., 1975, *Sea Ice.* Leningrad: Gidrometeoizdat Publishers. (Translated 1977, NSF TT75-52088.)

Duncan, J. H., 1981, "An experimental investigation of breaking waves produced by a towed hydrofoil," *Proc. Roy. Soc. London, A* **377,** 331.

Dunn, G. E., and Miller, B. I., 1964, *Atlantic Hurricanes.* Baton Rouge: Louisiana State University Press.

Duynkerke, P. G., and Driedonks, A. G. M., 1987, "A model for the turbulent structure of the stratocumulus-topped atmospheric boundary layer," *J. Atmos. Sci.* **44**, 43.

Dyer, A. J., and Bradley, E. F., 1982, "An alternative analysis of flux-gradient relationships at the 1976 ITCE," *Bound.-Layer Meteor.* **22**, 3.

Dyer, A. J., and Hicks, B. B., 1970, "Flux-gradient relationships in the constant flux layer," *Quart. J. Roy. Meteor. Soc.* **96**, 715.

Dyer, A. J., and Hicks, B. B., 1982, "Kolmogorov constants at the 1976 ITCE," *Bound.-Layer Meteor.* **22**, 137.

Eckart, C. H., 1960, *Hydrodynamics of Oceans and Atmospheres.* Oxford: Pergamon Press.

Ekman, V. W., 1904, "On dead water," *Sci. Res. Norw. North Polar Expedition 1893–96*, **5**.

Ekman, V. W., 1905, "On the influence of the earth's rotation on ocean currents," *Arkiv. Math. Astron. Fysik.* **2**, 11.

Elasser, W. M., and Culbertson, M. F., 1960, "Atmospheric radiation tables," *Met. Monogr.* **4**, 43.

Emanuel, K. A., 1987, "The dependence of hurricane intensity on climate," *Nature* **326**, 483.

Emanuel, K. A., 1988, "Towards a general theory of hurricanes," *Am. Scientist* **76**, 370.

Enfield, D. B., 1989, "El Niño, past and present," *Rev. Geophys.* **27**, 159.

Eriksson, E., 1959, "The yearly circulation of chloride and sulfur in nature; meteorological, geochemical and pedological implications," *Tellus* **11**, 376.

Etling, D., 1971, "The stability of Ekman boundary layer flow as influenced by temperature stratification," *Beitr. Phys. Atmos.* **44**, 168.

Etling, D., and Brown, R. A., 1993, "Roll vortices in the planetary boundary layer: A Review," *Bound.-Layer Meteor.* **65**, 215.

Fairall, C. W., Edson, J. B., Larsen, S. E., and Mestayer, P. G., 1990, "Inertial-dissipation air–sea flux measurements: a prototype system using real-time spectral computations," *J. Atmos. Oceanic. Tech.* **7**, 425.

Fairall, C. W., and Larsen, S. E., 1986, "Inertial dissipation methods and turbulent fluxes at the air ocean interface," *Bound.-Layer Meteor.* **34**, 287.

Faller, A. J., 1963, "An experimental study of the instability of the laminar Ekman boundary layer," *J. Fluid Mech.* **15**, 560.

Faller, A. J., 1965, "Large eddies in the atmospheric boundary layer and their possible role in the formation of cloud rows," *J. Atmos. Sci.* **22**, 176.

Faller, A. J., and Kaylor, R. E., 1966, "A numerical study of the instability of the laminar Ekman boundary layer," *J. Atmos. Sci.* **23**, 466.

Fandry, C. B., Leslie, L. M., and Steedman, R. K., 1984, "Kelvin-type coastal surges generated by tropical cyclones," *J. Phys. Oceanogr.* **14**, 582.

Farmer, D. M., 1975, "Penetrative convection in the absence of mean shear," *Quart. J. Roy. Meteor. Soc.* **101**, 869.

Farmer, D. M., and Lemon, D. D., 1984, "The influence of bubbles on ambient noise in the ocean at high wind speeds," *J. Phys. Oceanogr.* **14**, 674.

Favre, A., and Hasselmann, K. (eds.), 1978, *Turbulent Fluxes Through the Sea Surface, Wave Dynamics and Prediction.* New York: Plenum Press.

Fernando, H. J. S., 1991, "Turbulent mixing in stratified fluids," *Ann. Rev. Fluid Mech.* **23**, 455.

Fernando, H. J. S., Chen, R.-R., and Boyer, D. L., 1991, "Effects of rotation on convective turbulence," *J. Fluid Mech.* **228,** 513.

Fiedler, B. H., 1984, "An integral closure model for the vertical turbulent flux of a scalar in a mixed layer," *J. Atmos. Sci.* **41,** 674.

Fleagle, R. G., and Businger, J. A., 1980, *An Introduction to Atmospheric Physics* (second edition), *International Physics Series* (Volume 5). New York: Academic Press.

Frouin, R., Gautier, C., Katsaros, K. B., and Lind, R. J., 1988a, "A comparison of satellite and empirical formula techniques for estimating insolation over the oceans," *J. Clim. Appl. Meteor.* **97,** 1016.

Frouin, R., Gautier, C., and Morcrette, J.-J., 1988b, "Downard longwave irradiance at the ocean surface from satellite data: methodology and in-situ validation," *J. Geophys. Res.* **93,** 597.

Gadgil, S., Joseph, P. V., and Joshi, N. V., 1984, "Ocean–atmosphere coupling over monsoon regions," *Nature* **312,** 141.

Gargett, A. E., Hendricks, P. J., Sanford, T. B., Osborn, T. R., and Williams, A. J., III, 1981, "A composite spectrum of vertical shear in the upper ocean," *J. Phys. Oceanogr.* **11,** 1258.

Garratt, J. R., 1977, "Review of drag coefficients over oceans and continents," *Mon. Wea. Rev.* **105,** 915.

Garratt, J. R., 1992, *The Atmospheric Boundary Layer.* London and New York: Cambridge University Press.

Garrett, C. J. R., and Munk, W. H., 1972, "Space–time scales of internal waves," *Geophys. Fluid Dyn.* **3,** 225 (also: 1975, *J. Geophys. Res.* **80,** 281).

Garwood, R. W., Jr., 1977, "An oceanic mixed layer model capable of simulating cyclic states," *J. Phys. Oceanogr.* **7,** 455.

Gascard, J. C., 1978, "Mediteranean seep water formulation—baroclinic instabilities and oceanic eddies," *Oceanol. Acta* 315.

Gascard, J. C., and Clarke, R. A., 1983, "The formulation of Labrador sea water. II. Mesoscale and smaller-scale processes," *J. Phys. Oceanogr.* **13,** 1779.

Gaspar, P., 1988, "Modelling the seasonal cycle of the upper ocean," *J. Phys. Oceanogr.* **18,** 161.

Gaspar, P., Stull, R., and Boissier, C., 1988, "Long-term simulation of upper ocean vertical mixing using models of different types," in *Small-Scale Turbulence and Mixing in the Upper Ocean* (J. C. J. Nihoul and B. M. Jamart, eds.). *Elsevier Oceanogr. Ser.* **46,** 169.

Gautier, C., Diak, G., and Masse, S., 1980, "A simple physical model to estimate incident solar radiation at the surface from GOES satellite data," *J. Appl. Meteor.* **19,** 1005.

Gautier, C., and Katsaros, K. B., 1984, "Insolation during STREX. I. Comparison between surface measurements and satellite estimates," *J. Geophys. Res.* **39,** 11,778.

Geernaert, G. L., Larsen, S. E., and Hansen, F., 1987, "Measurements of the wind stress, heat flux, and turbulence intensity during storm conditions over the North Sea," *J. Geophys. Res.* **92,** 13,127.

Geisler, J. E., 1970, "Linear theory of the response of a two-layer ocean to a moving hurricane," *Geophys. Fluid Dyn.* **1,** 249.

Gerling, T. W., 1986, "Structure of the wind field from the Seasat SAR," *J. Geophys. Res.* **91,** 2308.

Gill, A. E., 1968, "Similarity theory and geostrophic adjustment," *Quart J. Roy. Meteor. Soc.* **94**, 586.

Gill, A. E., 1980, "Some simple solutions for heat induced tropical circulation," *Quart. J. Roy. Meteor. Soc.* **106**, 447.

Gill, A. E., 1982, *Atmosphere–Ocean Dynamics.* London and New York: Academic Press.

Gill, A. E., 1985, "Elements of coupled ocean-atmosphere models for the tropics," in *Coupled Ocean–Atmosphere Models* (J. C. J. Nihoul, ed.). New York: Elsevier, p. 303.

Gill, A. E., and Turner, J. S., 1976, "A comparison of seasonal thermocline models with observations," *Deep-Sea Res.* **23**, 391.

Gold, E., 1908, *Barometric Gradient and Wind Force.* London: Meteor. Off. M. O. 190, H. M. Stationary Office.

Goody, R. M., and Yung, Y. L., 1989, *Atmospheric Radiation* (second edition). New York, Oxford: Oxford University Press.

Greatbatch, R. J., 1983, "On the response of the ocean to a moving storm: the nonlinear dynamics," *J. Phys. Oceanogr.* **13**, 357.

Grenfell, T. C., and Maykut, G. A., 1977, "The optical properties of ice and snow in the Arctic Basin," *J. Glaciol.* **18**, 445.

Haney, R. L., 1971, "Surface thermal boundary conditions for ocean circulation models," *J. Phys. Oceanogr.* **1**, 241.

Hanna, S. R., 1969, "The formation of longitudinal sand dunes by large helical eddies in the atmosphere," *J. Appl. Meteor.* **8**, 874.

Hasse, L., 1986, "On Charnock's relation for the roughness at sea," in *Oceanic whitecaps* (E. C. Monahan and G. MacNiocaill, Eds.). Reidel, p. 49.

Hasselmann, K., 1962, "On the non-linear energy transfer in a gravity wave spectrum. I," *J. Fluid Mech.* **12**, 481.

Hasselmann, K., 1963a, "On the non-linear energy transfer in a gravity wave spectrum. II," *J. Fluid Mech.* **15**, 273.

Hasselmann, K., 1963b, "On the non-linear energy transfer in a gravity wave spectrum. III," *J. Fluid Mech.* **15**, 385.

Hasselman, K., 1967, "Nonlinear interactions treated by the methods of theoretical physics," *Proc. Roy. Soc.* London, A **299**, 77.

Hasselman, K., Barnett, T. T., Bouws, E., Carlson, H., Cartwright, D. E., Enke, K., Ewing, J. A., Gienapp, H., Hasselmann, D. E. Kruseman, P., Meerburg, A., Müller, P., Olbers, D. J., Richter, K., Sell, W., and Walden, H., 1973, "Measurements of wind-wave growth and swell decay during the Joint North Sea Wave Projection (JONSWAP)," *Deut. Hydrogr. Z., Suppl. A* **8**, #12.

Hasselmann, S., and Hasselmann, K., 1985, "Computations and parameterizations of the nonlinear energy transfer in a gravity wave spectrum. I. A new method for efficient computations of the exact nonlinear transfer integral," *J. Phys. Oceanogr.* **15**, 1369.

Hastenrath, S., 1982, "On meridional heat transports in the world ocean," *J. Phys. Oceanogr.* **12**, 922.

Hein, P. F., and Brown, R. A., 1988, "Observations of longitudinal rolls vortices during arctic cold air outbreaks over open water," *Bound.-Layer Meteor.* **45**, 177.

Hicks, B. B., and McMillen, R. T., 1984, "A simulation of the eddy accumulation method for measuring pollutant fluxes" *J. Climate Appl. Meteor.* **23**, 637.

Higbie, R., 1935, "The rate of absorption of a pure gas into a still liquid during short periods of exposure," *Trans. Am. Inst. Chem. Eng.* **31**, 365.

Hinze, J. O., 1975, *Turbulence* (second edition). New York: McGraw-Hill.

Hoeber, H., 1969, "Wind-, Temperatur- und Feuchteprofile in der wassernahen Luftschicht über dem äquatorialen Atlantik," *Sonderdruck aus Meteor. Forschungsergebnisse Reihe* **B**, Heft 3. Berlin, Stuttgart: Gebrüder Bornträger, p. 1.

Holland, J. Z., Chen, W., Almazon, J. A., and Elder, F. C., 1981, "Atmospheric boundary layer," in *IFYGL—the International Field Year for the Great Lakes* (E. J. Aubert and T. L. Richards, eds.). Dordrecht: Reidel, p. 147.

Holton, J. R., Wallace, J. M., Young, J. A., 1971, "On boundary layer dynamics and the ITCZ," *J. Atmos. Sci.* **28**, 275.

Hsiao, S. V., and Shemdin, O. H., 1983, "Measurements of wind velocity and pressure with a wave follower during MARSEN," *J. Geophys. Res.* **88** (C14), 9841.

Huang, N. D., 1979, "On surface drift currents in the ocean," *J. Fluid Mech.* **91**, 191.

Hurlbut, H. E., and Thompson, J. D., 1976, "A numerical model of the Somali current," *J. Phys. Oceanogr.* **6**, 646.

Hwang, P. A., Hsu, Y. H. L., and Wu, J., 1990, "Air bubbles produced by breaking wind waves: A laboratory study," *J. Phys. Oceanogr.* **20**, 19.

Janssen, P. A. E. M., 1989, "Wave-induced stress and the drag of air over sea waves," *J. Phys. Oceanogr.* **19**, 745.

Janssen, P. A. E. M., and Komen, G. J., 1985, "Effect of atmospheric stability on the growth of surface gravity waves," *Bound.-Layer Meteor.* **32**, 85.

Jeffreys, H., 1924, "On the formation of waves by wind," *Proc. Roy. Soc. London, Ser. A* **107**, 189.

Jeffreys, H., 1925, "On the formation of waves by wind. II," *Proc. Roy. Soc. London, Ser. A* **110**, 341.

Jeffreys, H., 1926, "The stability of a layer of fluid heated from below," *Phil. Mag. Ser. 2* **7**, 833.

Jelesnianski, C. P., 1967, "Numerical computations of storm surges with bottom stress," *Mon. Wea. Rev.* **95**, 740.

Jerlov, N. G., 1965, "Factors influencing the colour of the oceans," in *Studies on Oceanography*. Seattle: University of Washington Press, p. 260.

Jerlov, N. G., 1976, *Marine Optics*. New York: Elsevier.

Johnson, E. R., 1989, "Boundary currents, free currents and dissipation in the low-frequency scattering of shelf waves," *J. Phys. Oceanogr.* **19**, 1291.

Johnson, F. S., 1954, "The solar constant," *J. Meteor.* **11**, 431.

Johnson, R. D., and Cooke, R. C., 1979, "Bubble populations and spectra in coastal waters: a photographic approach," *J. Geophys. Res.* **84**, 3761.

Jones, W. L., Schroeder, L. C., Boggs, D. H., Bracalente, E. M., Brown, R. A., Dome, G. J., Pierson, W. J., and Wentz, F. J., 1982, "The geophysical evaluation of remotely sensed wind vectors over the ocean," *J. Geophys. Res.* **87**, 3297.

Jordan, C. L., 1958, "Mean soundings for the West Indies area," *J. Meteor.* **15**, 91.

Joseph, J. H., Wiscombe, W. J., and Weinman, J. A., 1976, "The Delta-Eddington approximation for radiative flux transfer," *J. Atmos. Sci.* **33**, 2452.

Kahma, K. K., 1981, "A study of the growth of the wave spectrum with fetch," *J. Phys. Oceanogr.* **11**, 1502.

Kahma, K. K., and Donelan, M. A., 1988, "A laboratory study of the minimum wind speed for wind wave generation," *J. Fluid Mech.* **192**, 339.

Kahn, P. H., and Businger, J. A., 1979, "The effect of radiative flux divergence on entrainment of saturated convective boundary layer," *Quart. J. Roy. Meteor. Soc.* **105**, 303.

Kaimal, J. C., 1975, "Sensors and techniques for direct measurement of turbulent fluxes and profiles in the atmospheric surface layer," in *Atmospheric Technology* (D. H. Lenshow, ed.). p. 7. Boulder, CO: NCAR.

Kaimal, J. C., and Businger, J. A., 1970, "Case studies of a convective plume and a dust devil," *J. Appl. Meteor.* **9**, 612.

Kaimal, J. C., Wyngaard, J. C., Izumi, Y., and Coté, O. R., 1972, "Spectral characteristics of surface layer turbulence," *Quart. J. Roy. Meteor. Soc.* **98**, 563.

Kajiura, K., 1962, "A note on the generation of boundary waves of the Kelvin type," *J. Oceanogr. Soc. Japan* **108**, 1212.

Kamenkovich, V. M., 1977, *Fundamentals of Ocean Dynamics.* Amsterdam: Elsevier.

Katsaros, K. B., 1977, "The sea surface temperature deviation at very low wind speeds: is there a limit?," *Tellus* **29**, 229.

Katsaros, K. B., 1980a, "Radiative sensing of sea surface temperatures," in *Instruments and Methods in Air–Sea Interaction* (F. Dobson, L. Hasse, and R. Davies, eds.). New York: Plenum Publ. Corp., p. 293.

Katsaros, K. B., 1980b, "The aqueous thermal boundary layer," *Bound.-Layer Meteor.* **18**, 107.

Katsaros, K. B., 1990, "Parameterization schemes and models for estimating the surface radiation budget," in *Surface Waves and Fluxes* (G. Geernaert and W. Plant, eds.). Boston: Kluwer Acad. Publ., p. 672.

Katsaros, K. B., and DeCosmo, J., 1990, "Evaporation at high wind speeds, sea surface temperature at low wind speeds: Examples of atmospheric regulation," *Proceedings of Workshop on Modelling the Fate and Influence of Marine Spray.* Luminy, Marseille, France.

Keller, J. B., and Watson, J. G., 1981, "Kelvin wave production," *J. Phys. Oceanogr.* **11**, 284.

Keller, W. C., Plant, W. J., and Weissman, D. E., 1985, "The dependence of X-band microwave sea return on atmospheric stability and sea state," *J. Geophys. Res.* **90**, 1019.

Kelly, R. D., 1984, "Horizontal roll and boundary-layer interrelationships observed over Lake Michigan," *J. Atmos. Sci.* **41**, 1816.

Kenyon, K. E., 1969, "Stokes drift for random gravity waves," *J. Geophys. Res.* **74**, 6991.

Khundzhua, G. G., and Andreyev, Ye. G., 1974, "An experimental study of heat exchange between the ocean and the atmosphere in small-scale interaction," *Izv. Atmos. Ocean. Phys.* **10**, 1110.

Killworth, P. D., 1976, "The mixing and spreading phase of MEDOC I," *Prog. Oceanogr.* **7**, 59.

Killworth, P. D., 1979, "On 'chimney' formations in the ocean," *J. Phys. Oceanogr.* **9**, 531.

Killworth, P. D., and Bigg, G. R., 1988, "An intercomparison of inverse methods using an eddy-resolving general circulation model," *J. Phys. Oceanogr.* **18**, 987.

Kimball, H. H., 1928, "Amount of solar radiation that reaches the surface of the earth on the land and on the sea and methods by which it is measured," *Mon. Wea. Rev.* **56,** 393.

Kitaigorodskii, S. A., Donelan, M. A., Lumley, J. L., and Terray, E. A., 1983, "Wave-turbulence interactions in the upper ocean. II. Statistical characteristics of wave and turbulent components of the random velocity field in the marine surface layer," *J. Phys. Oceanogr.* **13,** 1988.

Kitaigorodskii, S. A., and Lumley, J. L., 1983, "Wave-turbulence interactions in the upper ocean. I. The energy balance of the interacting fields of surface wind waves and wind-induced three dimensional turbulence," *J. Phys. Oceanogr.* **13,** 1977.

Klein, P., and Coantic, M., 1981, "A numerical study of turbulent processes in the marine upper layers," *J. Phys. Oceanogr.* **11,** 849.

Knox, R. A., and Anderson, D. L. T., 1985, "Recent advances in the study of the low-latitude ocean circulation," *Prog. Oceanogr.* **14,** 259.

Kolmogorov, A. N., 1941, "The local structure of turbulence in compressible turbulence for very large Reynolds numbers," *Compt. Rend. Akad. Nauk SSSR,* **30,** 301.

Kolovayev, P. A., 1976, "Investigation of the concentration and statistical size distribution of wind-produced bubbles in the near-surface ocean layer," *Oceanology* (Engl. transl.) **15,** 659.

Komen, G. J., Hasselmann, S., and Hasselmann, K., 1984, "On the existence of a fully-developed wind-sea spectrum," *J. Phys. Oceanogr.* **14,** 1271.

Kondo, J., 1975, "Air-sea bulk transfer coefficients in diabatic conditions," *Bound.-Layer Meteor.* **9,** 91.

Kondratjev, K. Ya., 1969, *Radiation in the Atmosphere.* New York: Academic Press.

Koračin, D., and Rogers, D. P., 1990, "Numerical simulations of the response of the marine atmosphere to ocean forcing," *J. Atmos. Sci.* **47,** 592.

Kraus, E. B., 1959, "The evaporation–precipitation cycle of the trades," *Tellus* **11,** 148.

Kraus, E. B., 1967, "Organized convection in the ocean surface layer resulting from slicks and wave radiation stress," *Physics of Fluids* **10,** Pt. 2, S294.

Kraus, E. B., 1972, *Atmosphere Ocean Interaction.* Oxford: Clarendon Press.

Kraus, E. B., 1977a, "Ocean surface drift velocities," *J. Phys. Oceanogr.* **7,** 606.

Kraus, E. B., 1987, "The torque and flux balances in the upper equatorial ocean," *J. Geophys. Res.* **92,** 14242.

Kraus, E. B., and Leslie, L. D., 1982, "The interactive evolution of the oceanic and atmospheric boundary layer of the trades," *J. Atmos. Sci.* **39,** 2760.

Kraus, E. B., and Morrison, R. E., 1966, "Local interactions between sea and air at monthly and annual time scales," *Quart. J. Roy. Meteor. Soc.* **92,** 114. (Also in *Selected Meterological Papers* No. 17, The Meteor. Soc. Japan, 1973.)

Kraus, E. B., Petersen, W. H., and Rooth, C. G., 1978, "The thermal evolution of the ocean," in *Proc. Evolution of Planetary Atmospheres and Climatology of the Earth.* Toulouse, France: Centre National d'Etudes Spatiales, p. 201.

Kraus, E. B., and Turner, J. S., 1967, "A one-dimensional model of the seasonal thermocline. II. The general theory and its consequences," *Tellus* **19,** 98.

Krauss, W., 1966, *Interne Wellen.* Berlin: Borntraeger.

Krishnamurti, Ruby, 1975, "On cellular cloud patterns," *J. Atmos. Sci.* **32,** 1373.

Kristensen, L., and Fitzjarrald, D. R., 1984, "The effect of line averaging on scalar flux measurements with a sonic anemometer near the surface," *J. Atmos. Oceanic Tech.* **1,** 138.

Kuettner, J., 1959, "The band structure of the atmosphere," *Tellus* **11,** 267.

Kuettner, J., 1967, "Cloudstreets, theory and observation," *Aero. Rev.* **42,** 52.

Landau, L. D., and Lifshitz, E. M., 1959, *Fluid Mechanics* (Translated from the Russian by J. B. Sykes and W. H. Reid). London: Pergamon Press.

Langmuir, I., 1938, "Surface motion of water induced by wind," *Science* **87,** 119.

Large, W. G., and Pond, S., 1981, "Open ocean momentum flux measurements in moderate to strong winds," *J. Phys. Oceanogr.* **11,** 324.

Larson, T. R., and Wright, J. W., 1975, "Wind-generated gravity-capillary waves: Laboratory measurements of temporal growth rates using microwave back-scatter," *J. Fluid Mech.* **70,** 417.

Latif, M., Flugel, M., and Xu, J. S., 1990, "An investigation of short-range climate predictability in the tropical Pacific," *Rep.* **52,** Max Plank Inst. für Meteor.

Lau, K.-M., 1981, "Oscillations in a simple equatorial climate system," *J. Atmos. Sci.* **38,** 248.

LeBlond, P. H., and Mysak, L. A., 1977, "Trapped coastal waves and their role in shelf dynamics," in *The Sea (Volume 6)* (E. D. Goldberg et al., eds.). New York: Wiley-Interscience.

Ledwell, J. R., 1984, "The variation of the gas transfer coefficient with molecular diffusivity," in *Gas Transfer at Water Surfaces* (W. Brutsaert and G. H. Jirka, eds.). Dordrecht: Reidel, P. 293.

Leetmaa, A., Quadfasel, D. R., and Wilson, D., 1982, "Development of the flow field during the onset of the Somali current, 1979," *J. Phys. Oceanogr.* **12,** 1,325.

Leibovich, S., 1983, "The form and dynamics of Langmuir circulations," *Ann. Rev. Fluid Mech.* **15,** 391.

Leibovich, S., Lele, S. K., and Moroz, I. M., 1989, "Nonlinear dynamics of Langmuir circulations and in thermosolutal convection," *J. Fluid Mech.* **198,** 471.

Leipper, D. F., 1967, "Observed ocean conditions and hurricane Hilda," *J. Atmos. Sci.* **24,** 182.

Lemasson, L., and Piton, B., 1968, "Anomalie dynamique de la surface de la mer le long de l'equateur dans l'Ocean Pacifique," *Cah. ORSTOM, Sér Océanogr.* **6,** 39.

LeMone, M. A., 1973, "The structure and dynamics of horizontal roll vortices in the planetary boundary layer," *J. Atmos. Sci.* **30,** 1077.

LeMone, M. A., 1976, "Modulation of turbulence energy by longitudinal rolls in an unstable planetary boundary layer," *J. Atmos. Sci.* **33,** 1308.

LeMone, M. A., and Meitin, R. J., 1984, "Three examples of fair-weather mesoscale boundary-layer convection in the tropics," *Mon. Wea. Rev.* **112,** 1985.

Lenschow, D. H., Wyngaard, J. C., and Pennell, W. T., 1980, "Mean-field and second-moment budgets in a baroclinic, convective boundary layer," *J. Atmos. Sci.* **37,** 1313.

Leslie, L. M., and Smith, R. K., 1970, "The surface boundary layer of a hurricane. II.," *Tellus* **22,** 288.

Lettau, H. H., 1962, "Theoretical wind spirals in the boundary layer of a barotropic atmosphere," *Beitr. Phys. Atmos.* **35,** 195.

Levich, V. G., 1962, *Physicochemical Hydrodynamics.* Englewood Cliffs, NJ: Prentice-Hall.

Li, F., Large, W., Shaw, W., Walsh, E. J., and Davidson, K., 1989, "Ocean radar backscatter relationship with near-surface winds: A case study during FASINEX," *J. Phys. Oceanogr.* **19,** 342.

Lighthill, M. J., 1962, "Physical interpretation of the mathematical theory of wave generation by wind," *J. Fluid Mech.* **14,** 385.

Lighthill, M. J., 1969, "Unsteady wind-driven ocean currents," *Quart. J. Roy. Meteor. Soc.* **95,** 675.

Lighthill, M. J., 1978, *Waves in Fluids.* Cambridge, New York: Cambridge University Press.

Lilly, D. K., 1966, "On the instability of the Ekman boundary flow," *J. Atmos. Sci.* **23,** 481.

Lilly, D. K., 1968, "Models of cloud-topped mixed-layers under a strong inversion," *Quart. J. Roy. Meteor. Soc.* **94,** 292.

Lind, R. J., and Katsaros, K. B., 1982, "A model of longwave irradiance for use with surface observations," *J. Appl. Meteor.* **12,** 1015.

Lind, R. J., and Katsaros, K. B., 1986, "Radiation measurements and model results from R/V *Oceanographer* during STREX 1980," *J. Geophys. Res.* **91,** 13,308.

Lind, R. J., and Katsaros, K. B., 1987, "Radiation measurements from R/V FLIP and R/V Acania during the mixed layer dynamics experiment (MILDEX)." Technical Report. University of Washington, Seattle, WA; Dept. of Atmospheric Sciences.

Liou, K. N., 1980, *An Introduction to Atmospheric Radiation.* New York: Academic Press.

Liss, P. S., 1983, "Gas transfer: experiments and geochemical implications," in *Air–Sea Exchange of Gases and Particles* (P. S. Liss and W. G. N. Slinn, eds.). Reidel, p. 241.

Liss, P. S., and Merlivat, L., 1986, "Air–sea gas exchange rates: Introduction and synthesis," in *The Role of Air–Sea Exchange in Geochemical Cycling* (P. Buat-Ménard, ed.). Dordrecht: Reidel, p. 113.

Liu, W. T., 1984, "Estimation of latent heat flux with Seasat-SMMR, a case study in N. Atlantic," in *Large-Scale Oceanographic Experiments and Satellites* (C. Gautier and M. Fieux, eds.). Hingham, MA: Reidel, p. 205.

Liu, W. T., 1989, "Moisture and latent heat flux variabilities in the tropical Pacific derived from satellite data," *J. Geophys. Res.* **93,** 6749.

Liu, W. T., and Businger, J. A., 1975 "Temperature profile in the molecular sublayer near the interface of a fluid in turbulent motion," *Geophys. Res. Lett.* **2,** 403.

Liu, W. T., Katsaros, K. B., and Businger, J. A., 1979, "Bulk parameterization of air–sea exchanges of heat and water vapor including the molecular constraints at the interface," *J. Atmos. Sci.* **36,** 1722.

Liu, W. T., and Large, W. G., 1981, "Determination of surface stress by Seasat-SASS: A case study with JASIN data," *J. Phys. Oceanogr.* **12,** 1603.

Longuet-Higgins, M. S., 1952, "On the statistical distribution of the heights of sea waves," *J. Mar. Res.* **11,** 245.

Longuet-Higgins, M. S., 1965, "The response of a stratified ocean to stationary or moving wind systems," *Deep-Sea Res.* **12,** 923.

Longuet-Higgins, M. S., 1969, "A non-linear mechanism for the generation of sea waves," *Proc. Roy. Soc. Ser. A* **311,** 529.

Lozovatsky, I. D., Ksenofontov, A. S., Erofeev, An Y., and Gibson, C. H., 1993, "Modelling of the evolution of vertical structure in the upper ocean by atmospheric forcing and intermittent turbulence in the pycnocline," *J. Mar. Sys.* **4,** 263.

Lueck, R. G., Crawford, W. R., and Osborn, T. R., 1983, "Turbulent dissipation over the continental slope off Vancouver Island," *J. Phys. Oceanogr.* **13,** 1809.

Lumb, F. E., 1964, "The influence of cloud on hourly amount or total solar radiation at the sea surface," *Quart. J. Roy. Meteor. Soc.* **90,** 43.

Lumley, J. L., and Panofsky, H. A., 1964, *The Structure of Atmospheric Turbulence.* New York: Interscience.

Luther, M. E., O'Brien, J. J., and Meng, A. H., 1985, "Morphology of Somali current system during the southwestern monsoon," In *Coupled Ocean–Atmosphere Models* (J. C. J. Nihoul, ed.). New York: Elsevier, p. 405.

Madden, R. A., and Robitaille, F. E., 1970, "A comparison of the equivalent potential temperature and the static energy," *J. Atmos. Sci.* **27,** 327.

Malkus, J. S., 1962, "Large scale interactions," in *The Sea* (Volume I) (M. N. Hill, ed.). New York, N.Y.: Wiley-Interscience, p. 88.

Malkus, J. S., and Riehl, H. 1960, "On the dynamics and energy transformation in steady-state hurricanes," *Tellus* **12,** 1.

Malmgren, F., 1927, "On the properties of sea ice," *Sci. Results; Norwegian North Polar Exped. with the Maud 1918–1925* **1,** 1.

Manabe, S., and Stouffer, R. J., 1988, "Two stable equilibria of a coupled ocean–atmosphere model," *J. Climate* **1,** 841.

Marks, R., and Monahan, E. C., 1989, "Relationship between marine aerosols, whitecaps and low-elevation winds observed during the HEXMAX experiment in the North Sea," in *Climate and Health Implications of Bubble-Mediated Sea–Air Exchange* (E. C. Monahan and M. A. VanPatten, eds.). Connecticut Sea Grant College Progr. Publ. CT-SG-89-06, p. 83.

Martin, M., and Berdahl, P., 1984, "Characteristics of infrared sky radiation in the United States," *Solar Energy* **33,** 321.

Mason, P. J., 1989, "Large-eddy simulation of the convective atmospheric boundary layer," *J. Atmos. Sci.* **46,** 1492.

Mason, P. J., and Thompson, D. J., 1987, "Large-eddy simulations of the neutral-static stability planetary boundary layer," *Quart. J. Roy. Meteor. Soc.* **113,** 413.

Matsuno, T., 1966, "Quasi-geostrophic motions in the equatorial area," *J. Meteor. Soc. Japan* **44,** 25.

Matunobu, Y., 1966, "Motion of a deformed drop in Stokes flow," *J. Phys. Soc. Japan* **21,** 1596.

Maxworthy, T. and Narimousa, S., 1994, "Unsteady, turbulent convection into a homogeneous rotating fluid," *J. Phys. Oceanogr.* **24,** (in press).

Maykut, G. A., 1985, "An introduction to ice in the polar oceans," *Applied Physics Lab./U. of Washington Publ.* (APL-UW 8510).

Maykut, G. A., and Perovich, D. K., 1987, "The role of shortwave radiation in the summer decay of a sea ice cover," *J. Geophys. Res.* **92,** 7032.

Maykut, G. A., and Untersteiner, N., 1971, "Some results from a time-dependent, thermodynamic model of sea ice," *J. Geophys. Res.* **76,** 1550.

McAlister, E. D., and McLeish, W., 1970, "A radiometer system for measurement of the total heat flux from the sea," *Appl. Optics* **9,** 2697.

McCreary, J. P., 1981, "A linear stratified model of the equatorial undercurrent," *Phil. Trans. Roy. Soc. London, Ser. A* **298,** 603.

McCreary, J. P. Jr., 1983, "A model of tropical ocean–atmosphere interaction," *Mon. Wea. Rev.* **111,** 370.

McCreary, J. P., Jr., and Anderson, D. L. T., 1985, "Simple models of El Niño and the Southern Oscillation," in *Coupled Ocean–Atmosphere Models* (J. C. J. Nihoul, ed.). New York: Elsevier, p. 345.

McCreary, J. P., and Anderson, D. L. T., 1991, "An overview of coupled ocean–atmosphere models of El Nino and the Southern Oscillation," *J. Geophys. Res.* **96,** 3125.

McDougall, T. J., 1987, "Thermobaricity, Cabbeling and Water Mass Conversion," *J. Geophys. Res.* **92,** 5448.

Mellor, G., and Yamada, T., 1974, "A hierarchy of turbulence closure models for planetary boundary layers," *J. Atmos. Sci.* **31,** 1791.

Middleton, J. F., and Wright, D. G., 1988, "Shelf wave shattering due to a longshore jump in topography," *J. Phys. Oceanogr.* **18,** 230.

Mie, Gustav, 1908, "Beitrage zur Optik trüber Medien, speziell kolloidaler Metallösungen," *Ann. Physik* **25,** 377.

Miles, J. W., 1957, "On the generation of surface waves by shear flows. I," *J. Fluid Mechanics* **3,** 185.

Miles, J. W., 1959a, "On the generation of surface waves by shear flows. II," *J. Fluid Mechanics* **6,** 568.

Miles, J. W., 1959b, "On the generation of surface waves by shear flows. III," *J. Fluid Mechanics* **6,** 583.

Millero, F. J., 1978, "Freezing point of sea water," in *UNESCO Tech. Pap. Mar. Sci. No. 28, Annex 6.* Paris; UNESCO.

Millero, F. J., Gonzales, A., and Ward, G. K., 1976, "The density of seawater solutions at one atmosphere as a function of temperature and salinity," *J. Mar. Res.* **34,** 61.

Mitsuyasu, H., Tasai, F., Suhara, T., Mizuno, S., Ohkuso, M., Honda, T., and Rikiishi, K., 1975, "Observations of the directional spectrum of ocean waves using a clover leaf buoy," *J. Phys. Oceanogr.* **5,** 750.

Moeng, C. H., 1984, "A large eddy-simulation model for the study of planetary boundary layer turbulence," *J. Atmos. Sci.* **41,** 2052.

Moeng, C. H., and Wyngaard, J. C., 1984, "Statistics of conservative scalars in the convective boundary layer," *J. Atmos. Sci.* **41,** 3161.

Monahan, E. C., 1968, "Sea spray as a function of low elevation windspeed," *J. Geophys. Res.* **73,** 1127.

Monahan, E. C., 1989, "From the laboratory tank to the global ocean," in *Climate and Health Implications of Bubble-Mediated Sea–Air Exchange* (E. C. Monahan and M. A. VanPatten, eds.). Connecticut Sea Grant College Progr. Publ. CT-SG-89-06, p. 43.

Monahan, E. C., and O'Muircheartaigh, I. G., 1980, "Optical power-law description of oceanic whitecap coverage dependence on wind speed," *J. Phys. Oceanogr.* **10,** 2094.

Monahan, E. C., and O'Muircheartaigh, I. G., 1987, "Comments on 'Albedos and glitter patterns of a wind-roughened sea surface'," *J. Phys. Oceanogr.* **17,** 549.

Monin, A. S., and Obukhov, A. M., 1954, "Basic laws of turbulent mixing in the surface layer of the atmosphere," *Tr. Akad. Nauk SSSR, Geofiz Inst.* **24,** 163.

Montgomery, R. B., and Palmén, E., 1940, "Contribution to the question of the Equatorial Countercurrent," *J. Mar. Res.* **3,** 112.

Moore, R. K., and Fung, A. K., 1979, "Radar determination of winds at sea," *Proc. IEEE* **67,** 1504.

Morel, A., 1965, "Resultats experimentaux concernant la penetration de la lumière du jour dans les eaux Mediterranéan," *Cah. Oceanogr.* **17,** 177.

Morgan, G. W., 1956, "On the wind-driven ocean circulation," *Tellus* **8,** 301.

Moum, J. N., and Osborne, T. R., 1986, "Mixing in the main thermocline," *J. Phys. Oceanogr.* **16,** 1250.

Mourad, P. D., and Brown, R. A., 1990, "Multiscale large eddy states in weakly stratified planetary boundary layers," *J. Atmos. Sci.* **47,** 414.

Muira, Y., 1986, "Aspect ratios of longitudinal rolls and convection cells observed during cold air outbreaks," *J. Atmos. Sci.* **43,** 26.

Munk, W. H., 1950, "On the wind-driven ocean circulation," *J. Meteor.* **7,** 79.

Munk, W. H., 1981, "Internal waves and small-scale processes," in *Evolution of Physical Oceanography* (B. A. Warren and C. Wunsch, eds.). Cambridge, MA: MIT Press.

Munk, W. H., Snodgrass, F. E., and Carrier, G., 1956, "Edge waves on a continental shelf," *Science* **123,** 127.

Meelin, J. D., Latif, M., Alaart, M. A. F., Cane, M. A., Cubash, U., Gates, W. L., Ghent, P. R., Ghil, M., Gordon, C., Lau, N. C., Mechoso, C. R., Mehl, G. A., Oberhuber, J. M., Philander, S. G. H., Shopf, P. S., Sperber, K. R., Sterl, A., Tokioka, T., Tribbia, J., and Zebiak, S. E., 1992, "Tropical air–sea interaction in general circulation models," *Clim. Dyn.* **7,** 73.

Nicholls, S., Brummer, B., Fiedler, A., Grant, A., Hauf, T., Jenkins, G., Readings, C., and Shaw, W., 1983, "The structure of the turbulent atmospheric boundary layer," *Phil. Trans. Roy. Soc. London A* **308,** 291.

Nieuwstadt, F. T. M., and Businger, J. A., 1984, "Radiative cooling near the top of a cloudy mixed layer," *Quart. J. Roy. Meteor. Soc.* **110,** 1073.

Nihoul, J. C. J. (ed.), 1993, "Sub-mesoscale air–sea interactions," *J. Mar. Sys.* **4,** Special Issue No 2/3.

Nihoul, J. C. J. (ed.), 1985, *Coupled Ocean–Atmosphere Models. Elsevier Oceanography Series 40.* Amsterdam, New York: Elsevier.

Niiler, P. P., and Kraus, E. B., 1977, "One-dimensional models of the upper ocean," in *Modelling and Prediction of the Upper Layers of the Ocean* (E. B. Kraus, ed.). Oxford: Pergamon Press, p. 143.

Nikolayeva, Y. I., and Tsimring, L. S., 1986, "Kinetic model of the wind generation of waves by a turbulent wind," *Izv. Acad. Sci. USSR, Atmos. Ocean Phys.* **22,** 102.

Nikuradse, J., 1933, "Strömungsgesetze in rauhen Rohren," *V.D.I. Forschungsheft* **361,** 1.

Nordeng, T. E., 1991, "On the wave age dependent drag coefficient and roughness length at sea," *J. Geophys. Res.* **96,** 7167.

Oakey, N. S., 1985, "Statistics of mixing parameters in the upper ocean during JASIN phase 2," *J. Phys. Oceanogr.* **15,** 1662.

O'Brien, J. J., and Reid, R. O., 1967, "The non-linear response of a two-layer, baroclinic ocean to a stationary, axially-symmetric hurricane. I," *J. Atmos. Sci.* **24**, 197.

Obukhov, A. M., 1946, "Turbulence in an atmosphere with a non-uniform temperature," *Tr. Akad. Nauk. USSR, Inst. Teoret. Geofiz.* **1**. Translated and published in *Bound.-Layer Meteor.* **2**, 7, 1971.

Oncley, S. P., 1989, *Flux parameterization Techniques in the Atmospheric Surface Layer.* Ph.D. Thesis, University of California, Irvine, Dept. Mech. Eng.

Oncley, S. P., Delaney, A. C., Horst, T. W., and Tans, P. P., 1993, "Verification of flux measurements using conditional sampling," *Atmos. Environ.* **27A**, 2417.

Ono, N., 1967, "Specific heat and heat of fusion of sea ice," in *Physics of Snow and Ice* (H. Oura, ed.). Hokkaido, Japan: Inst. Low Temp. Sci., p. 599.

Ooyama, K., 1969, "Numerical simulation of the life-cycle of tropical cyclones," *J. Atmos. Sci.* **26**, 3.

Ostlund, H. G., 1967, "Hurricane tritium. I. Preliminary results on Hilda 1964 and Betsy 1965," *Geophys. Mongr. No.* **11**. Washington D.C.: Am. Geophys. Union.

Panofsky, W. K. H., and Phillips, M., 1962, *Classical Electricity and Magnetism* (second edition). Reading, MA: Addison-Wesley.

Paulson, C. A., 1970, "The mathematical representation of wind speed and temperature profiles in the unstable atmospheric surface layer," *J. Appl. Meteor.* **9**, 857.

Paulson, C. A., and Parker, T. W., 1972, "Cooling of a water surface by evaporation, radiation, and heat transfer," *J. Geophys. Res.* **77**, 491.

Payne, R. E., 1972, "Albedo of the sea surface," *J. Atmos. Sci.* **29**, 959.

Pedlosky, J., 1965, "A study of the time dependent ocean circulation," *J. Atmos. Sci.* **22**, 267.

Pellew, A., and Southwell, R. W., 1940, "On maintained convection in a fluid heated from below," *Proc. Roy. Soc. London, Ser. A.* **125**, 312.

Peng, T.-H., Broecker, W. S., Mathieu, G. G., Li, Y.-H., and Bainbridge, A. E., 1979, "Radon evasion rates in the Atlantic and Pacific oceans as determined during the GEOSECS program," *J. Geophys. Res.* **84**, 2471.

Petersen, W. H., 1978, *A Steady Thermo-Haline Convection Model.* Ph.D. Thesis, University of Miami.

Philander, S. G. H., 1990, *El Niño, La Niña and the Southern Oscillation.* San Diego, CA: Academic Press.

Phillips, O. M., 1957, "On the generation of waves by turbulent wind," *J. Fluid Mech.* **2**, 417.

Phillips, O. M., 1958, "The equilibrium range in the spectrum of wind-generated waves," *J. Fluid Mech.* **4**, 426.

Phillips, O. M., 1960, "On the dynamics of unsteady gravity waves of finite amplitude. I," *J. Fluid Mech.* **9**, 193.

Phillips, O. M., 1961, "On the dynamics of unsteady gravity waves of finite amplitude. II," *J. Fluid Mech.* **11**, 143.

Phillips, O. M., 1977, *The Dynamics of the Upper Ocean* (second edition). London and New York: Cambridge University Press.

Pierson, W. J., and Moskowitz, L., 1964, "A proposed spectral form for fully developed wind seas based on the similarity theory of S. A. Kitaigorodskii," *J. Geophys. Res.* **69**, 5181.

Pike, A. C., 1971, "The intertropical convergence zone studied with an interacting atmosphere and ocean model," *Mon. Wea. Rev.* **99**, 469.

Plant, W. J., 1982, "A relationship between wind stress and wave slope," *J. Geophys. Res.* **87**, 1961.

Platzman, G. W., 1968, "The Rossby wave," *Quart. J. Roy. Meteor. Soc.* **94**, 225.

Pollard, R. T., 1970, "On the generation by winds of inertial waves in the ocean," *Deep-Sea Res.* **17**, 795.

Pollard, R. T., Rhines, P. B., and Thompson, R. O. R. Y., 1973, "The deepening of the wind-mixed layer," *Geophys. Fluid Dyn.* **3**, 381.

Preobrazhenskii, L. Y., 1973, "Estimate of the content of spray-drops in the near-water layer of the atmosphere," *Fluid Mech. Sov. Res.* **2**, 95.

Price, J. F., 1981, "Upper ocean response to a hurricane," *J. Phys. Oceanogr.* **11**, 153.

Price, J. F., Weller, R. A., and Pinkel, R., 1986, "Diurnal cycling: observations and models of the upper ocean response to diurnal heating, cooling and wind mixing," *J. Geophys. Res. (Oceans)* **91**, 8411.

Quinn, J. A., and Otto, N. C., 1971, "Carbon dioxide exchange at the air–sea interface: flux augmentation by chemical reaction," *J. Geophys. Res.* **76**, 1539.

Randall, D. A., 1980a, "Conditional instability of the first kind upside down," *J. Atmos. Sci.* **37**, 125.

Randall, D. A., 1980b, "Entrainment into a stratocumulus layer with distributed radiative cooling," *J. Atmos. Sci.* **37**, 148.

Reichardt, H., 1940, "Die Wärme Übertragung in turbulenten Reibungsschichten," *Z. Angew. Math. Mech.* **20**, 297.

Reid, R. O., 1958, "Effect of Coriolis force on edge waves. I. Investigation of normal modes," *J. Mar. Res.* **16**, 109.

Resnyanskiy, Yu. D., 1975, "Parameterization of the integral turbulent energy dissipation in the upper quasi-homogeneous layer of the ocean," *Izv. Atm. Ocean. Phys.* **11**, 453.

Rhines, P. B., 1977, "The dynamics of unsteady currents," in *The Sea* (Volume 6) (E. D. Goldberg et al., eds.). New York: Wiley-Interscience.

Richardson, L. F., 1920, "The supply of energy from and to atmospheric eddies," *Proc. Roy. Soc. Lond A* **97**, 354.

Richter, F. M., 1973, "Convection and the large-scale circulation of the mantle," *J. Geophys. Res.* **78**, 8735.

Riehl, H., 1963, "On the origin and possible modification of hurricanes," *Science* **141**, 1001.

Riehl, H., Yeh, T. C., Malkus, J. S., and LaSeur, N. E., 1951, "The north-east trade of the Pacific Ocean," *Quart. J. Roy. Meteor. Soc.* **77**, 598.

Riley, D. S., Donelan, M. A., and Hui, W. H., 1982, "An extended Miles' theory for wave generation by wind," *Bound.-Layer Meteor.* **22**, 209.

Riley, J. P., and Skirrow, J. (eds.), 1965, *Chemical Oceanography* (Volume 1). New York: Academic Press.

Roach, W. T., and Slingo, A., 1979, "A high resolution infrared radiative transfer scheme to study the interaction of radiation with cloud," *Quart. J. Roy Meteor. Soc.* **105**, 603.

Roether, W., and Kromer, B., 1984, "Optimum application of the radon deficit method to obtain air-sea gas exchange rates," in *Gas Transfer at Water Surfaces* (W. Brutsaert and G. H. Jirka, eds.). Dordrecht: Reidel, p. 447.

Rogers, D. P., and Koračin, D., 1992, "Radiative transfer and turbulence in the cloud-topped marine atmospheric boundary layer," *J. Atmos. Sci.* **49**, 1473.

Rooth, C., 1982, "Hydrology and ocean circulation," *Prog. Oceanogr.* **11**, 131.

Rooth, C., and Xie, L.-A., 1992, "Air–sea boundary layer dynamics in the presence of mesoscale currents," *J. Geophys. Res.* **97**, 14431.

Rossiter, J. R., 1954, "The North Sea surge of 31 January and 1 February 1953," *Phil. Trans. Roy. Soc. London, Ser, A* **246**, 371.

Rotunno, R., and Emanuel, K. A., 1987, "An air–sea interaction theory for tropical cyclones. II. Evolutionary study using a nonhydrostatic axisymmetric numerical model," *J. Atmos. Sci.* **44**, 542.

Sanford, T. B., Black, P. G., Feeney, J. R., Forristall, J. W., George, Z., and Price, J. F., 1987, "Ocean response to a hurricane. I. Observations," *J. Phys. Oceanogr.* **17**, 2065.

Sarachik, E. S., 1978, "Boundary layers on both sides of the tropical ocean surface," *Review papers on equatorial oceanography. Proc. FINE Workshop*, Ft. Lauderdale: Nova/N.Y.I.T. Univ. Press.

Saunders, P. M., 1964, "Sea smoke and steam fog," *Quart. J. Roy. Meteor. Soc.* **90**, 156.

Saunders, P. M., 1968, "Radiance of sea and sky in the infrared window 800–1200 cm$^{-1}$," *J. Opt. Soc. Am.* **58**, 645.

Saunders, P. M., 1973, "The instability of a baroclinic vortex," *J. Phys. Oceanogr.* **3**, 61.

Schmetz, J., and Raschke, E., 1981, "An approximate computation of infrared radiative fluxes in a cloudy atmosphere," *Pure Appl. Geophys.* **119**, 248.

Schmetz, J., Slingo, A., Nicholls, S., and Raschke, E., 1983, "Case studies of radiation in the cloud-capped atmospheric boundary layer," *Phil. Trans. Roy. Soc. London, Ser. A* **308**, 377.

Schmetz, P., Schmetz, J., and Raschke, E., 1986, "Estimation of daytime downward longwave radiation at the surface from satellite and grid point data," *Theor. Appl. Climatol.* **37**, 136.

Schopf, P. S., and Cane, M. A., 1983, "On equatorial dynamics, mixed layer physics and sea surface temperature," *J. Phys. Oceanogr.* **13**, 917.

Schroeder, L. C., Boggs, D. H., Dome, G. J., Halberstam. I. M., Jones, W. L., Pierson, W. J., and Wentz, F. W., 1982, "The relationship between wind vector and normalized radar crossection used to derive Seasat-A satellite scatterometer winds," *J. Geophys. Res.* **87**, 3318.

Schulejkin, W. W., 1960, *Theory der Meereswellen.* Berlin: Akademie-Verlag.

Schwerdtfeger, P., 1963, "The thermal properties of sea ice," *J. Glaciol.* **4**, 789.

Seckel, G. R., and Beaudry, F. H., 1973, "The radiation from the sun and sky over the North Pacific Ocean (abstr.)," *Trans. Am. Geophys. Union* **54**, 1114.

Sellers, W. D., 1969, "A climate based on the energy balance of the earth–atmosphere system," *J. Appl. Meteor.* **8**, 392.

Shakleton, N. J., 1978, "Evolution of the earth's climate during the Tertiary Era," in *Proc. Evolution of Planetary Atmospheres and Climatology of the Earth.* Toulouse, France: Centre National d'Etudes Spatiales, p. 49.

Shay, L. K., and Elsberry, R. L., 1987, "Near-inertial ocean current response to hurricane Frederic," *J. Phys. Oceanogr.* **17**, 1249.

Shay, T. J., and Gregg, M. C., 1984, "Turbulence in an oceanic convective mixed layer," *Nature* **310,** 282.

Shemdin, O. H., and Hsu, E.-Y., 1967, "Direct measurement of aerodynamic pressure above a simple progressive gravity wave," *J. Fluid Mech.* **30,** 403.

Shirer, H. N., 1987, *Nonlinear Hydrodynamic Modeling. A Mathematical Introduction. Lecture Notes in Physics.* Berlin: Springer-Verlag, p. 271.

Slingo, A., Brown, R., and Wench, C. L., 1982a, "A field study of nocturnal stratocumulus. III. High resolution radiative and microphysical observations," *Quart. J. Roy. Meteor. Soc.* **108,** 145.

Slingo, A., Nicholls, S., and Schmetz, J., 1982b, "Aircraft observations of marine stratocumulus during JASIN," *Quart. J. Roy. Meteor. Soc.* **108,** 833.

Slingo, A., and Schrecker, H. M., 1982, "On the shortwave radiative properties of stratiform water clouds," *Quart. J. Roy. Meteor. Soc.* **108,** 407.

Smith, R. C., and Tyler, J. E., 1967, "Optical properties of clear natural water," *J. Opt. Soc. Am.* **57,** 589.

Smith, S. D., 1981, "Coefficients for sea-surface wind stress and heat exchange," Rep. BI-R-81-19, Dartmouth, N.S.: Bedford Inst. Oceanogr.

Smith, S. D., 1988, "Coefficients for sea-surface wind stress, heatflux, and wind profiles as a function of windspeed and temperature," *J. Geophys. Res.* **93,** 15,467.

Smith, S. D., Anderson, R. J., Jones, E. P., Desjardins, R. L., Moore, R. M., Hertzman, O., and Johnson, B. D., 1991, "A new measurement of $CO_2$ eddy flux in the nearshore atmospheric surface layer," *J. Geophys. Res.* **96,** 8881.

Smith, S. D., and Banke, E. G., 1975, "Variation of the sea surface drag coefficients with windspeed," *Quart. J. Roy. Meteor. Soc.* **101,** 665.

Smithsonian Institution, 1971, *Smithsonian Meteorological Tables* (sixth revised edition, prepared by R. J. List). Washington, D.C.

Snyder, R. L., Dobson, F. W., Elliott, J. A., and Long R. B., 1981, "Array measurements of atmospheric pressure fluctuations over surface gravity waves," *J. Fluid Mech.* **102,** 1.

Soloviev, A. V., 1982, "On vertical structure of the thin surface layer of the ocean at weak winds," *Fiz. Atmos. Okeana* **18,** 751.

Soloviev, A. V., and Vershinsky, V. N., 1982, "The vertical structure of the thin surface layer of the ocean under conditions of low windspeed," *Deep Sea-Res.* **29,** 1437.

Stage, S. A., and Businger, J. A., 1981, "A model for entrainment into a cloud-topped marine boundary layer. II. Discussion of model behavior and comparison with other cloud models," *J. Atmos. Sci.* **38,** 2230.

Stocker, T. F., and Mysak, L. A., 1992, "Climate fluctuations on the century time-scale: A review of high-resolution proxy data and possible mechanisms," *Climatic Change* **20,** 227.

Stokes, G. G., 1847, "On the theory of oscillatory waves," *Trans. Cambridge Philos. Soc.* **8,** 441

Stommel, H., 1948, "The westward intensification of wind-driven ocean currents," *Trans. Am. Geophys. Union* **99,** 202.

Stommel, H., 1961, "Thermohaline convection with two stable regimes of flow," *Tellus* **13,** 224.

Stommel, H., 1962, "On the smallness of sinking regions in the ocean," *Proc. of the Ntl. Acad. Sci. U.S.A.* **48,** 766.

Stommel, H., 1965, *The Gulf Stream* (Second Edition). Berkeley: University of California Press.

Stull, R. B., 1976, "Energetics of entrainment across a density interface," *J. Atmos. Sci.* **33**, 1260.

Stull, R. B., 1984, "Transilient turbulence theory. I. The concept of eddy mixing across finite distances," *J. Atmos. Sci.* **41**, 3351.

Stull, R. B., 1986, "Transilient turbulence theory, III. Bulk dispersion rate and numerical stability," *J. Atmos. Sci.* **43**, 50.

Stull, R. B., and Kraus, E. B., 1987, "The transilient model of the upper ocean," *J. Geophys. Res.* **92**, 10745.

Sverdrup, H. U., 1945, *Oceanography for Meteorologists.* London: Allen and Unwin Ltd.

Sverdrup, H. U., Johnson, M. W., and Fleming, R. H., 1942, *The Oceans; Their Physics, Chemistry and General Biology.* Englewood Cliffs, NJ: Prentice Hall.

Sverdrup, H. U., and Munk, W. H., 1947, "Wind, sea and swell; Theory of relations for forecasting," *Hydrographic Office Publ. No.* 601, U.S. Department of the Navy.

SWAMP (Sea Wave Modelling Project) Group, 1985, "An intercomparison study of wind wave prediction models, in *Ocean Wave Modelling.* Plenum, p. 1.

Swinbank, W. C., 1963, "Long-wave radiation from clear skies," *Quart. J. Roy. Meteor. Soc.* **89**, 339.

Tatro, P. R., and Mollo-Christensen, E. L., 1967, "Experiments on Ekman layer instability," *J. Fluid Mech.* **28**, 531.

Taylor, T. D., and Acrivos, A., 1964, "On the deformation and drag of a falling viscous drop at low Reynolds number," *J. Fluid. Mech.* **18**, 466.

Taylor, P. K., and Guymer, T. H., 1983, "The structure of an atmospheric warm front and its interaction with the boundary layer," *Phil. Trans. Roy. Soc. London, Ser. A* **308**, 341.

Tennekes, H., 1970, "Free convection in the turbulent Ekman layer of the atmosphere," *J. Atmos. Sci.* **27**, 1027.

Tennekes, H., 1973, "A model for the dynamics of the inversion above a convective boundary layer," *J. Atmos. Sci.* **30**, 558.

Tennekes, H., and Driedonks, A. G. M., 1981, "Basic entrainment equations for the atmospheric boundary layer," *Bound.-Layer Meteor.* **20**, 515.

Tennekes, H., and Lumley, J. L., 1972, *A First Course in Turbulence.* Cambridge, MA: MIT Press.

Thompson, B. (Count Rumford), 1800, "The propagation of heat in fluids," in *Essays Political, Economical and Philosophical* (T. Cadell and W. Davies, eds.). London, p. 197. (Also in *Collected Works,* S. C. Brown, ed. Harvard University Press.)

Thompson, R. E., 1970, "On the generation of Kelvin-type waves by atmospheric disturbances," *J. Fluid Mech.* **42**, 657.

Thorpe, S. A., 1982, "On the clouds of bubbles formed by breaking wind-waves in deep water, and their role in air-sea gas transfer," *Phil. Trans. Roy. Soc. London, Ser. A* **304**, 155.

Thorpe, S. A., 1992, "Bubble clouds and the dynamics of the upper ocean," *Quart. J. Roy. Meteor. Soc.* **118**, 1.

Thorpe, S. A., Cure, M. S., Graham, A., and Hall, A. J., 1994, "Sonar observations of Langmuir circulation and estimation of dispersion of floating particles," *J. Atmos. & Oceanic Tech.* (submitted).

Toba, Y., 1973, "Local balance in the air-sea boundary processes, III. On the spectrum of wind waves," *J. Oceanogr. Soc. Japan* **29**, 209.

Turner, J. S., and Kraus, E. B., 1967, "A one-dimensional model of the seasonal thermocline. I. A laboratory experiment and its interpretation," *Tellus* **19**, 88.

Tyler, J. E., and Preisendorfer, R. W., 1962, "Light," in *The Sea* (Volume I) (M. N. Hill, ed.). New York: Interscience, p. 397.

Untersteiner, N., 1961, "On the mass and heat budget of arctic sea ice," *Arch. Meteor. Geophys. Bioklimatol.* **A12**, 151.

Untersteiner, N., 1964, "Calculations of temperature regime and heat budget of sea ice in the central Arctic," *J. Geophys. Res.* **69**, 4755.

Untersteiner, N., 1968, "Natural desalination and equilibrium salinity profile of perennial sea ice," *J. Geophys. Res.* **73**, 1251.

van de Hulst, H. C., 1957, *Light Scattering by Small Particles.* New York: Wiley.

Veronis, G., 1981, "Dynamics of large-scale ocean circulation," in *Evolution of Physical Oceanography* (B. A. Warren and C. Wunsch, eds.). Cambridge, MA: MIT Press.

Veronis, G., and Stommel, H., 1956, "The action of variable wind stresses on a stratified ocean," *J. Mar. Res.* **15**, 43.

Walker, G. T., 1928, "World weather," *Quart. J. Roy. Meteor. Soc.* **54**, 79.

Walmsley, J. L., 1988, "On theoretical wind speed and temperature profiles over the sea with applications to data from Sable Island, Nova Scotia," *Atmos.-Ocean* **26**, 203.

Walter, B. A., and Overland, J. E., 1984, "Observations of longitudinal rolls in a near-neutral atmosphere," *Mon. Wea. Rev.* **112**, 200.

Watson, K. M., West, B. J., and Cohen, B. I., 1976, "Coupling of surface and internal gravity waves,: A Hamiltonian model," *J. Fluid Mech.* **77**, 185.

Weaver, A. J., and Hughes, T. M. C., 1992, "Stability and variability of thermohaline circulation and its link to climate," *Trends in Phys. Oceanogr.* Trivandrum, India.

Webb, E. K., Pearman, G. I., and Leuning, R., 1980, "Correction of flux measurements for density effects due to heat and vapor transfer," *Quart. J. Roy. Meteor. Soc.* **106**, 85.

Welander, P., 1963, "On the generation of wind streaks on the sea surface by action of a surface film," *Tellus* **15**, 67.

Welander, P., 1982, "A simple heat-salt oscillator," *Dyn. Atm. Oceans* **6**, 233.

Welander, P., 1986, "Thermohaline effects in the ocean circulation and related simple models," in *Large-Scale Transport Processes in Oceans and Atmosphere* (J. Willebrand and D. L. T. Anderson, eds.). Dordrecht: Reidel.

Weller, R. A., and Price, J. F., 1988, "Langmuir circulation within the ocean mixed layer," *Deep-Sea Res.* **35**, 711.

Wells, N. C., 1979, "A coupled ocean-atmosphere experiment: the ocean response," *Quart. J. Roy. Meteor. Soc.* **105**, 355.

Whitham, G. B., 1974, *Linear and Non-Linear Waves.* New York: Wiley-Interscience.

Wilczak, J. M., 1984, "Large-scale eddies in the unstably stratified atmospheric surface layer. I. Velocity and temperature structure," *J. Atmos. Sci.,* **41**, 3537.

Wilczak, J. M., and Businger, J. A., 1983, "Thermally indirect motions in the convective atmospheric boundary layer," *J. Atmos. Sci.* **40**, 343.

Wilkin, J. L., and Chapman, D. C., 1987, "Scattering of continental shelf waves at a discontinuity in shelf width," *J. Phys. Oceanogr.* **17**, 713.

Williams, R. M., and Paulson, C. A., 1977, "Microscale temperature and velocity spectra in the atmospheric boundary layer," *J. Fluid Mech.* **83,** 547.

Willson, R. C., 1984, "Measurements of solar total irradiance and its variability," *Space Sci. Rev.* **38,** 203.

Winton, M., and Sarachik, E. S., 1993, "Thermohaline oscillations induced by strong steady salinity forcing of ocean general circulation models," *J. Phys. Oceanogr.* **23,** 1389.

Wipperman, F. D., Etling, D., and Kirstein, H. J., 1978, "On the instability of a PBL with Rossby number similarity," *Bound.-Layer Meteor.* **15,** 301.

Woodcock, A. H., 1940, "Convection and soaring over the open sea," *J. Mar. Res.* **3,** 248.

Woodcock, A. H., and Stommel, H., 1947, "Temperatures observed near the surface of a fresh water pond at night," *J. Meteor.* **4,** 102.

Woods, J., and Barkmann, W., 1986, "The response of the upper ocean to solar heating, I. The mixed layer," *Quart. J. Roy. Meteor. Soc.* **112,** 1.

Worthem, S., and Mellor, G., 1979, "Turbulence closure model applied to the upper tropical ocean," *Deep-Sea Res., GATE Suppl.* **26,** 237.

Worthington, L. V., 1976, *On the North Atlantic Circulation.* Baltimore: The Johns Hopkins University Press.

Wu, J., 1975, "Wind-induced drift currents," *J. Fluid Mech.* **68,** 49.

Wu, J., 1980, "Wind stress coefficients over the sea surface near neutral conditions—A revisit, *J. Phys. Oceanogr.* **10,** 727.

Wu, J., 1983, "Sea-surface drift currents induced by wind and waves," *J. Phys. Oceanogr.* **13,** 1441.

Wu, J., 1988, "Variations of whitecap coverage with wind stress and water temperature," *J. Phys. Oceanogr.* **18,** 1448.

Wu, J., 1990, "On parameterization of sea spray," *J. Geophys. Res.* **95,** 18, 269.

Wu, J., 1992, "Bubble flux and marine aerosol spectra under various wind velocities," *J. Geophys. Res.* **97,** 2327.

Wunsch, C., and Gill, A. E., 1976, "Observations of equatorially trapped waves in Pacific sea level variations," *Deep-Sea Res.* **23,** 371.

Wyngaard, J. C., and Brost, R. A., 1984, "Top-down and bottom-up diffusion of a scalar in the convective boundary layer," *J. Atmos. Sci.* **41,** 102.

Wyngaard, J. C., and Coté, O. R., 1971, "The budgets of turbulent kinetic energy and temperature variance in the atmospheric surface layer," *J. Atmos. Sci.* **28,** 190.

Yamada, T., 1983, "Simulation of nocturnal drainage flows by a $q^2$-1 turbulence closure model," *J. Atmos. Sci.* **40,** 91.

Yoshida, K., 1959, "A theory of the Cromwell current (the equatorial undercurrent) and of the equatorial upwelling—an interpretation in a similarity to a coastal circulation," *J. Oceanogr. Soc. Japan* **15,** 159.

Zedel, L., and Farmer, D. M., 1991, "Organized structures in subsurface bubble clouds: Langmuir circulations in the open ocean," *J. Geophys. Res.* **96,** 8889.

Zilitinkevich, S. S., Leichtmann, D. L., and Monin, A. S., 1967, "Dynamics of the atmospheric boundary layer (English edition)," *Izv. Atmos. Ocean. Phys.* **3,** 297.

# AUTHOR INDEX

# SUBJECT INDEX

Absorptance, 78
Absorption, 76
Absorption coefficient, monochromatic, 80–81
Absorption spectra
of atmospheric gases, 83
Adiabatic changes of state, 52, 53
Advection, 15
Affinity of vaporization, 42
Ageostrophic transport method, 179
Air Mass Transformation Experiment (AMTEX), 223
Albedo, 84
of sea surface, 84
of sea ice, 85
Alternating tensor, 3
Antarctic Bottom Water (AABW), 300
Aspect ratio, 28
Atlantic countercurrent, 287
Atmospheric forcing, 239–40, 259–62
Auto-correlation function, 19, 24
Available potential energy (APE), 12, 17, 275, 292

Baroclinic deformation radius, 248
Baroclinic mode, 245, 249, 257, 264
Baroclinic wake, 266
Barotropic mode, 245, 249, 264
Barotropy, 245
Beer's law, 81
Bernoulli's equation, 9
Bjerknes, V., 183, 246
Bjerknes circulation theorem, 266
Black body, 78
radiance, 80
irradiance, 79–80
Boundary conditions
dynamic, 8
kinematic, 8

Boundary Layer Group of Air Force Cambridge Research Laboratories (AFCRL), 151, 174
Bottom water formation, 298–300
Boussinesq approximation, 10, 104, 213
Brunt-Vaisala frequency, 13, 53, 110
Bubbles, 58–69
environmental effects, 67–69
equilibrium pressure in, 59–61
generation of, 58–59
Reynolds numbers for, 62
size spectra of, 62–65
terminal velocity of, 61–62
Bulk temperature, 167
Bulk transfer coefficients, 180–81
Buoyancy
acceleration, 10
flux, 17, 197–98, 214, 217, 239
force, 17
frequency, 13

$^{14}$C-method, 166
Capillary waves. *See* Waves
Carnot cycle, 310
Charnock's relation, 145, 157, 180
Chemical potential, 41–42, 45
specific, 13
Chimneys. *See* Convective plumes
Clausius Clapeyron equation, 40
Cloud clusters, 306
Cloudstreets, 192–93, 198
Cloud-topped mixed-layer, 226–37
$CO_2$, 46, 48, 300
Cold water tongue, 313
Coherent structures, 189–201
Colligative properties, 42–43
Concentration, 43–46
equilibrium, 43
Conditional sampling method, 174–76

356